T0310439

Advances in Materials Science for Environmental and Nuclear Technology II

Advances in Materials Science for Environmental and Nuclear Technology II

Ceramic Transactions, Volume 227

Edited by
S. K. Sundaram
Kevin Fox
Tatsuki Ohji
Elizabeth Hoffman

A John Wiley & Sons, Inc., Publication

Published by John Wiley & Sons, Inc., Hoboken, New Jersey.
Published simultaneously in Canada.

For general information on our other products and services or for technical support, please contact our
Customer Care Department within the United States at (800) 762-2974, outside the United States at
(317) 572-3993 or fax (317) 572-4002.

Wiley also publishes its books in a variety of electronic formats. Some content that appears in print may
not be available in electronic formats. For more information about Wiley products, visit our web site at
www.wiley.com.

Library of Congress Cataloging-in-Publication Data is available.

ISBN: 978-1-118-06000-1
ISSN: 1042-1122

oBook ISBN: 978-1-118-14452-7
ePDF ISBN: 978-1-118-14449-7

Printed in the United States of America.

10 9 8 7 6 5 4 3 2 1

Contents

MATERIALS SOLUTIONS FOR THE NUCLEAR RENAISSANCE

GREEN TECHNOLOGIES FOR MATERIALS MANUFACTURING AND PROCESSING

Preface

The Materials Science and Technology 2010 Conference and Exhibition (MS&T'10) was held October 17-21, 2010, in Houston, Texas. A major theme of the conference was Environmental and Energy Issues. Papers from three of the symposia held under that theme are included in this volume. These symposia include Clean Energy: Fuel Cells, Batteries, Renewables – Materials, Processing and Manufacturing; Materials Solutions for the Nuclear Renaissance; and Green Technologies for Materials Manufacturing and Processing. These symposia included a variety of presentations with sessions focused on Advanced Nuclear Fuels, Materials Performance in Extreme Environments, Immobilization of Nuclear Wastes, Irradiation and Corrosion Effects, Modeling, Solid Oxide Fuel Cells, Batteries, Clean Fuel Combustion, and Solar Energy.

The success of these symposia and the publication of the proceedings could not have been possible without the support of The American Ceramic Society and the organizers of the above mentioned symposia. The symposia organizers included Willam E. Lee, Josef Matyas, Ramana G. Reddy, Kumar Sridharan, Zhenguo (Gary) Yang, Prabhakar Singh, Prashant Kumta, Ayyakkannu Manivannan, Abdul-Majeed Azad, Colleen Legzdins, Arumugam Manthiram, Donald W. Collins, Mrityunjay Singh, Richard Sisson, and Allen Apblett. Their assistance, along with that of the session chairs, was invaluable in ensuring the creation of this volume.

S. K. Sundaram, *Alfred University, USA*
Kevin Fox, *Savannah River National Laboratory, USA*
Tatsuki Ohji, *AIST, JAPAN*
Elizabeth Hoffman, *Savannah River National Laboratory, USA*

Clean Energy: Materials, Processing, and Manufacturing

SLAG CHARACTERIZATION FOR THE DEVELOPMENT OF NEW AND IMPROVED SERVICE LIFE MATERIALS IN GASIFIERS USING FLEXIBLE CARBON FEEDSTOCK

James Bennett[1], Seetharaman Sridhar[1,2], Jinichiro Nakano[1], Kyei-Sing Kwong[1], Tom Lam[1,3], Tetsuya Kaneko[1,2], Laura Fernandez[1,2], Piyamanee Komolwit[1,2], Hugh Thomas[1], Rick Krabbe[1]

[1] US Department of Energy National Energy Technology Laboratory, 1450 Queen Ave., Albany, OR 97321 USA
[2] Department of Materials Science and Engineering, Carnegie Mellon University, 5000 Forbes Ave., Pittsburgh, PA 15213 USA
[3] Alfred University, 1 Saxon Dr., Alfred, NY 14802 USA

ABSTRACT

In modern gasifiers, the carbon feedstock (coal, petcoke and/or biomass) is determined largely by carbon content, cost, availability, and environmental concerns. Ash impurities in the carbon feedstock vary widely in quantity and chemistry, impacting gasifier operation. Ash from mineral impurities in the feedstock liquefy at the elevated temperatures of gasification; impacting slag chemistry, viscosity, melting temperature, surface and interfacial tension – ultimately determining gasifier operating temperature and refractory service life. The slag itself experiences wide variations in the relative fraction and state of crystalline material (oxides, sulfides and metallic), non-crystalline (glass) material, or gas phases formed from feedstock ash. It is these variations that have a critical impact gasifier operation, determining slag fluidity along the walls and the chemical and physical stability (wear) of the refractory liner. In this paper, two aspects of joint research between NETL and CMU on slag and slag/refractory interactions will be discussed. The first area is researching phase formation in synthetic petcoke/coal slag (SiO_2-Al_2O_3-Fe_2O_3-CaO-V_2O_3) under simulated gasification conditions ($1500°C$ and 10^{-8} atm oxygen partial pressure). The second area focuses on interactions between coal and petcoke slags with commercial refractory currently used (high chrome oxide) or having the potential for use as a gasifier liner (high alumina). Refractory materials studied in the simulated gasifier environment were fired brick of the following compositions: $90wt\%Cr_2O_3$-$10wt\%Al_2O_3$ and $100wt\%Al_2O_3$. Information from this research is being used to improve the performance of or to develop new refractory liner materials for gasifiers, and to understand mixed feedstock slag behavior under gasification conditions.

INTRODUCTION

Gasification as a modern industrial process converts carbon feedstock (typically coal and petcoke), water, and oxygen (oxygen shortage - reducing concentration) at elevated temperatures and high pressure into CO and H_2; with excess heat, gases (typically CO_2, H_2S, CH_4, NH_3, HCN, N_2, and Ar), slag (from mineral impurities in the carbon feedstock), and carbon (excess added to ensure maximum CO production) produced as process by-products. The CO/H_2 (syngas) product from a gasifier is used for power generation, as a feedstock for the chemical industry (used in the production of chemicals including fertilizer or Fischer Tropsch liquids), and is considered a leading candidate for H_2 production in a hydrogen based economy. Gasification is being explored as a critical technology in the success of DOE's Near Zero Emissions Advanced Fossil Fuel Power Plants, and could play a key role in defining the nation's long-term energy security in both power and liquid fuels. With a DOE goal of 90 pct CO_2 capture, the closed loop of high pressure gasification is easily adapted to supply a "pure" CO_2 stream for reuse or sequestration, giving it a cost advantage over many other processes. It thought that if biomass is used as a part of a carbon feedstock in gasification, approximately 30 pct of the total carbon additions as biomass will make the process carbon neutral.

Several types of gasifiers exist, with this research focused on the entrained bed air cooled slagging gasifier, which produces a molten slag from mineral impurities in the carbon feedstock. These gasifiers, shown in figure 1, are lined with high chrome oxide refractory materials that lasts anywhere from 3 months to 30 months, a life determined by operating conditions and the severity of the gasifier environment. The operating conditions of slagging gasifiers involves temperatures between 1325°-1575°C, a reducing atmosphere (about 10^{-8} oxygen partial pressure), and pressures between 2.07 – 6.89 MPa. In that environment, the service life of the high chrome oxide liner does not meet the performance needs of gasifier users, and limits achieving an on line gasifier availability of 85-95% targeted for utility applications and more than 95% in applications such as chemical feedstock production[1]. Failure to meet these goals has created a potential roadblock to widespread acceptance and commercialization of advanced gasification technology, and is the reason refractory liners were identified as a key barrier to widespread commercialization of gasification technology[2].

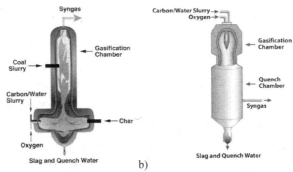

Figure 1. Two types of slagging gasifies used by industry; a) two stage slagging gasifier, and b) single stage gasifier.

Failure of the hot face gasifier lining is expensive, both in terms of material replacement costs (as high as $1,000,000) and in lost production (due to the 1-3 week gasifier shutdown for liner repair/replacement and the resulting lost syngas production). Besides current performance issues, questions also exist concerning high chrome oxide linings and fuel flexibility. High alkali and alkaline earth levels found in new carbon sources being considered for gasification, such as biomass or western coals, will change refractory liner wear and may cause the formation of hexavalent chrome oxide in the refractory liner; resulting in their possible classification as a hazardous waste. Non-chrome oxide materials, which were evaluated for their potential as a gasifier liner material in the mid 70's through the mid 80's by USDOE, EPRI, and industry sponsored research, were found to have inferior performance when compared to high chrome oxide refractories[3-8]. Chrome oxide levels above 75 pct have been determined to survive the severe service environment of gasification[9]. Current additives to chrome oxide that contribute to their superior corrosion performance are alumina and zirconia. Since the early research, improvements in raw material purity, grain structure, and in refractory fabrication technology have occurred that merit non-chrome oxide materials to be re-evaluated as gasifier containment materials, especially with regard to mixed feedstock environments. The practice of mixed feedstock with make it highly unlikely that any one brick composition will have adequate performance in all slag environments – the impact of potential slag compositions must be evaluated and used to determine liner materials. Additional benefits of no or low chrome oxide refractories could include lower cost (due to the use of raw materials with lower cost than chrome oxide), lower brick density (leading to lighter brick weight per unit size), an increased number of refractory material suppliers (creating an environment for increased innovation and price competition), and faster installation of linings (non-chrome oxide materials have the potential for installation techniques not possible with high chrome oxide liner materials).

Carbon feedstock currently used in slagging gasifiers includes coal, petroleum coke, or combinations of them. Because of the rising cost of carbon feedstock, concerns over CO_2 emissions, feedstock availability, as well as other drivers; fuel flexibility (the use of mixed or varying carbon feedstock in gasification) is currently practiced or considered by many gasifier operators. One drawback to fuel flexibility is the unknown effect impurities present in them have on the liner service life and the gasification process. Mineral impurities in carbon feedstock can range up to 10 wt pct or higher for coal, 1 wt pct for petcoke, and from 1 to 20 wt pct for biomass. The impurities of concern include oxides of Si, Al, Ca, Mg, Fe, K, Na, and V; which may form aggressive molten slags at the high temperature of gasification, slags that dissolve or penetrate the refractory liner and lead to their premature failure. Besides gasifier temperature, oxygen partial pressure, and the quantity of ash associated with the carbon feedstock will also impact liner wear. In addition to lining wear, slag must flow from a gasifier. Additives to control slag flow and the temperature of gasification may be made, but have an unknown impact on refractory wear. Past refractory research focused on developing highly slag resistant refractory materials, but failed to consider other properties of slags from carbon feedstock, and how modification of them might reduce refractory wear and impart good slag flow properties. Slag chemistry can be controlled to minimize interactions with the refractory liner or its penetration into the porous refractory. Known wear mechanisms of refractories caused by slag include structural spalling (brought about by slag penetration into a refractory), chemical spalling (caused by slag/refractory interactions on the refractory surface), and/or corrosion (chemical dissolution of the refractory in the slag)[10]. Instead of considering slag impact on lining wear, the gasifier is typically operated at a temperature where slag will flow it as a liquid, yet will produce the desired syngas composition. If a gasifier is operated at too low of a temperature, slag will not flow from it, instead clogging the exit and leading to premature system shutdown. If the gasifier is operated at too high a temperature, accelerated wear of the refractory lining will occur, a known consequence that is not controlled.

To obtain a fundamental understanding of the interactions of mixed carbon feedstock with current and proposed refractory liner materials, the National Energy Technology Laboratory (NETL) and Carnegie Mellon University (CMU) have partnered to evaluate slag phases formed under gasification conditions and to study the slag/refractory interactions of those mixed feedstock materials. This paper summarizes the joint research that has been conducted.

COAL AND PETCOKE SLAG CHEMISTRY

A range of coal ash chemical compositions exist in the U.S.; with gasifier use dependent on the price, carbon content, availability, shipping distance, ash and moisture content, and other factors. Petcoke used in gasification also vary greatly in ash chemistry; with its feedstock use dependent on factors similar to coal. Because of the wide variations in coal chemistry, an average based on approximately 300 analyses[11] was used as a base for compositions in this research, and an average of nine for petcoke[12-14]. Those chemistries are listed in table 1; with Al, Si, Fe, Ca, and V the most critical for slag viscosity and refractory/slag interactions. Oxides of these compounds have a large impact on slag viscosity, refractory dissolution, and the ability of slag to wet and penetrate refractory surfaces; properties which influence refractory wear.

Of the coal and petcoke ashes generated in a slagging gasifier, the behavior of petcoke slag under gasification conditions is the least understood; with limited thermodynamic data available on vanadium in software program (such as FactSage™) at low oxygen partial pressures and elevated temperatures encountered during gasification. Because of the lack of this fundamental information, slag behavior and materials interactions of mixed feedstock are difficult to predict, and are the focus of reported research. Mixed feedstock containing biomass will be evaluated at a later date, with elements of concern in biomass including Al, Si, Fe, Ca, K, Mg, and Na.

Table 1. Average coal[11] and petcoke[12-14] slag compositions used to evaluate mixed feedstock slag properties.

Ash Source	Percent Ash	Wt Pct Oxide								
		Al_2O_3	SiO_2	Fe_2O_3	CaO	MgO	Na_2O	K_2O	V_2O_3	NiO
Avg. Coal	10.0	27.0	46.6	16.4	6.2	1.3	1.0	1.5	NA	NA
Avg. Petcoke	1.0	5.4	16.0	7.4	6.1	1.1	0.9	0.6	53.1	9.5

NA = Not analyzed

KINETIC STUDIES OF SYNTHETIC COAL/PETCOKE CRYSTAL FORMATION

Research Objective/Test Description

The study of coal/petcoke mixture crystallization under gasification conditions is discussed below, which was previously reported[15]. A total of six average coal and petcoke mixtures (based on the chemical compositions in table 1) were ball milled in 500 ml Nalgene® plastic bottles containing ethanol as a grinding/mixing procedure for 48 hours using reagent grade oxide powders (starting particle size below 44 microns). The chemistry of the mixed slags was analyzed by ICP, with the results given in table 2. A Confocal Scanning Laser Microscope (CSLM – Lasertec 1LM21H) was used to melt samples, with a figure of the furnace and sample holder assembly shown in figure 2. Slag samples were placed in a cylindrical platinum crucible (99.99 pct Pt) prior to the melting and quenching experiment. Mixtures were heated in air to 1500°C at rate of 77°C/sec and "premelted" to produce liquid slag, after which the furnace was cooled to a specific test temperature, held for 2 minutes, then the furnace environment changed to a CO/CO_2 gas mixture to create an oxygen partial pressure of 10^{-8} atm – similar to that existing in a commercial gasifier at that temperature. Initial heating was done in air because thermodynamic data did not exist on V_2O_3 liquid formation at 1500°C, while the starting vanadium compound of V_2O_5 was known to completely liquefy in air. After samples were held at the desired test temperature for 2 minutes, a CO/CO_2 gas mixture was introduced to create an oxygen partial pressure of approximately 10^{-8} atm and the time for initial crystallization to occur on the slag surface recorded. When the atmosphere was changed from air to CO/CO_2 (creating an O_2 partial pressure of approximately 10^{-8}), vanadium oxide present in the mixture shifted in composition to V_2O_3, and crystallization of supersaturated vanadium solutions occurred.

Table 2. Chemistry of petcoke/coal slag mixtures used in time temperature transformation studies.

Petcoke/Coal Ash Wt Ratio	Oxide Chemistry (wt pct)					
	SiO_2	Al_2O_3	CaO	Fe_2O_3	K_2O	V_2O
0/100	52.0	24.0	6.4	14.5	2.9	0
10/90	49.3	22.4	6.8	14.1	2.7	4.7
30/70	43.9	19.2	7.5	13.2	2.4	13.7
50/50	38.4	15.9	8.3	12.3	2.0	23.1
70/30	32.9	12.6	9.0	11.5	1.7	32.5
100/0	25.0	7.8	10.1	10.2	1.2	46.0

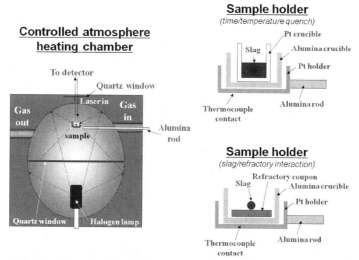

Figure 2 Confocal Scanning Laser Microscope used for heating test samples of slag in Pt crucibles or slag on refractory coupons. Slag and refractory materials are shown in the sample chamber.

Results and Conclusions

The crystallization sequence at different times for a petcoke slag at 1500°C is shown in figure 3. Note that crystallization of V_2O_3 did not occur until about 50 seconds after the atmosphere change. The crystallization in figure 3 occurred at the pre-melting temperature of the slag mixture in air, and indicated the impact oxygen partial pressure has on the phases of vanadium present in the liquid slag.

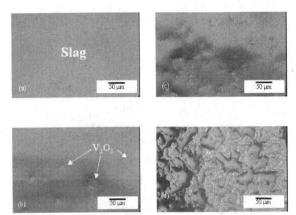

Figure 3. Crystallization sequence for a 100 pct petcoke mixture at 1500°C and an oxygen partial pressure of 10^{-8} atm. Times to crystallization in the images are as follows: a) 47 sec; b) 52 sec; c) 60 sec; and d) 138 sec. Images b, c, and d show the crystallization of V_2O_3 (karelianite).

Crystallization as a time temperature transformation diagram occurring at an oxygen partial pressure of 10^{-8} atm is shown in figure 4. At temperatures above 1500°C, karelianite (V_2O_3) was the primary phase identified. At lower temperatures tested (below about 1350°C), phases of karelianite, spinels (VFe_2O_4, FeV_2O_4, and $FeAl_2O_4$), and anorthorite (CaO-Al_2O_3-$2SiO_2$) were found to precipitate. Slag compositions containing 10 wt pct or less petcoke did not have any phases that precipitated above about 1375°C, while compositions containing 100 wt pct petcoke had no phases other than karelianite that precipitated. Above 1350°C, only karelianite was found to precipitate in all compositions. It is of interest that a thermodynamic program, FactSage™, suggested mullite may also be stable at temperatures above 1350°C, a phase not found in these rapid quench studies, but a phase identified in long term coal/petcoke phase equilibrium studies presented in the next section.

Phases that precipitated from a slag influence overall system viscosity, remaining liquid viscosity (because of chemistry changes in the liquid), the ability of the altered liquid slag to penetrate pores, and how the liquid slag interacts with the refractory. As mentioned, greater detail of the study is found in the original work[15].

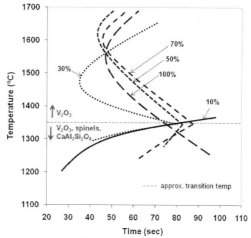

Figure 4. Time temperature transformation diagram showing crystallization of different petcoke/coal mixtures at an oxygen partial pressure of 10^{-8} atm.

THERMODYNAMIC STUDIES OF PHASE EQUILIBRIA IN SYNTHETIC COAL/PETCOKE MIXED SLAGS

Research Objective/Test Description

The phase equilibrium of synthetic coal/petcoke mixtures composed of five oxide components (SiO_2-Al_2O_3-FeO-CaO-V_2O_3) were studied as part of an effort to establish phase equilibrium conditions to use in thermodynamic databases. This work is on-going, and the research presented represents preliminary data. The five oxides considered in this study are the major components in coal and petcoke slags impacting viscosity and wear behavior. Phase information obtained from the study will be used to predict equilibrium conditions existing in gasifiers. The average coal and petcoke compositions used in the study are listed in table 3, and have their origin in the petcoke and compositions listed in table 1. The ratios of petcoke and coal evaluated corresponded to petcoke to coal additions of 0-72 pct on a carbon feedstock basis. Because the amount of V_2O_3 increases as petcoke additions were made and the amount of FeO and CaO stayed fairly constant, FeO and CaO were fixed at constant values in the study, while SiO_2, Al_2O_3, and V_2O_3 were varied, with ratios of Al_2O_3 and SiO_2 ranging from 1/4 to 1/1.

Exposure conditions for the petcoke/coal compositions were 1500°C at an oxygen partial pressure of 10^{-8} atm (obtained by using a 64% CO/36% CO_2 gas exposure environment). Samples were held at temperature for 72 hours to obtain equilibrium. Oxides used in the study were reagent grade materials with particle sizes less than 44 microns. Powders were dry mixed in 150 ml Nalgene® plastic bottles at 55 rpm, with mixing stopped and manual stirring done on a periodic basis to prevent agglomerates from forming and to improve mixing. A total of 450 mgs of each mixed powder was placed individually in Pt crucibles (99.99 pct pure) that were approximately 10mm diameter and 7.5 mm high. The Pt crucibles containing test material was then placed in an alumina trays for handling and quenching (as shown in figure 5). Test materials was heated to 1200°C at a rate of 191°C/hr, followed by a heating rate of 120°C/hr to 1500°C. After the timed sample soak at 1500°C, samples were quenched in water to "freeze" phases present, after which they were analyzed for crystalline phases and chemistry using XRD (Rigaku Ultima III), SEM (JEOL-7000F), and TEM (Phillips CM200).

Table 3. Chemistry and phases formed in petcoke/coal compositions after a 72 hr exposure at 1500°C in a 10^{-8} atm oxygen partial pressure.

Sample	Chemical Composition (Wt Pct)					Phase Identified
	Al_2O_3	CaO	FeO	SiO_2	V_2O_3	
1	14.7	6.9	12.8	65.6	0	Slag
2	18.8	6.9	12.8	61.4	0	Slag
3	14.7	6.7	12.8	64.2	1.7	Slag
4	36.5	6.6	15.8	41	0	Slag + Mullite
5	25.1	6.6	12.7	55.7	0	Slag + Mullite
6	38.9	7	13.5	39	1.6	Slag + Mullite
7	35.4	7	14.3	39.8	3.5	Slag + Mullite
8	33.6	7.5	14	38.9	6	Slag + Mullite
9	32.8	6.9	13.7	44.4	2.1	Slag + Mullite
10	11.7	6.4	12.3	60.9	8.7	Slag + Karelianite
11	20.8	6.6	15.9	49.6	7	Slag + Karelianite
12	14.2	6.8	14	62	3	Slag + Karelianite
13	15	7	12.1	60.9	4.9	Slag + Karelianite
14	33.9	7.7	13.8	36.6	8	Slag + Karelianite + Mullite
15	25.1	6.9	14.2	46.4	7.4	Slag + Karelianite +Mullite
16	28.7	7.6	13.1	42.8	7.7	Slag + Karelianite + Mullite

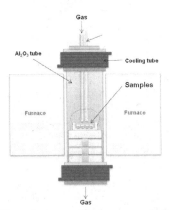

Figure 5. Furnace and experimental setup used to expose synthetic petcoke/coal slag samples at 1500°C to an oxygen partial pressure of 10^{-8} atm.

Results and Conclusions

The following phases were noted in the petcoke/coal slag composition listed in table 3: amorphous slag, mullite ($3Al_2O_3$-$2SiO_2$), and karelianite (V_2O_3). Some substitution of vanadium for aluminum in the mullite structure, and of aluminum for vanadium in the karelianite structure was noted, and is being evaluated further. A preliminary isothermal phase diagram based on the data obtained to date is shown in figure 6, along with SEM images of the phases formed at 1500°C. Phase equilibrium data is being obtained for under other temperatures and oxygen partial pressures conditions, and will be used to revise this proposed phase diagram.

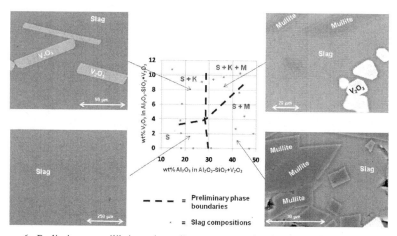

Figure 6. Preliminary equilibrium phase diagram versus chemistry of petcoke/slag synthetic mixtures with 13.5 wt pct FeO and 7 wt pct CaO and variations of SiO_2, Al_2O_3, and V_2O_3 when they were exposed to a simulated gasifier environment of 1500°C, and oxygen partial pressure of 10^{-8} atm. Phases are S = Slag, K = Karelianite, M = Mullite

Four phase areas are included in the proposed isothermal phase diagram (figure 6) and include slag, slag plus mullite, slag plus karelianite, and slag plus mullite and karelianite. Typical coal/petcoke carbon feedstock and the corresponding slags formed from them should fall within the proposed diagram; and indicate system saturation and changes in mullite with respect to changes in Al_2O_3 and SiO_2. Slags high in petcoke may become saturated with V_2O_3 or V_2O_3 and mullite, depending on the amount of Al_2O_3 and vanadium present and the system temperature and oxygen partial pressure.

INTERACTIONS OF REFRACTORY MATERIALS (Al_2O_3 and Cr_2O_3) WITH COAL AND PETCOKE SLAGS

Research Objective/Test Description

Interactions between coal and petcoke slags and refractory materials currently used (high chrome oxide) or with potential for use as gasifier liner materials (high alumina in this study) were previously evaluated[16] to understand interface reactions occurring at the surface and are summarized in this section. Synthetic coal and petcoke slag compositions were used, with the compositions listed in table 4. Note that both petcoke and coal slag compositions are based on compositions previously described and listed in table 1. Two refractory materials were evaluated, a high chrome oxide commercial material of approximate composition 90 wt pct Cr_2O_3/10 wt pct Al_2O_3; and a silica bonded high alumina commercial refractory that was 99 pct Al_2O_3. Chemical and physical properties of the refractory materials are also listed in table 4.

Table 4. Chemical and physical properties of synthetic slag (petcoke and coal) and refractory compositions used to evaluate interface interactions.

	Slag		Refractory Material*	
	Petcoke	Coal	High Chrome	High Alumina
Chemistry (wt %) - SiO_2	25	52	0.2	0.2
- Al_2O_3	7.8	24	10.2	99.6
- CaO	10.1	6.4	0.3	Trace
- MgO	NA	NA	0.1	Trace
- Fe_2O_3	10.2	14.5	0.1	0.1
- K_2O	1.2	2.9	NA	NA
- $Na_2O + K_2O$	-	-	NA	0.1
- V_2O_5	46	NA	-	-
- Cr_2O_3	-	-	89.0	NA
- TiO_2	-	-	0.2	Trace
Apparent Porosity (%)	-	-	16.7	19.4

NA = Not analyzed   * = Data from product data sheet

Synthetic slag mixtures listed in table 4 were pre-melted in carbon crucibles by heating above their melting temperature in high purity Ar for two hours. Slag samples then were reheated a second time for 2 hours at 1500°C in an atmosphere of CO/CO_2 to create stable phases existing in that environment prior to testing. Coal or petcoke slags particles were then placed on specific test coupon areas (grains or matrix) for timed exposures at 1500°C of 0 or 10 minutes. Reactions of the premelted slag on the refractory coupon were evaluated as sessile drop interactions using the CSLM described earlier and shown in figure 2. The heating rate for samples tested was approximately 77°C/sec to the test temperature of 1500°C. Slag particles were approximately 300-500 microns in diameter and weighed between 0.02 and 0.04 mg. Refractory coupons were approximate 3 mm square by 1 mm thick, with a surface finish of 1 micron. Testing was done in a CO/CO_2 atmosphere corresponding to an oxygen partial pressure of 10^{-8} atm. After heating and holding at 1500°C for the designated time, test samples were He quenched at an average cooling rate from 99°C/sec from 1500 to 900°C, than cooled at 77°C/sec to room temperature. Exposed surface and side profile microstructures were evaluated by SEM to determine changes occurring at the slag/refractory interaction. A total of 16 experiments were conducted.

Results and Conclusions
Slag and high chrome, high alumina refractory surface microstructures (after 10 minutes of exposure at 1500°C in a CO/CO_2 atmosphere corresponding to an oxygen partial pressure of 10^{-8} atm) are shown in figures 7 and 8 respectively. Significant surface changes were not noted on either the coal or petcoke slags that contacted high chrome grains, although when petcoke

contacted grains of high chrome oxide, it wet and spread; and it appeared to wet and interact with grains of alumina, with crystallization of V_2O_3 occurring on the surface of alumina grains.

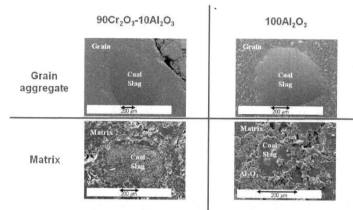

Figure 7. Surface interactions between coal slag and commercial refractory samples (high chrome oxide and alumina) after 10 minutes of exposure at 1500°C in a CO/CO₂ atmosphere corresponding to an oxygen partial pressure of 10^{-8} atm.

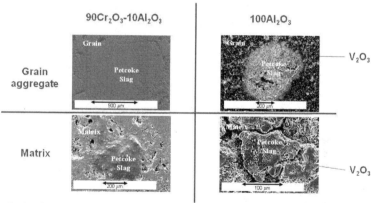

Figure 8. Surface interactions between petcoke slag and commercial refractory samples (high chrome oxide and alumina) after 10 minutes of exposure at 1500°C in a CO/CO₂ atmosphere corresponding to an oxygen partial pressure of 10^{-8} atm.

When interactions between coal and petcoke slags and the high chrome and alumina matrix materials were evaluated, coal slag wet and penetrated the high chrome oxide matrix, while the high alumina matrix interacted with the slag and had high alumina grains that were released from the contact area, remaining on the surface of the coupon (figure 7). Petcoke slags penetrated the surface of the high chrome oxide refractory, and had excess slag containing particles (figure 8). When the petcoke slag interacted with the high alumina material, a coarser surface resulted, probably from material dissolution into the slag and the release of some alumina grains.

The greater interaction of petcoke slags with high alumina materials would indicate the potential for greater refractory wear during service and thus, a shorter service life. The potential for greater wear in high chrome oxide refractory materials was also indicated when exposed to petcoke slags. A cross section image of petcoke slag/high chrome oxide matrix material exposed for 10 minutes (figure 9) showed possible chemical spalling of a Cr/V compound at the slag/refractory interface. This type wear mechanism is known to occur in high chrome oxide refractories exposed to coal slag [17].

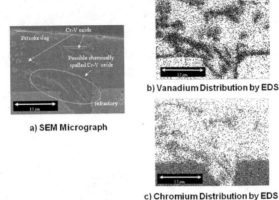

b) Vanadium Distribution by EDS

a) SEM Micrograph

c) Chromium Distribution by EDS

Figure 9. Cross section of petcoke slag/high chrome oxide refractory interface evaluated by SEM after 10 minutes of exposure at 1500°C in a CO/CO_2 atmosphere corresponding to an oxygen partial pressure of 10^{-8} atm. a) slag/refractory cross section, b) EDS imaging indicating vanadium, and c) EDS imaging for chrome. Note the possible chemical spalling on the surface of Cr/V compound.

CONCLUSIONS

Gasification is used to produce CO and H_2 (syngas), with two types of feedstock (coal and petcoke) used individually or in mixtures as carbon sources. Syngas generated by gasification is used in the production of power and chemicals, and is a possible source of H_2 in a hydrogen based economy. Gasification as an industrial process is limited by the on line availability of the process, which is directly impacted by refractory liner failure. Mineral impurities in the carbon feedstock create ash; which at the high temperature (1325-1575°C) and low oxygen partial pressures (10^{-8} atm) of gasification, melt to produce molten slag. The molten slag interacts with the Cr_2O_3 refractory liner currently used to protect the gasifier, causing refractory failure primarily by chemical corrosion or spalling. The properties of slag can be controlled to reduce refractory wear or to develop improved performance refractory materials, but are not well understood. NETL and CMU have conducted joint research using synthetic coal/petcoke slags to study: 1) slag properties under gasification conditions, and 2) slag/refractory interactions. The following conclusions concerning slags and refractory/slag interactions under gasification conditions can be drawn:

- Synthetic coal/petcoke slags containing $SiO_2-Al_2O_3-FeO-CaO-V_2O_3$ were heated to 1500°C in a Confocal Scanning Laser Microscope (CSLM), than cooled at different temperatures for up to 2 minutes. Karelianite (V_2O_3) was formed as the primary phase above 1350°C; and karelianite, spinels (VFe_2O_4, FeV_2O_4, and $FeAl_2O_4$), and anorthorite ($CaO-Al_2O_3-2SiO_2$) were formed below 1350°C. Synthetic slag containing 100 pct petcoke formed only karelianite at all test temperatures. Synthetic slag containing 0 and 10 pct petcoke formed no phase other than slag above 1375°C.
- Different synthetic coal/petcoke slag compositions containing $SiO_2-Al_2O_3-FeO-CaO-V_2O_5$ were heated at 1500°C for 72 hrs in an oxygen partial pressure of 10^{-8} atm. Based on phase formation four regions are proposed in a preliminary phase diagram: slag; slag plus karelianite; slag plus mullite ($3Al_2O_3-2SiO_2$); or slag, mullite and karelianite. The type of phases formed was dependent on slag composition.
- Interactions between mixtures of synthetic coal/petcoke slags and commercial high chrome and alumina refractories evaluated at 1500°C for 10 minutes in an oxygen partial pressure of 10^{-8} atm using a CSLM indicated the greatest interactions between coal and petcoke slags and alumina refractory materials. Chemical spalling was noted on high chrome oxide refractories and petcoke slags.

Knowledge of the properties of coal and petcoke slags under gasification conditions is being used to develop improved performance refractory materials, or to improve the performance of existing gasifier materials through the possible control of slag properties.

REFERENCES
[1] G.Stiegel, and S. Clayton, "DOE Gasification Industry R& D Survey: A Perspective of Long Term Market Trends and R&D Needs"; Proceedings of the Gasification Technologies 2001 Annual Meeting; San Francisco, CA.
[2] Gasification Markets and Technologies – Present and Future – An Industry Perspective; US DOE/FE Report 0447; US DOE; (July, 2002); pp. 1-53.
[3] J.A. Bonar, C.R. Kennedy, and R.B. Swaroop; "Coal-Ash Slag Attack and Corrosion of Refractories"; Amer. Ceram. Soc. Bul.; Vol. 59, No. 4, 1980; pp 473-478.
[4] W.T. Bakker, Greenberg, M. Trondt, and U. Gerhardus; "Refractory Practice in Slagging Gasifiers"; Amer. Ceram. Soc. Bul.; Vol. 63, No. 7, 1984; pp 870-876.
[5] S. Greenberg and R.B. Poeppel, "The Corrosion of Ceramic Refractories Exposed to Synthetic Coal Slags by Means of the Rotation-Cylinder Technique: Final Report," Research Report ANL/FE—85-15, research sponsored by USDOE/FE and EPRI, April 1986, 66 pp.
[6] A.P. Starzacher, "Picrochromite Brick - A Qualified Material for Texaco Slagging Gasifiers," Radex-Rundschau, Vol. 1, 1988, pp. 491-501.
[7] R.E. Dial, "Refractories for Coal Gasification and Liquefaction," Amer. Ceram. Soc. Bul., Vol. 54, No. 7 (1975), pp 640-43.
[8] M.S. Crowley, "Refractory Problems in Coal Gasification Reactors," Amer. Ceram. Soc. Bul.; Vol. 54, No. 12 (1975), pp 1072-74.
[9] W.T. Bakker, "Refractories for Present and Future Electric Power Plants," Key Engineering

Materials, Trans Tech Publications, (1993), Vol. 88, pp. 41-70.

[10] J.P. Bennett, J. Kwong, B. Riggs (Eastman Chemical, Kingsport, TN); Identification and Elimination of Refractory Failure in an Air Cooled Slagging Gasifier; Proceedings of the 44th Annual Ref. Sym.; St. Louis, MO; March 26-27, 2008.

[11] W.A. Selvig and F.H. Gibson; "Analysis of Ash from United States Coals"; USBM Bulletin, Pub. 567; 1956; 33 pp.

[12] S.V. Vassilev et al; "Low Cost Catalytic Sorbents for NOx Reduction. 1. Preparation and Characterization of Coal Char Impregnated with Model Vanadium Components and Petroleum Coke Ash"; Fuel; Vol 81, 2002; pp 1281-1296.

[13] R.E. Conn; "Laboratory Techniques for Evaluating Ash Agglomeration Potential in Petroleum Coke Fired Circulating Fluidized Bed Combustors"; Fuel Pro.Tech.; Vol 44, 1995; pp 95-103.

[14] R.W. Bryers; "Utilization of Petroleum Coke and Petroleum Coke/Coal Blends as a Means of Steam Raising"; Fuel Proc. Tech.; Vol. 44, 1995; pp 121-141.

[15] J. Nakano, S. Sridhar, T. Moss, J. Bennett, and K.S. Kwong; "Crystallization of Synthetic Coal-Petcoke Slag Mixtures Simulating Those Encountered in Entrained Bed Slagging Gasifiers"; Energy and Fuels; 23, 2009; pp 4723-4733.

[16] J. Nakano, S. Sridhar, J. Bennett, K.S. Kwong, and T. Moss; "Interactions of Refractory Materials With Molten Gasifier Slags"; Inter. J. of Hydrogen Energy ; 2010; doi:10.1016/j.ijhydene.2010.04.117.

[17] J.P. Bennett, K.S. Kwong, H. Thomas, and R. Krabbe; "Post-Mortem Analysis of High Chrome Oxide Refractory Liners from Slagging Gasifiers Using Analytical Tools/Thermodynamics"; presented and published in CD proceedings of UNITECR 2009; Oct, 2009; Salvador, Brazil; 4 pp.

CHARACTERIZATION OF ELECTROCHEMICAL CYCLING INDUCED GRAPHITE ELECTRODE DAMAGE IN LITHIUM-ION CELLS

Sandeep Bhattacharya, A. Reza Riahi, Ahmet T. Alpas
Department of Mechanical, Automotive and Materials Engineering
University of Windsor, 401 Sunset Avenue,
Windsor, Ontario, Canada N9B 3P4

ABSTRACT
Electrodes made of high purity graphite were subjected to electrochemical tests to investigate graphite surface degradation using in-situ observations by means of a digital microscope with large depth of field. The fraction of graphite removed from the surface increased with decreasing voltage scan rate. Cross-sectional SEM investigations of samples prepared using focused ion-beam (FIB) milling indicated that SEI deposits were formed preferentially inside the cavities on graphite surface. Subsurface damage features of graphite included partial delamination of the graphite layers and fragmentation of graphite into fine particles.

INTRODUCTION
The high reversible lithium capacity (372 mAh/g) as well as good electronic and ionic conductivity of graphite has made it the preferred anode material for rechargeable lithium-ion batteries[1]. However, an impediment of the use of graphite as negative electrode in lithium-ion cells is the loss of capacity during the first charge/discharge cycle. Yazami et al.[2] showed that lithium can be reversibly inserted into graphite at room temperature[3]. It has been shown that the spacing between the graphite layers increased from 3.35 A° to 3.37 A° upon lithium intercalation[4-6]. This expansion in graphite d-spacing may lead to build-up of internal mechanical stresses within graphite layers. Grunes et al.[7] studied distortions of the graphite band structure upon intercalation of alkali metals using electron energy-loss spectroscopy. Christensen and Newman's mathematical model[8] predicted that carbon particle surface was most likely to fracture during de-insertion of lithium-ion. Deshpande et al.[9] developed a model for computing diffusion-induced stresses in electrodes and concluded that the driving force (and probability) of crack formation during de-lithiation and lithiation were different. Results of an experimental study conducted using acoustic emission technique by Ohzuku et al.[10] and Sawai et al.[11] indicated that the rate of fracture of the anode material (MnO_2) was high during the initial cycle. Using in-situ optical microscopy, Bhattacharya et al.[12] observed occurrence of extensive graphite particle removal during the de-lithiation stage and showed that the potential difference exerted on graphite negative electrode acted as the driving force for graphite surface damage. The damage was arrested when the operating voltage reached a critical value. The magnitude of damage in lithiation stage was much less compared to de-lithiation, which was attributed to the formation of a solid electrolyte layer (SEI).
The structure of the SEI layer is not well known but it was suggested that the SEI film has a mosaic-type structure consisting of polyheteromicrophases[13,14], where inorganic compounds may predominate in the inner part and organic compounds are usually formed in the outer part[15]. Raman microscopy and AFM studies by Kostecki et al.[16] monitored structural changes of graphite electrodes in lithium-ion cells cycled at room temperature. Degradation events observed at the electrode surface were more pronounced than those observed in the bulk.

Degradation promoted anode surface reactivity with the electrolyte and inorganic decomposition products were found to deposit preferentially at the degraded regions. Zhang et al.[17] suggested that damage in graphite may be due to gas production from electrolyte solution decomposition and subsequent releasing. As proposed by Aurbach et al.[18,19], ethylene gas could be formed by the decomposition of ethylene carbonate in the electrolyte. Therefore, it is plausible that electrolyte solutions that diffused into cracks inside graphite may decompose and release gases, which might cause an increase of internal pressure causing damage. Focused ion beam microscopy (FIB) has been employed to investigate the formation and evolution of the SEI film on natural graphite spheres[17], which revealed that initially the thickness range of SEI was 450 - 980 nm when it has a rough, non-uniform morphology, and attained a thickness of approximately 1,600 nm after 24 cycles.

This article focuses on characterization of damage that occurred on graphite electrode surfaces when they were electrochemically cycled between different values of applied potential differences. *In-situ* optical microscopy was used to sequentially observe graphite surface morphology changes. The role of voltage scan rate on the degradation process has been discussed. . Cross-sectional microstructures, obtained using FIB milling, were used to investigate the subsurface damage features.

EXPERIMENTAL

Cylindrical working (negative) electrodes, of 5 mm diameter, made from high purity graphite containing traces of Al, Fe, Mg and Si, were subjected to cyclic voltammetry tests. The electrode surface was prepared using standard metallographic techniques to a final polish using 1200 grit SiC papers and cleaned ultrasonically (to remove the fine graphite particles from the natural cavities present in graphite) using acetone and dried at 100°C for 1 hour. The density of this particular grade of graphite was calculated as 1.6 g/cc.

For in-situ observation of surface degradation of graphite electrodes, a cell with an optical grade quartz glass cover at the top was constructed using chemically modified PTFE having exceptional chemical and permeation resistance. The graphite electrode was placed at the center of the cell. A 99.99% pure lithium wire (containing ~0.01% Na as impurity and a resistivity of 9.446 μΩ-cm at 20°C) of diameter 3.2 mm, was used as the counter electrode. The reference electrode was also made of a 3.2 mm diameter 99.99% pure lithium wire. The electrolyte solution was formed using a 1 M LiClO$_4$ (electrolyte salt) in a 1:1 (by vol.) mixture of ethylene carbonate and 1,2-dimethoxy ethane.

The cell components were assembled, and the electrolyte was injected in the cell, inside a glovebox operating in an Argon atmosphere. Cyclic voltammetry tests were performed using a Solartron Modulab Potentiostat/Galvanostat between a base voltage of 0.00 V and different peak voltages of 0.50 V, 1.00 V, 2.00 V and 3.00 V, at scan rates of 500 μV/s, 2 mV/s, 5 mV/s and 10 mV/s. The open-circuit potential of the completed lithium-ion cells was 3.20 V.

Surface morphologies of cycled graphite electrodes were observed using an optical surface profilometer WYKO NT 1100 that utilizes white-light interferometry and a FEI Quanta 200 FEG Scanning Electron Microscope (SEM). Optical profilometer data was collected over a cumulative area of 225×296 μm^2 using an average of 5 selected regions on graphite surface. Damaged subsurface microstructures were obtained from cross-sections using a Zeiss NVISION 40 dual beam SEM/FIB. Cross-sectional trenches were milled at specific locations on cycled graphite at an accelerating voltage of 30 kV. Final milling was conducted at low ion-beam currents ranging from 80 - 40 pA.

RESULTS AND DISCUSSIONS

In-situ Optical Microscopy

Cyclic voltammetry experiments were conducted by increasing the voltage from the base value of 0.00 V to a peak voltage of 3.00 V (V = 0.00 V → 3.00 V). The initial high slope observed in the current density curve in **Fig.1 (a)** is an indication of the presence of lithium ions that were already diffused inside the graphite from the electrolytic solution before the start of the experiment. The fall of the current density values after 1.53 V marked the completion of the de-lithiation of the graphite electrode[20].

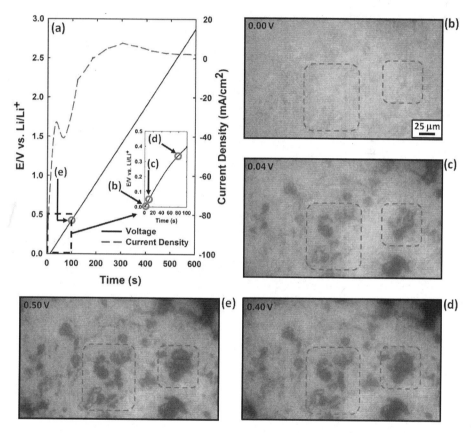

Fig.1. (a) Graphite electrode cycled from V = 0.00 V → 1.00 V at 5.00 mV/s scan rate. *In-situ* digital microscopy images at 1000× show graphite surface at **(b)** initial condition (at 0.00 V), and changes after **(c)** 8 s, **(d)** 80 s and **(e)** 100 s. Typical damage features are highlighted in **Figs. 1(c) - (e)**.

Thus, graphite damage during de-lithiation stage was studied during the voltage increase from the base value, and high magnification optical images (at 1000×) were recorded simultaneously at regular time intervals as shown in **Figs.1 (b) - (e)**. The emergence of regions with dark appearance observed in the sequence of the optical microscope images in **Figs.1 (b) - (e)** was due to loss of material in the from graphite electrode surface resulting in formation of cavities of sizes between 5 - 50 μm.

The surface cavity formation process continued with increasing the voltage; both the number and the size of the cavities increased rapidly until the potential reached 0.40 V [**Fig.1 (d)**]. Interestingly, almost all damage on graphite surface occurred within a short period of time (< 100 s) after the start of the experiment. Then, damage continued at lesser rate until the operating voltage reached 0.50 V [**Fig.1 (e)**], after which almost no new cavity formation was observed up to the peak voltage of 3.00 V.

Cyclic voltammetry tests conducted at different linear scan rates of 500 μV/s, 2.00 mV/s and 10.00 mV/s between 0.00 and 3.00 V (V = 0.00 V → 3.00 V) were used for quantification of graphite surface damage. High magnification optical images (at 1000×), were used to calculate the percentage area fractions of the damaged regions on graphite surface. The area fractions of the surface cavities were monitored for the initial 200 s of each experiment. It was observed that at low scan rates (e.g. 500 μV/s), cavity formation process was severe and the area fraction increased abruptly to more than 30% within less than 25 s. For a high scan rate of 10.00 mV/s, only 15% area fraction was removed within the first 25 s. Thus, damage to graphite occurred during the initial stage of the experiment, and depended on the voltage scan rate. After this high damage stage, almost no additional damage occurred regardless of the scan rate.

Surface Morphology of Cycled Graphite Electrodes

The average surface roughness, R_a, of a mechanically polished graphite surface was 118 ± 5 nm, according to optical surface profilometer (WYKO NT1100) data. A 3-dimensional optical profilometry image of polished graphite electrode is shown in **Fig.2 (a)**. Here, the initial roughness observed was due to the presence of natural cavities and defects present on the graphite surface as shown in the SEM image of the polished surface [**Fig.2 (b)**].

Surface roughness values were recorded at the end of the 10th cycle when the tests were run from 0 to peak voltages of 0.50 V, 1.00 V, 2.00 V and 3.00 V, using a scan rate of 2.00 mV/s and plotted against the peak voltages in **Fig.2 (c)**. It was observed that the surface morphology was non-uniform (R_a = 337 ± 48 nm) at the end of test run to a peak voltage value of 0.50 V. On the other hand, the electrode surface attained a more uniform roughness of R_a = 221 ± 12 nm when the tests were run to a high peak voltage of 3.00 V, exposing graphite to a large potential difference; this could be attributed to the formation of a surface deposits (SEI) that filled the cavities that were formed. Evidence for deposition of SEI inside the surface cavities is examined in the next section.

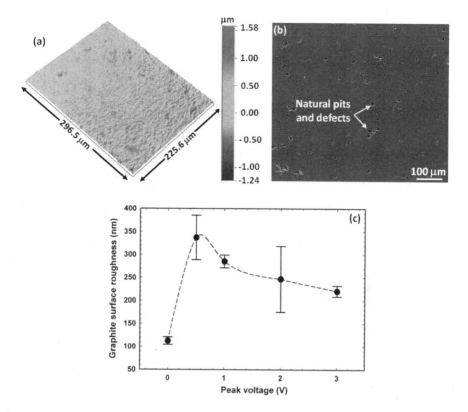

Fig.2. (a) 3-Dimensional surface profilometry image of an untreated graphite electrode. **(b)** SE-SEM image of a polished graphite surface at 400× showing the presence of natural cavities in graphite. **(c)** Variations of the average surface roughness with respect to different peak voltages. Roughness data was measured from 5 selected regions of (225×296) μm^2 area of cycled graphite electrodes.

Surface and Subsurface Damage in Graphite

Fig.3 (a) shows a typical example of a cavity observed on graphite electrode surface after cyclic voltammetry experiments were run between 0.00 and 1.00 V for 1 cycle. These cavities appeared as the dark areas in **Fig.1**. SEI was found to form inside the cavity.

A high magnification SE-SEM image [**Fig.3 (b)**] reveals the spherical morphology of the SEI particles having diameters in the range of 250 - 750 nm. The EDS spectrum of an SEI is shown in **Fig.3 (c)**, indicating the presence of oxygen and chlorine, which was present as a component of the electrolyte ($LiClO_4$).

Fig.4 (a) shows a FIB-milled cross-sectional view of the subsurface region of a graphite electrode cycled for V = 0.00 V → 1.00 V at a linear scan rate of 500.0 μV/s. The light grey regions observed at the subsurface region indicate typical features of graphite damage inflicted during cycling. The damaged region extended to a depth of about 25 μm. The region below the damaged graphite consisted of undisturbed graphite noted by its dark grey contrast.

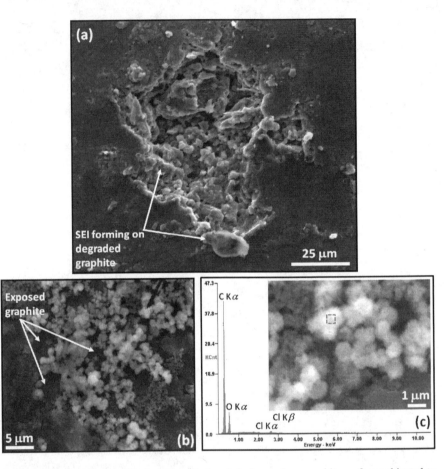

Fig.3. (a) SE-SEM image of a degraded site (cavity) observed on graphite surface subjected to cyclic voltammetry experiments cycled between 0.00 V and 1.00 V at a linear voltage rate of 500 μV/s. **(b)** SEI particles with spherical morphology and **(c)** EDS spectrum taken from the spot on an SEI particle (indicated in the inset).

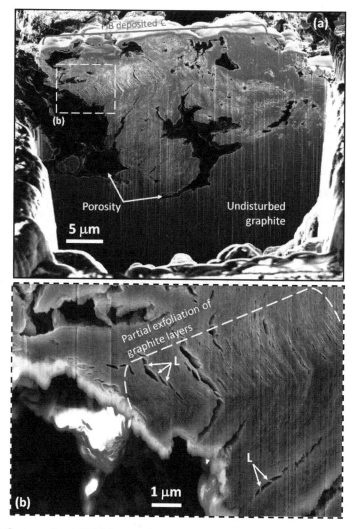

Fig.4. (a) Cross-sectional SEM image of a FIB-milled region on graphite electrode surface, cycled for V = 0.00 V → 1.00 V at a linear scan rate of 500.0 µV/s. **(b)** High magnification image of the region indicated as (b) in **(a)**, showing the existence of delaminated graphite layers. Delaminated graphite layers are seen to be held by ligaments (marked by L) connecting the layers at certain locations.

A particular region of interest is marked in **Figs.4 (a)** as **(b)** and shown in **Fig.4 (b)** at a higher magnification. In this region, damage to graphite layers manifested itself by formation of interlayer cracks giving evidence for local delamination of graphite layers during electrochemical tests. The separation width between the delaminated graphite layers was highly variable. In this particular location, the typical separation distance was between 150 - 200 nm. Many delaminated graphite layers were still held by ligaments (marked by L) connecting the layers at certain locations as shown in **Fig.4 (b)**. Interestingly, the delamination process has shown some ductility rather than the typical cleavage-type fracture[21] normally observed at the basal planes of graphite. The partially exfoliated graphite layers were randomly distributed in the damaged region.

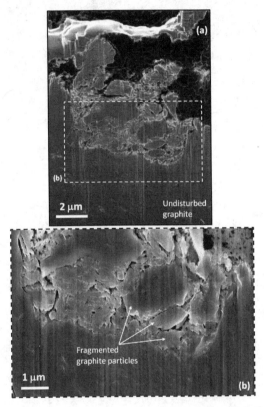

Fig.5. (a) SEM image of a subsurface region milled using FIB applied to a graphite electrode surface, cycled from V = 0.0 V → 1.0 V at a linear scan rate of 500.0 μV/s. Image shows formation of fragmented graphite particles due to electrochemical cycling. **(b)** High magnification image of the regions indicated as (b) in **(a)**, showing the existence of fragmented graphite particles.

Apart from partial exfoliation of the graphite layers, damage also occurred during cycling by fragmentation of graphite particles [**Figs.5 (a)** and **(b)**] into smaller ones, with sizes ranging widely from 2 μm to 250 nm. Fragmentation could cause formation of cavities observed using optical microscopy in **Fig.1**. The exact of mechanism of graphite fragmentation is not known. However, two mechanisms could be suggested; coalescence of delaminated graphite layers and formation of instable cracks and fracture as a result of evolution of gases formed by the electrolytic reduction[17]. It has been suggested that the cause of this damage could be the formation of ethylene gas[18,19] generated due to decomposition of ethylene carbonate. This may have caused an increase in the internal pressure causing almost instantaneous damage as observed in **Fig.5 (b)**. On the other hand, damage in the form of delamination of graphite layers is probably more gradual. Coalescence of such delaminated layers could also lead to graphite fragmentation. More detailed analytical TEM studies are needed to determine the micro-mechanisms of graphite damage.

CONCLUSIONS

Damage on the graphite surface due to electrochemical cycling was investigated using a combination of in-situ optical microscopy, profilometry, and cross-sectional FIB, SEM, EDS techniques. The main conclusions can be summarized as follows:

1. Evidence of graphite damage manifested itself by loss of graphite particles, causing formation of cavities of 5 - 50 μm width by 25 μm depth on the graphite surface. The extent of damage was high at lower voltage scan rates and eventually subsided as the operating voltage reached the peak voltage.
2. Evidence of graphite micro-damage was found inside the cavities formed on the graphite surface as SEI deposits were preferentially nucleated inside such degraded sites. Cross-sectional SEM results indicated presence of partially delaminated graphite layers and fragmentation of surface graphite layer into graphite particles. Graphite fragmentation is thought to cause cavity formation either by fracture due to the internal pressure increase as a result of electrolytic decomposition induced gas formation, or by coalescence of delaminated graphite layers.

ACKNOWLEDGEMENTS

The authors would like to thank the Natural Sciences and Engineering Research Council of Canada (NSERC) for providing financial support. Canadian Center of Electron Microscopy, operated by the Brockhouse Institute for Materials Research at McMaster University, Hamilton, Canada, is gratefully acknowledged for their assistance with work carried out using the Zeiss NVISION 40 SEM/FIB.

REFERENCES

[1]J.R. Dahn, A.K. Sleigh, H. Shi, B.M. Way, W.J. Weydanz, J.N. Reimers, Q. Zhong, and U. VonSacken, Lithium Batteries: New Materials, Developments and Prospecives, 1 (1994).
[2]R. Yazami, and P. Touzain, A reversible graphite-lithium negative electrode for electrochemical generators, *J.Power Sources*, **9**, 365-71 (1983).
[3]T. Ohzuku, Y. Iwakoshi, and K. Sawai, Formation of lithium-graphite intercalation compounds in nonaqueous electrolytes and their application as a negative electrode for a lithium ion (shuttlecock) cell, *J. Electrochem. Soc.*, **140**, 2490-97 (1993).

[4]M. S. Dresselhaus, and G. Dresselhaus, Intercalation compounds of graphite, *Adv. Phys.*, **51**, 1-186 (2002).

[5]S. A. Solin, The Nature and Structural Properties of Graphite Intercalation Compounds, *Adv. Chem. Phys.*, **49**, 455-532 (1982).

[6]T. Enoki, M. Suzuki, and M. Endo, Graphite Intercalation Compounds and Applications, Oxford University Press, New York (2003).

[7]L. A. Grunes, I. P. Gates, J. J. Ritsko, E. J. Mele, D. P. DiVincenzo, M. E. Preil, and J. E. Fische, Valence and core electronic excitations in LiC_6, *Phys. Rev. B*, **28**, 6681-86 (1983).

[8]J. Christensen, and J. Newman, Stress generation and fracture in lithium insertion materials, *J. Solid State Electrochem.*, **10**, 293-319 (2006).

[9]R. Deshpande, Y. Qi, and Y-T Cheng, Effects of concentration-dependent elastic modulus on diffusion-induced stresses for battery applications, *J. Electrochem. Soc.*, **157**, A967-71 (2010).

[10]T. Ohzuku, H. Tomura, and K. Sawai, Monitoring of particle fracture by acoustic emission during charge and discharge of Li/MnO_2 cells, *J. Electrochem. Soc.*, **144**, 3496-500 (1997).

[11]K. Sawai, K. Yoshikawa, H. Tomura, and T. Ohzuku, Progress in Batteries & Battery Material, **17**, 201 (1998).

[12]S. Bhattacharya, A.R. Riahi, and A.T. Alpas, In-situ observations of lithiation/de-lithiation induced graphite damage during electrochemical cycling, *Scripta Mater.* doi: 10.1016/j.scriptamat.2010.09.035, (2010).

[13]E. Peled, D. Golodnitsky, and G. Ardel, Advanced model for solid electrolyte interphase electrodes in liquid and polymer electrolytes, *J.Electrochem.Soc.*, **144**, L208-10 (1997).

[14]E. Peled, D. Golodnitsky, G. Ardel, and V. Eshkenazy, The sei model-application to lithium-polymer electrolyte batteries, *Electrochim.Acta*, **40**, 2197-204 (1995).

[15]J.O. Besenhard, Handbook of Battery Materials, (1999).

[16]R. Kostecki, and F. McLarnon, Microprobe study of the effect of Li intercalation on the structure of graphite, *J.Power Sources*, **119-121**, 550-54 (2003).

[17]H. Zhang, F. Li, C. Liu, J. Tan, and H. Cheng, New insight into the solid electrolyte interphase with use of a focused ion beam, *J. Phys. Chem. B*, **109**, 22205-11 (2005).

[18]D. Aurbach, B. Markovsky, A. Shechter, Y. Ein-Eli, and H. Cohen, A comparative study of synthetic graphite and Li electrodes in electrolyte solutions based on ethylene carbonate-Dimethyl carbonate mixtures, *J.Electrochem.Soc.*, **143**, 3809-20 (1996).

[19]D. Aurbach, B. Markovsky, I. Weissman, E. Levi, and Y. Ein-Eli, On the correlation between surface chemistry and performance of graphite negative electrodes for Li ion batteries, *Electrochim. Acta*, **45**, 67-86 (1999).

[20]Mark W. Verbrugge, and Brian J. Koch, Lithium intercalation of carbon-fiber microelectrodes, *J.Electrochem.Soc.*, **143**, 24-31(1996).

[21]I. Ioka, S. Yoda, and T. Konishi, Behavior of acoustic emission caused by microfracture in polycrystalline graphites, *Carbon*, **286**, 879-85 (1990).

TITANIUM-DIOXIDE-COATED SILICA MICROSPHERES FOR HIGH-EFFICIENCY DYE-SENSITIZED SOLAR CELL

Devender[1]* and Ajay Dangi[2]

[1]Department of Ceramic Engineering, Institute of Technology, Banaras Hindu University, Varanasi 221005, India

[2]Department of Mechanical Engineering, Institute of Technology, Banaras Hindu University, Varanasi 221005, India

ABSTRACT

In order to improve the efficiency of a dye-sensitized solar cell (DSSC) by using silica microspheres coated with titanium dioxide (TiO_2), silica microspheres were prepared by the sol-gel method using tetraethyl orthosilicate as the starting material and hydrochloric acid to catalyze the reaction. Silica microspheres thus formed were coated with TiO_2. The presence of TiO_2 improved the photocatalytic properties of silica, which is responsible for a significant increase in the decomposition of the dye molecules. Enhancement of this decomposition was ascertained by measuring the difference in initial and final concentrations of dye in a solution containing the coated microspheres and the die. The use of the microspheres can boost the efficiency of DSSCs.

INTRODUCTION

The development of renewable, clean energy sources is imperative in the 21st century. Among many available sources, solar energy is the most appropriate one, being completely renewable, safe and clean. Sunlight can be directly converted to electricity by solar cells. Although single- or poly-crystalline silicon solar cells are the most economical ones for now, difficulties in their cost reduction still stand in the way of popularization especially in developing countries[1].

Dye-sensitized solar cell (DSSCs) developed by Gratzel and coworkers[2, 3] have achieved efficiencies a little over 7% on small areas[4]. Because of their expected inexpensive production[3, 5], extensive efforts are being increasingly undertaken in scaling up their manufacturing process. Possible applications—such as transparent windows and stand-alone power generator[6]—and the ease of manufacturing has attracted commercial interest in DSSCs[3]. Commercial manufacture of

DSSCs is still infeasible and more research is required on several aspects such as technical performance and manufacturability, costs, design, and long-term stability. During last few years, all of these parameters have been investigated extensively. Since then, despite several reports on the performance of small-size DSSCs have been published, but their efficiency has not been enhanced significantly[7]. Currently, the highest efficiency of a small-area DSSC using an ionic liquid electrolyte is 7.3%[8] and 11.1% for acetonitrile-based DSSC[9], still about two times lower than of silicon solar cells[1].

The redox system in a DSSC consists of two electrode coated with a transparent conductive oxide (TCO) glass, with one electrode having a nanoporous TiO_2 coating which is sensitized with dyes for harvesting visible light, and the other electrode coated with platinum[2, 3, 5, 7]. The iodide/tri-iodide (I^-/I_3^-) redox couple acts as the electrolyte in the gap between the two. Under illumination, the dye molecules are excited and initial charge separation occurs by injection of an electron from the dye into the conduction band of the TiO_2. This electron travels to the external load by way of the nanoporous TiO_2 and the TCO glass and returns via the I^-/I_3^- electrolyte. Nano-structured TiO_2 has been used extensively in DSSCs as it provides a high surface area and an efficient route for the movement of electrons to the external load[7].

In this paper, improved photocatalytic property of microspheres of silica coated with TiO_2 has been demonstrated.

EXPERIMENTAL

The first step is the preparation of silica microspheres using the sol-gel method, followed by soaking of microspheres in TiO_2 solution leading to the formation of TiO_2-coated silica microspheres (TCSMs).

Preparation of sol: Tetraethyl orthosilicate ($Si(OC_2H_5)_4$, Fluka Chemicals, Switzerland) and hydrochloric acid (HCl) were mixed to start the hydrolysis and subsequent polymerization of alkoxides. HCl catalyzes the hydrolysis of $Si(OC_2H_5)_4$ which otherwise is very slow because of low susceptibility to attack due to four oxygen atoms ($p\Pi$-$d\Pi$ bonding). In the presence of an acid, an alkoxide group is protonated, as a result of which electron density is reduced in silicon, which is then susceptible to attack by water. A heavy phase—the sol—formed at the bottom of the beaker and was separated using a pipette.

Preparation of TCSMs: Neutral surfactant, tween 80 and CCl$_4$ (Fluka Chemicals, Switzerland) were mixed using a magnetic stirrer and the heavy phase obtained was added to this solution. The partially hydrolyzed products of Si(OC$_2$H$_5$)$_4$ reacted with either unhydrolyzed or partially hydrolyzed Si(OC$_2$H$_5$)$_4$, forming -Si-O-Si- linkages through a condensation reaction. The hydrolysis and the subsequent polymerization reaction generated polymeric particles. The particles grew in time due to the continuous reaction, thereby leading to an increase in the viscosity of the sol. Methanol (S-D Fine-Chem, Mumbai) was added in the solution which after some time led to the formation of spheres. These spheres settled down and were separated from the solution. The microspheres obtained were cleaned with distilled water, dried in an oven, heated for sintering in a furnace (Nisabjee, Kolkata), and then soaked in a 2% solution of TiO$_2$. After soaking, the TCSMs were removed and then completely dried.

RESULTS AND DISCUSSION

The TCSMs were kept on a glass slide and viewed under an optical microscope. An optical image is presented in Fig. 1. The diameters of the TCSMs range from 60 to 250 μm.

The infrared transmission spectrum (Cary 50 scan, Varian) of the uncoated silica microspheres, taken after drying, is shown in Figure 2. The minimums in the spectrum correspond to the different bonds present in the network. Thus, the minimum at 1644 cm^{-1} is due to vibration of physically adsorbed water in the gel network, the minimum at 1086 cm^{-1} is due to vibration of the Si-O-Si bond, and the one at 954 cm^{-1} is due to vibration of the Si-O-H bond. Typical vibration band corresponding to the Si-O-Si bond clearly suggests the formation of Si-O-Si network.

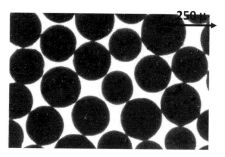

Figure1.Photograph of TCSMs

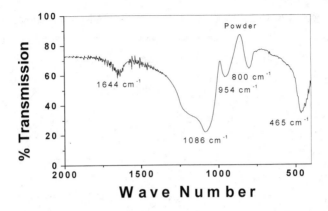

Figure2. IR transmission spectrum of a powder of uncoated silica microspheres.

Now, to check the effect on the photocatalytic behavior of the silica microspheres after coating with TiO$_2$, R6G dye was used. This dye decomposes in the presence of ultraviolet (UV) light and silica microspheres. It also decomposes under the photocatalytic action of TiO$_2$ and UV exposure. The decomposition of dye by the uncoated microspheres was first carried out, and then by the TCSMs.

The concentration of the R6G dye was calculated by UV-Visible transmission spectra of different solutions shown in Fig. 3. Peaks of absorption spectra of different solutions were compared with calibrated values i.e. absorption peaks of different known concentrations of R6G dye solution. The data on concentrations are presented in Table 1. From Table 1, a marked increase in UV decomposition of R6G dye by TCSMs is evident.

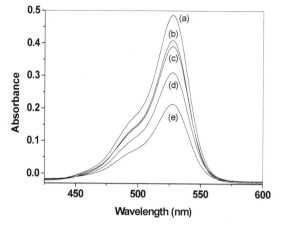

a) 6.3×10^{-6} M -R6G aq. Solution

b) Dye-365 nm UV-2h

c) Dye Microsphere without UV-2h

d) Dye-Microsphere- UV-365nm-2h

e) Dye-TiO$_2$ coated microsphere-
    UV 365 nm-2h

Figure 3. Variation of absorbance peaks with wavelength for different solutions.

Table1. Decomposition of dye observed indifferent environments.

Sample	Initial concentration of dye	Final concentration (after decomposition)	Amount of dye decomposed
TCSMs in presence of UV light	6.3×10^{-6}M	2.83×10^{-6}M	3.47×10^{-6}M
Uncoated silica microspheres in presence of UV light	6.3×10^{-6}M	3.99×10^{-6}M	2.31×10^{-6}M
Silica microspheres without UV light	6.3×10^{-6}M	4.9×10^{-6}M	1.4×10^{-6}M
UV light only	6.3×10^{-6}M	5.16×10^{-6}M	1.14×10^{-6}M
R6G dye alone	6.3×10^{-6}M	6.3×10^{-6}M	0

CONCLUDING REMARKS

Our conclusion from the foregoing experiments is that TiO_2-coated silica microspheres show superior photo catalytic properties than uncoated silica microspheres and their use in dye-sensitized solar cells can improve the efficiency of DSSCs. These coated microspheres can also have wide applications in areas like water purification and waste water treatment.

REFERENCES

1. R. Hezel, Progress in manufacturable high-efficiency silicon solar cells, Advances in Solid State Physics, 44, 1205-1206, (2004).

2. B. O'Regan, M. Gratzel, A low-cost high-efficiency solar cell based on dye-sensitized colloidal TiO_2 films, Nature, 353, 737–740, (1991).

3. M. Gratzel, Perspectives for dye-sensitized nanocrystalline solar cells, Prog. Photovoltaics, 8, 171–185, (2000).

4. M.A. Green, K. Emery, K. Bucher, D.L. King, S. Igari, Solar cell efficiency tables (version 11), Prog. Photovoltaics, 6, 35–42, (1998).

5. H. Otaka, M. Kira, K. Yano, S. Ito, H. Mitekura, T. Kawata, F.Matsui, Multi-colored dye-sensitized solar cells, J. Photochem. Photobiol.A—Chem., 164, 67–73, (2004).

6. G. Phani, M.P.J. Bertoz, J. Hopkins, I.L. Skryabin, G.E. Tulloch, Electric window modules based on nanocrystalline titania solar cells, Australia and New Zealand Solar Energy Society, Proceedings of Solar, (1997)

7. S. Dai, J.Weng, Y. Sui, C. Shi, Y. Huang, S. Chen, X. Pan, X. Fang, L. Hu, F. Kong, K. Wang, Dye-sensitized solar cells, from cell to module, Sol. Energy Mater. Sol. Cells, 84, 125–133, (2004).

8. D. Kuang, P. Wang, S. Ito, et al., Stable mesoscopic dye-sensitized solar cells based on tetracyanoborate ionic liquid electrolyte, J. Am. Chem. Soc., 128, 7732–7733, (2006).

9. M. Green, K. Emery, Y. Hishikawa, W. Warta, Solar cell efficiency tables, Prog. Photovolt: Res. Appl., 17, 85–94, (2009).

EFFECT OF TITANIUM AND IRON ADDITIONS ON THE TRANSPORT PROPERTIES OF MANGANESE COBALT SPINEL OXIDE

Jeffrey W. Fergus, Kangli Wang and Yingjia Liu
Auburn University
Auburn, AL, USA

ABSTRACT

The transport properties of manganese cobalt oxides are important for their application as coatings for interconnect alloys in solid oxide fuel cells (SOFCs). In particular, the transport of chromium should be low to reduce chromium volatilization, while the transport of electrons should be high to minimize ohmic resistance. These transport properties can be changed by doping with ions of different valences and/or site (octahedral/tetrahedral) preferences. In this work, manganese cobalt spinel oxides have been doped with titanium or iron to improve coating performance. In particular, the effects of the dopants on the reaction with chromia and on the electrical conductivity will be presented.

INTRODUCTION

The fuel flexibility resulting from the high operating temperatures of solid oxide fuel cells (SOFCs) expands the range of potential applications[1]. The high operating temperatures, however, also create materials challenges[2] by increasing the rates of reaction between components and with the surrounding gases. One such materials challenge is the development of interconnects[3-5]. Chromia-forming alloys exhibit reasonable oxidation resistance in intermediate temperature SOFC conditions, but chromia is oxidized in air to form volatile species that deposit on the cathode and degrade cell performance. The amount of volatilization can be reduced by alloying, such as with the addition of manganese that leads to the formation of a spinel phase on the scale with reduced chromia activity (relative to Cr_2O_3) and thus reduced chromium volatilization. However, long operation times require further reduction in the amount of chromium volatilization, which can be accomplished with ceramic coatings[6]. One promising coating material system is the spinel $(Mn,Co)_3O_4$[7-10], which has been shown to reduce chromium volatilization[11,12]. The purpose of this work is to evaluate the effect of iron and titanium additions on the reactivity and properties of $(Mn,Co)_3O_4$.

EXPERIMENTAL

Pellets of $Mn_{1.5}Co_{1.5}O_4$, $MnCo_2O_4$, $MnCo_{1.66}Fe_{0.34}O_4$ and $MnCo_{1.66}Ti_{0.34}O_4$ were prepared by solid-state synthesis from MnO, Co_3O_4, Fe_3O_4 and TiO_2 powders. The powders were mixed with deionized water, ball milled for 48 hours, dried overnight, pressed into pellets and then sintered in air at 1200°C for 24 hours.

The interaction of the prepared $(Mn,Co,Ti,Fe)_3O_4$ samples with Cr_2O_3 was evaluated by placing pellets of the materials in contact and heating in air at 900°C. The amount of reaction was determined using scanning electron microscopy (SEM) with energy dispersive x-ray spectroscopy (EDS).

The conductivities of coating materials were determined using 4-point dc conductivity measurements on rectangular bars (13-14 mm x 5-6 mm x 1.5-2 mm) that were sectioned from sintered pellets (25 mm diameter, 1.5-2 thickness). The coefficients of thermal expansion were measured using dilatometry between room temperature and 1000°C at a heating rate of 10°C/min with die-pressed and sintered cylindrical samples (12 mm diameter, 14-15 mm height).

RESULTS AND DISCUSSION

The effect of iron and titanium doping on the reaction between the manganese cobalt oxide spinel and chromia is illustrated in Figure 1, which shows the chromium concentration gradient in the spinel oxide after being in contact with Cr_2O_3 in air at 900°C for 144 hours. As previously reported[13] for the reaction between $(Mn,Co)_3O_4$ andh Cr_2O_3, a two-layer reaction zone is formed at the interface. The layer in contact with Cr_2O_3 (to the right in Figure 1) has relatively constant chromium concentration at a value of approximately two chromium ions per formula unit (i.e. $(Mn,Co)Cr_2O_4$). Neutron diffraction results[14] indicate that chromium occupies the octahedrals sites, of which there are two per formula unit in the spinel structure. This constant concentration, along with platinum marker experiments and the reaction layer morphology, indicate that this layer grows by the diffusion of manganese and cobalt from the manganese cobalt oxide spinel to the Cr_2O_3. At the same time, chromium diffuses into the spinel which results in the decreasing chromium concentration towards the left in Figure 1.

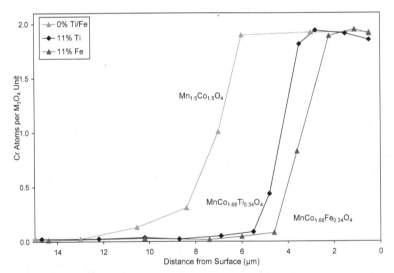

Figure 1. Chromium contents in undoped, Ti-doped and Fe-doped $(Mn,Co)_3O_4$ after being in contact with Cr_2O_3 for 144 hours in air at 900°C.

The results in Figure 1 show that the extent of reaction decreases significantly with the addition or iron or titanium. The thickness of the layer with constant chromium content in the reaction layer in $Mn_{1.5}Co_{1.5}O_4$ is 1.5-2.5 times larger than that in the iron- or titanium-doped samples. In addition, the chromium gradient in the other reaction layer is steeper with iron- and titanium-doping. Thus, iron and titanium are beneficial for reducing the reaction layer thickness, but their effect on other coating properties is also important.

One critical property is electrical conductivity, since the cell current passes through the interconnect coating during fuel cell operation. The measured conductivities for $(Mn,Co)_3O_4$ are shown in Figure 2 along with results from the literature[15,16]. The composition with a Mn:Co ratio of 1:1 ($Mn_{1.5}Co_{1.5}O_4$) has the highest conductivity. These conductivities are compared with the conductivities of the iron- and titanium-doped compositions in Figure 3. The conductivity of $MnCo_{1.66}Fe_{0.34}O_4$ is about the same as that of $Mn_{1.5}Co_{1.5}O_4$. Although the conductivity of $MnCo_{1.66}Ti_{0.34}O_4$ is lower than that of $Mn_{1.5}Co_{1.5}O_4$, it is higher than that of $MnCo_2O_4$.

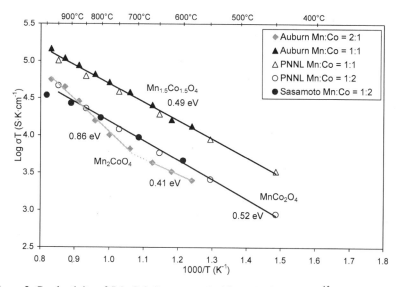

Figure 2. Conductivity of $(Mn,Co)_3O_4$ compared with results from PNNL[15] and Sasamoto et al.[16].

Another important coating property is the coefficient of thermal expansion (CTE), since differences in thermal expansion can generate stresses during thermal cycling, which can lead to cracking. Table I shows that the CTEs of $MnCo_{1.66}Fe_{0.34}O_4$ and $MnCo_{1.66}Fe_{0.34}O_4$ are a little higher than that of $Mn_{1.5}Co_{1.5}O_4$, but they are lower than that of $MnCo_2O_4$, which suggests that the dopants may partially offset the increase in CTE associated with increasing cobalt content.

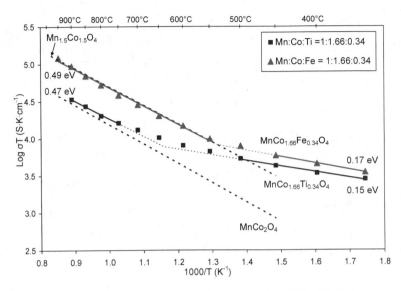

Figure 3. Effect of iron and titanium on the conductivity of (Mn,Co)$_3$O$_4$ in air.

Table I. Coefficients of Thermal Expansion (CTE) for 25-1000°C.

Composition	CTE (°C^{-1})
Mn$_{1.5}$Co$_{1.5}$O$_4$	10.6 x 10^{-6}
MnCo$_2$O$_4$	14.1 x 10^{-6}
MnCo$_{1.66}$Fe$_{0.34}$O$_4$	11.2 x 10^{-6}
MnCo$_{1.66}$Ti$_{0.34}$O$_4$	12.0 x 10^{-6}

CONCLUSIONS

The addition of iron or titanium to (Mn,Co)$_3$O$_4$ decreases the rate of reaction with Cr$_2$O$_3$, with little or no decrease in the electrical conductivity. Of the two compositions, the iron-doped composition, MnCo$_{1.66}$Fe$_{0.34}$O$_4$, exhibits less reaction, a higher conductivity and a better CTE match with other SOFC components. However, additional work is needed to characterize the properties of the reaction layers formed to identify any potential detrimental effects on coating performance.

ACKNOWLEDGMENTS

Financial support from the Department of Energy through the Building EPSCoR-State/National Laboratory Partnerships Program (Timothy Fitzsimmons, Program Officer) and the National Energy Technology Laboratory (NETL) (Briggs White, Program Officer) is gratefully acknowledged.

REFERENCES

[1] E.D. Wachsman and S.C. Singhal, Solid Oxide Fuel Cell Commmercialization, Research and Challenges, *ECS Interface*, **18**[3], 38-43 (2009).

[2] A.J. Jacobson, "Materials for Solid Oxide Fuel Cells, *Chem. Mater.*, **22**, 660-674 (2010).

[3] W.J. Quadakkers, J. Piron-Abellan, V. Shemet and L. Singheiser, Metallic Interconnectors for Solid Oxide Fuel Cells – A Review, *Materials at High Temperatures*, **20**, 115-127 (2003)..

[4] J.W. Fergus, Metallic Interconnects for Solid Oxide Fuel Cells, *Mater. Sci. Eng. A*, **397**, 271-283 (2005).

[5] Z. Yang, K.S. Weil, D.M. Paxton and J.W. Stevenson, Selection and Evaluation of Heat-Resistant Alloys for SOFC Interconnect Applications, *J. Electrochem. Soc.*, **150**, A1188-A1201 (2003).

[6] N. Shaigan, W. Qu, D.G. Ivey and W. Chen, A Review of Recent Progress in Coatings, Surface Modifications and Alloy Developments for Solid Oxide Fuel Cell Ferritic Stainless Steel Interconnects, *J. Power Sources*, **195**, 1529-1542 (2010).

[7] Y. Larring and T. Norby, Spinel and Perovskite Functional Layers between Plansee Metallic Interconnect (Cr-5 wt% Fe-1 wt% Y_2O_3) and Ceramic $(La_{0.85}Sr_{0.15})_{0.91}MnO_3$ Cathode Materials for Solid Oxide Fuel Cells, *J. Electrochem. Soc.* **147**, 3251-3256, (2000).

[8] Z. Yang, G.-G. Xia, G.D. Maupin and J.W. Stevenson, Conductive Protection Layers on Oxidation Resistance Alloys for SOFC Interconnect Applications, *Surf. Coating Tech.*, **201**, 4476-4483 (2006).

[9] M.R. Bateni, P. Wei, X. Deng and A. Petric, Spinel Coatings for UNS 430 Stainless Steel Interconnects, *Surf. Coating Tech.* **201**, 4677-4684 (2007).

[10] M.J. Garcia-Vargas, M. Zahid, F. Tietz and A. Aslanides, Use of SOFC Metallic Interconnect Coated with Spinel Protective Layers using the APS Technology, *ECS Trans.* **7**, 2399-2405 (2007).

[11] H. Kurokawa, C.P. Jacobson, L.C. DeJonghe and S.J. Visco, Chromium Vaporization of Bare and of Coated Iron-Chromium Alloys at 1073 K, *Solid State Ionics*, **178**, 287-296 (2007).

[12] C. Collins, J. Lucas, T.L. Buchanan, M. Kopczyk, A. Kayani, P.E. Gannon, M.C. Deibert, R.J. Smith, D.-S. Choi and V.I. Gorokhovsky, Chromium Volatility of Coated and Uncoated Steel Interconnects for SOFCs, *Surf. Coating Tech.* **201**, 4467-4470 (2006).

[13] J. Fergus, K. Wang and Y. Liu, Interactions between $(Mn,Co)_3O_4$ SOFC Interconnect Coating Materials and Chromia, in *Supplemental Proceedings: Volume 2: Materials Characterization, Computation, Modeling and Energy*, 473-480 (The Minerals, Metals & Materials Society, 2010).

[14] A. Purwanto, A. Fajar, H. Mugirahardjo, J.W. Fergus and K. Wang, Cation Distribution in Spinel $(Mn,Co,Cr)_3O_4$ at Room Temperature, *J. Appl. Crystallography*, **43**, 394-400 (2010).

[15] Z. Yang, G.-G. Xia, H.-H. Li and J.W. Stevenson, $(Mn,Co)_3O_4$ Spinel Coatings on Ferritic Stainless Steels for SOFC Interconnect Applications, *Int. J. Hydrogen Energy*, **32**, 3648-3654 (2007).

[16] T. Sasamoto, N. Sumi, A. Shimaji, O. Yamamoto and Y. Abe, High Temperature Electrical Properties of the Spinel-Type Oxide $MnCo_2O_4$-$NiMn_2O_4$, *J. Mater. Sci. Soc. Jpn.*, **33**, 32-37 (1996).

EFFECT OF HYDROGEN ON BENDING FATIGUE LIFE FOR MATERIALS USED IN
HYDROGEN CONTAINMENT SYSTEMS

Patrick Ferro
Gonzaga University
Spokane, WA, USA

ABSTRACT
    Hydrogen containment systems must be engineered for multiple charge/discharge cycles.
Challenges from possible hydrogen embrittlement, pressure, temperature, cost and other
parameters affect storage design decisions. The scope of the present work is a study of recent
hydrogen storage designs and the effect of hydrogen exposure on bending fatigue life. A brief
review of recent designs in portable hydrogen storage preliminarily indicates that aluminum
alloys are widely used for container materials. Other frequently used materials for containers
and other hydrogen hardware include austenitic stainlesses, brass and high molecular weight
polymers. Preliminary bending fatigue results for cold-rolled steel and 316 stainless steel are
presented.

INTRODUCTION
    The effects of hydrogen on metals has been well researched. The effects on steel and
aluminum alloys, for example, have been well-reported and summarized[1-4]. The effect of
hydrogen on metals that may affect design is embrittlement from exposure to hydrogen. The
effects of hydrogen exposure time, pressure, alloy system, alloy content and other variables are
factors which may affect the embrittlement phenomenon.
    A recent review by San Marchi and Somerday[5] summarizes previous mechanical
properties effects of high pressure hydrogen on structural metals. The data given by San Marchi
and Somerday provides critical fracture toughness and yield strength data for commonly used
engineering materials including steel, precipitation hardened stainless steel and aluminum. The
work underscores the importance of alloy chemistry on design, among other parameters. The
design of many components in the present day hydrogen infrastructure is shown to rely on
commonly-available engineering materials. Materials for hydrogen gas service include
austenitic stainless steel (including Types 304, 316 and A-286), carbon steels (including ASTM
A106 Grade B, for example), low alloy (pressure vessel) steels (containing chromium and
molybdenum) and aluminum alloys[5].
    In recent years, the need for development of alternatives to petroleum-based energy has
led to advances in hydrogen-based technologies. In addition to hydrogen-powered vehicles,
there have been advances in the realm of other products and technologies that utilize hydrogen.
Most of these products require a steady supply of hydrogen. The growing infrastructure
requirements indicate a need for more materials design data. Sandia National Laboratories is the
lead DOE laboratory for the study of gaseous hydrogen embrittlement of structural metals, and
offers a comprehensive resource for engineering data particularly materials and mechanical
properties data[6].
    Aoki et al.[7] have reported the effects of fatigue on austenitic stainless in hydrogen.
Aoki reports a retardation in the onset of fatigue failure in hydrogen, possibly due to the absence
of water and oxygen in the controlled hydrogen environment. The data also suggests that once a
crack is initiated, crack propagation is accelerated in a hydrogen environment[7].

Tartaglia, Lazzari, Hui and Hayreynen[8] provide mechanical properties results from hydrogen embrittlement experiments on 4340 steel. The work of Tartaglia et al. shows the effect of austempering, compared to quenched and tempered, on the resistance of 4340 to hydrogen embrittlement.

The present work seeks to illuminate some aspects of the present hydrogen enbrittlement design challenge. The present work includes a bending fatigue investigation on possible candidate materials for hydrogen storage containers.

BACKGROUND

At least three different companies are currently offering commercially-available portable hydrogen storage containers. Sizes for portable containers that are readily available range in size from 20 to 900 standard liters of hydrogen (sl). Companies that currently offer portable containers include Ovonic Hydrogen (Rochester Hills MI), Solid H (Bailey CO) and MH (through fuelcellstore.com)[9].

Ovonics offers four sizes of canisters, ranging from 75 sl to 900 sl. Ovonic's standard metal hydride alloy allows for charging a container in air in under two hours with 1.7 MPa (250 psig) applied hydrogen. Besides the standard alloy, two other lower plateau pressure alloys are offered by Ovonics, for international shipment or for lower pressure design. Canisters that are ordered with the two non-standard alloys will not store the same amount of hydrogen as the canisters with the standard alloy. The material from which the Ovonic canisters are made is DOT 3AL, a specified variant of 6061-T6.

Solid H offers several canisters in sizes ranging between 30 and 370 sl. The metal hydride alloy type used by Solid H are AB type alloys, in three different chemistries. Solid H Alloys L, M and H have respective room temperature plateau pressures of 2 atm, 5 atm and 10 atm. The designs offered by Solid H appear to be two distinct geometric container shapes. The CL series of Solid H's designs appear to be offered in only one size (370 sl) and the BL series appears to include four sizes. The container material for the BL series is stainless steel, and is unreported for the CL series container. The CL series container appears to be spun aluminum or spun stainless.

MH offers a small cylinder that stores 20 sl of hydrogen. The mass of hydride is 0.13 kg, and the alloy chemistry is 65.6 Ni, 33.7 Al and small percentages of Rh and Mn. The reported container material is an aluminum alloy.

Table I summarizes the reported information on several of the commercially-available hydrogen storage containers. The cost on a price per reversible stored gram of hydrogen basis is reported based on prices reported in July 2010[9].

Table I.   Summary of commercially available metal hydride hydrogen storage canisters[9]

Product name	Container material	Reported metal hydride data	Reversible hydrogen volume	Cost/gram of stored H_2 ($/g)
MH20	aluminum alloy	65.6 Ni, 33.7 Rh, 33.7 Al, 0.18 Mn; 0.13 kg	20 sl (1.8g)	80.6
Solid H CL370	not reported	AB type (Alloy L, 2 atm at RT; Alloy M, 5 atm at RT; Alloy H, 10 atm at RT)	370 sl	15.2
Ovonics (several)	DOT 3AL	charges in <2h at 250 psig H_2 or two other alloys with lower RT pressures for international shipping	75 sl, 100 sl, 275 sl, 900 sl	11.9 - 56.3
Solid H BL series (several)	stainless steel	AB type, Alloy A or international shipping, use Alloy L, 2 atm at RT; Alloy M, 5 atm at RT; Alloy H, 10 atm at RT	30 sl, 60 sl, 120 sl, 250 sl	102 - 407

Fig. 1 shows a graph of the commercially available canisters ranked on a milligrams of reversible hydrogen stored per dollar cost. The parameter ranges from less than 10 to more than 80 mg/USD.

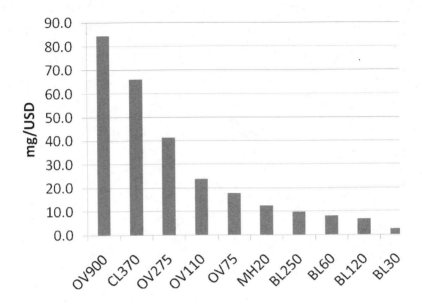

Fig. 1. Histogram of commercially-available portable hydrogen storage container ranked as a function of mass of reversibly stored hydrogen per dollar cost of product using July 2010 prices. The parameter, mg/USD ranges from less than ten to more than eighty.

Besides the commercial market for portable hydrogen storage containers, another industry that is developing hydrogen storage containers is the hydrogen-powered vehicle market. For example, Honda, Hyundai, Nissan and Daimler have appeared to have increased hydrogen vehicle development efforts in 2009. By contrast, Ford, Nissan-Renault and GM have appeared to remained at the same level or possibly decreased hydrogen-powered vehicle development and testing in 2009[10]. Possible reasons for slowing hydrogen-powered vehicle development research for somer manufacturers may be due to increased attention on battery-powered electric vehicle research.

For the companies that have pursued continual hydrogen development, fuel cell powered vehicles (as opposed to hydrogen internal combustion vehicles) appears to be the dominant development effort based on numbers of development vehicles. Honda introduced the fuel-cell powered FCX in 1999 and has since offered a second generation FCX Clarity. The FCX Clarity has been available to the public in the US and Japan since 2008[10]. Hyundai has announced the intention of producing 500 fuel-cell powered vehicles in 2010[10]. Hyundai's eventual production goal is 100,000 fuel-cell vehicles by 2013.

The Honda FCX Clarity uses a high-pressure compressed gas tank to store hydrogen on-board[11]. The maximum pressure in the tank is 34.5 MPa (5000 psig) and the tank volume is 173 liters (45.7 gallons). The Honda website does not specify the materials or construction of the hydrogen tank[11].

Quantum Technologies appears to be one of the leaders in carbon-fiber reinforced composite tanks for storing hydrogen. The basic construction of these tanks, which are capable of a maximum pressure of 70 MPa (10,000 psig), is a internal liner made of high molecular weight polymer surrounded by a carbon fiber-epoxy composite and an outer shell for impact damage protection. The internal liner is designed for hydrogen impermeability, and the carbon fiber-epoxy outer casing is designed to bear the stresses caused by the high internal gas pressure[12].

Besides using high molecular weight liners, other designs have used fiber-wound composite tank with an aluminum liner and composite tanks with austenitic stainless steel liners. For example, Dynetek Industries (Calgary, AB) offers several composite hydrogen storage tanks, including a thin-wall aluminum liner with a carbon fiber overwrap[13].

Based on reviewing the current information regarding commercially-available portable hydrogen storage containers and potential means of storing hydrogen for vehicle on-board applications, it appears that the metallic material choices for working with hydrogen are variations of aluminum alloys and stainless steel (e.g. 304, 316, 321). Brass is used in compression fittings and other hardware associated with hydrogen storage and fuel cells. For example, pressure relief devices on the top of portable canisters may be made from brass and brass nuts are used for connecting high-pressure lines to gas bottles. The metallic materials that will be initially studied in a bending fatigue study will include aluminum, austenitic stainless steel and low carbon steel. Other materials may be studied as the investigation progresses.

PROCEDURE

The bending fatigue tests that were performed were using a VSS-40H bending fatigue tester from Fatigue Dynamics (Walled Lake MI) with a shutdown controller. Bending fatigue experiments were performed by mounting a specimen into the VSS-40H tester. Cycling was started with the machine in 'bypass' mode. The shutdown threshhold percentage was set while the specimen was initially cycling at a preset cycling rate. The shutdown threshhold percentages that were used for the experiments were 80% or 90% of the full initial applied bending stress. After the threshhold percentage (as a function of the full gain) was set on the shutdown controller, the gain was changed to 100% while the specimen was still in its full stress during initial cycling. The last step in starting an experiment was to switch the shutdown controller from 'bypass' to 'run' mode. The number of cycles experienced by a specimen during the threshhold setup in bypass mode was 200 cycles or less.

Once the machine was switched to 'run' mode, the bending fatigue tester cycled the specimen until failure. The cycling frequency ranged between 200 and 400 cycles per minute. Failure was defined when the stress necessary to sustain bending deformation dropped to a percentage of the initial applied bending load.

After the specimen failed, it was removed and the location of the failure was identified by measuring the distance from the start of the specimen radius to the middle of the failure location on specimen.

Prior to installing a specimen into the machine, the displacement of the rotating arm on the fatigue tester was set for a given displacement of the fatigue specimen. The rotating arm was adjusted to a position on the machine spindle and noted. The position of the rotating arm on the machine spindle affects the cantilevered displacement of the fatigue specimen. Fig. 2 shows a photo of the VSS-40H fatigue tester, with a view of the rotating arm.

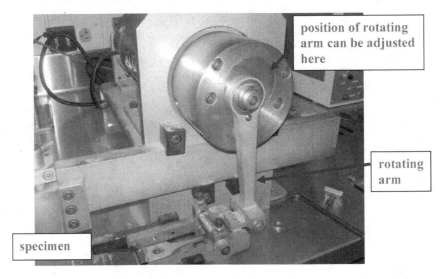

Fig. 2. Photo of rotating arm on VSS-40H fatigue tester. A specimen is shown in the bottom of the photo prior to installation.

The specimens were manufactured according to the dimensions shown in Fig. 3. Specimens were CNC plasma cut from 30.5 cm x 30.5 cm sheets, and tested in the as-plasma-cut condition.

The materials that were studied included A366 cold-rolled steel and Type 316 stainless steel. The sheets of material were obtained from Coeur d'Alene Metals, Spokane WA. Specimen thicknesses were 0.6 mm (0.025 inches) for 316 stainless and 0.9 mm (0.035 inches) for cold rolled steel. Fig. 4 shows a photo of a sheet of specimens cut from a sheet of A366 cold-rolled steel using CNC plasma cutting.

Hydrogen was applied by generating hydrogen gas at nominally one atmosphere pressure from reacting zinc pellets with 1M hydrochloric acid. The hydrogen gas from the reaction was sent into a flask with bending fatigue specimens at room temperature. Specimens stayed in the hydrogen flask for one to two hours. After hydrogen exposure, samples were removed from the flask and installed into the bending fatigue tester.

Bending stress was estimtaed by using elasticity equations for a cantilevered beam. The specimen was approximated as a beam of rectangular cross-section with an average width of 13.5 mm. The bending stress was calculated at a distance of 1 cm from the clamped end of the specimen, which was the average location of failure.

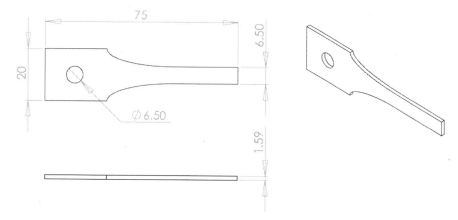

Fig. 3. Sketch of the bending fatigue specimen design that was used for the experiments. Dimensions shown are in mm. The thickness of the specimen shown above is 1.59 mm (0.064") but thicknesses for the specimens used in the present experiments were 0.6 mm for 316 stainless and 0.9 mm for cold-rolled steel.

Fig. 4. Sheet of specimens in the as-plasma cut condition. The specimens shown are from a sheet of A366 cold rolled steel, 0.9 mm thick. Hardness of specimens in the as-plasma cut condition was 38 HRC.

RESULTS

Table II shows data for fatigue tests that were performed on A366 cold rolled steel. The steel sheet thickness from which specimens were CNC plasma cut was 0.9 mm (0.035 inch). The hardness of the sheet in the as-plasma cut condition was 38 HRC. The last column in the table gives the failure location from the clamped end of the beam. The average failure location was 10 mm.

Table II.   Fatigue results, cold rolled steel, 0.9 mm, 38 HRC

Maximum bending stress	Rolling orientation	Cycling frequency	Cycles to failure	Failure location
380 MPa	parallel	250 cycles/min	69100	12.4 mm
380 MPa	parallel	400 cycles/min	13500	11.7 mm
380 MPa	parallel	400 cycles/min	10400	13.3 mm
380 MPa	parallel	400 cycles/min	16400	8.3 mm
380 MPa	transverse	250 cycles/min	47400	8.1 mm
380 MPa	transverse	250 cycles/min	59300	11.3 mm
380 MPa	transverse	400 cycles/min	14000	12.9 mm
380 MPa	transverse	400 cycles/min	21300	10.5 mm
380 MPa	transverse	400 cycles/min	12600	11 mm
470 MPa	parallel	400 cycles/min	5700	8.9 mm
470 MPa	parallel	300 cycles/min	10000	9.7 mm
470 MPa	transverse	400 cycles/min	12700	9.0 mm
470 MPa	transverse	400 cycles/min	9000	11.7 mm
470 MPa	transverse	300 cycles/min	11000	9.2 mm

Fig. 5 shows cycles to failure as a function of cycling frequency for A366 cold rolled steel specimens. The data indicates that cycling frequency appears to have a possible effect on cycles to failure. For the range of frequencies tested, it appears that higher cycling frequency is associated with shorter fatigue life. Rolling orientation did not appear to affect fatigue life in the preliminary data.

Fig. 6 shows cycles to failure for samples of 316 stainless steel. Samples were 0.6 mm thick, and were tested in the as-plasma cut condition. The data shows the fatigue results for samples, some of which were exposed to 1 atm hydrogen for one hour immediately prior to testing. Based on the preliminary results shown in fig. 6, an effect on fatigue life from one hour exposure to 1 atm hydrogen on 316 stainless fatigue samples is not apparent.

The data presented in Table II and in figs. 5 and 6 are examples of the type of preliminary work that is underway for additonal materials including steels, aluminum alloys and austenitic stainless steel. For each of the materials, fatigue tests will be performed in the unexposed and exposed to hydrogen conditions. Additional work includes increasing the hydrogen gas exposure times and pressures, as well as increasing the temperature of the samples during hydrogen exposure.

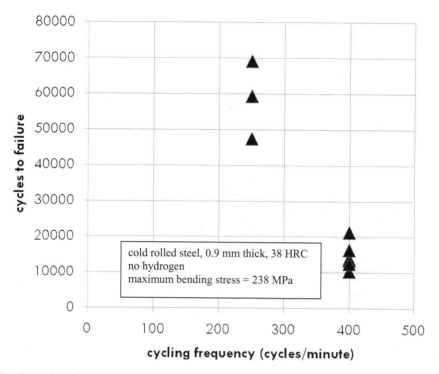

Fig. 5. Cycles to failure as a function of cycling frequency for cold rolled mild steel (0.9 mm 385 HRC) at an alternating bending stress of 238 MPa.

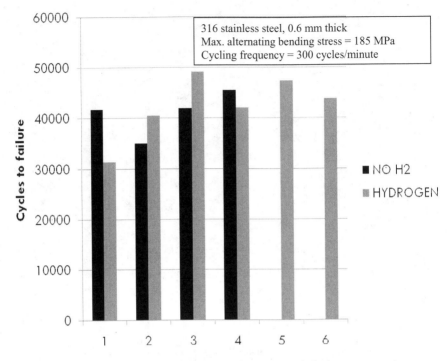

Fig. 6. Cycles to failure for 316 stainless steel for samples some of which were exposed to hydrogen before fatigue cycling. The maximum alternating bending stress during testing was 185 MPa. The dark bars represent data for samples which were not exposed to hydrogen. The gray bars represent data for samples whihc were exposed to 1 atm hydrogen for a minimum of one hour immediately prior to installation in the bending fatigue tester.

DISCUSSION

One of the challenges to the investigation will be in the application of hydrogen to the specimens. In the present work, exposure to hydrogen for specimens was achieved by placing specimens in a flask and passing hydrogen gas from the reaction of zinc and hydrochloric acid through the flask. The pressure of hydrogen is nominally only one atmosphere with this method, and the exposure takes place before the fatigue cycling. Future work may include an investigation into application of higher pressure hydrogen (e.g. 1.5 MPa), on specimens before the fatigue experiments are started, and at higher temperatures. Also in the design stage is a method for performing the fatigue cycling in a hydrogen atmosphere. For the latter experimental apparatus, hydrogen pressures of up to 0.3 MPa may be employed.

By exposing samples to hydrogen prior to testing, the effects of internal hydrogen embrittlement (IHE) are being investigated, rather than hydrogen exposure embrittlement (HEE). Precharging the specimens may be a valid method for studying the effects of hydrogen in austenitic stainlesses, for example, since the solubility of hydrogen in austenite is relatively high and because the diffusivity of hydrogen in austenite is relatively low[4,5]. The precharging effect may be may be presumed to have a possible embrittling effect if the time between hydrogen exposure and fatigue testing is relatively short. In the present investigation, the precharge time, temperature and pressure are all relatively low (1 h at 1 atm, at room temperature). Increasing the precharge time, temperature or pressure will saturate the austenitic microstructure with more hydrogen, and possibly show an effect on properties from hydrogen. For comparison, other investigators perform precharging at temperatures and pressures that range up to 300°C and 135 MPa respectively[5].

Studying the effect of precharging on cold-rolled steel, for example, may not yield show an effect on properties from hydrogen, since the diffusivity of hydrogen in ferritics is relatively high. The hydrogen may outgas from the precharged steel specimen during cycling and thus yield difficult-to-interpret results. However, if the time from hydrogen expsosure to testing is kept short and if the cycling experimentation is also relatively short (by using high fatigue stresses, for example) it may be possible to discern an effect.

One of the current experimental challenges is to study the effects of hydrogen environment embrittlement. Building a fatigue tester with an in situ hydrogen capability will require considerations for hydrogen ventilation and safety, among other concerns.

Another approach that may be investigated in future fatigue work may involve alternating precharging and cycling until failure. For example, a specimen may be precharged and then cycled for 1000 cycles and then back to precharging, etc., until failure occurs. This approach may simulate a real world hydrogen exposure and mechanical cycling process.

Besides bending fatigue, other mechanical properties tests that will be investigated for the types of materials discussed includes tensile fatigue, tensile testing and load relaxation testing. Each of the experiments will be performed on specimens in the unexposed and exposed to hydrogen conditions.

CONCLUSIONS

Austenitic stainlesses and aluminum alloys are frequently specified for hydrogen storage container materials. Additionally, steels and brass components are often exposed to hydrogen in the hydrogen infrastructure. The present work is focused on developing bending fatigue test methods and other mechanical properties tests to study the effect of hydrogen exposure on common structural materials. Preliminary results indicate possible effects due to bending fatigue cycling rate. Rolling orientation for the mild steel sheets has not shown to be a factor in fatigue cycling for the range of fatigue tests that have been performed as of this writing. Hydrogen exposure effects on fatigue properties will continue to be investigated in future work.

ACKNOWLEDGMENTS
The author acknowledges John Cadwell, Larry Shockey, Rob Hardie and Patrick Nowacki of the Gonzaga University Engineering Department for their help with instrumentation, equipment assistance and materials preparation.

REFERENCES

1.   R. Gibala, R.F. Hehemann, eds., Hydrogen Embrittlement and Stress Corrosion Cracking, Americal Society for Metals (1984)

2.   R.A. Oriani, Hydrogen Embrittlement of Steels, Am. Rev. Mater. Sci. 3:327-57 (1978)

3.   J.P. Hirth, Effects of Hydrogen on the Properties of Iron and Steel, Met. Trans. A, v.11A June 1980 861-890 (1980)

4.   H.G. Nelson, Hydrogen Embrittlement, Treatise on Materials Sci. and Tech., v. 25, Academic Press, pp. 275-359 (1983)

5.   C. San Marchi, B.P. Somerday, Effects of High-Pressure Gaseous Hydrogen on Structural Metals, 2007-01-0433

6.   LC Sandia National Laboratories, Technical Reference on Hydrogen Compatibility of Materials, http://www.ca.sandia.gov/matlsTechRef/, accessed May 25, 2010

7.   Y. Aoki, K. Kawamoto, Y. Oda, H. Noguchi, K. Higashida, Fatigue Characteristics of a Type 304 Austenitic Stainless Steel in a Hydrogen Gas Environment, Int. J. Fracture (2005) 133:277-288

8.   J.M. Tartaglia, K.A. Lazzari, G.P. Hui, K.L. Hayrynen, A Comparison of Mechanical Properties and Hydrogen Embrittlement Resistance of Austempered vs Quenched and Tempered 4340 Steel, Met. and Mat. Trans. A, v39A, March 2008, pp. 559-576

9.   www.fuelcellstore.com, accessed 26 July 2010

10.   Wikipedia, http://en.wikipedia.org/wiki/Hydrogen_vehicle, accessed May 14, 2010

11.   Honda website, http://corporate.honda.com/environment/fuel_cells.aspx?id=fuel_cells_fcx, accessed May 14, 2010

12.   US DOE website, http://www1.eere.energy.gov/hydrogenandfuelcells/storage/hydrogen_storage.html, accessed May 14, 2010

13.   Dynetek Industries, http://www.dynetek.com/products.php, accessed 31 May 2010

INVESTIGATION OF SECONDARY PHASES FORMATION DUE TO PH₃ INTERACTION WITH SOFC ANODE

Huang Guo, Gulfam Iqbal, and Bruce Kang
Mechanical and Aerospace Engineering Department, West Virginia University

ABSTRACT

Nickel Yttria-Stabilized Zirconia (Ni-YSZ) is a common choice for solid oxide fuel cell (SOFCs) anode material because of its excellent performance in hydrogen and simulated syngas. However trace amount of phosphine (PH_3), commonly present in coal-derived syngas, adversely interacts with the anode material and degrades its electrochemical performance and material properties. This degradation is attributed to the formation of secondary phases such as nickel phosphate, nickel phosphite etc. In this research, the poisoning effects of PH_3 are investigated in hydrogen with and without steam. Post-test analyses of the cell indicate the presence of Ni_3P in both dry and wet hydrogen that does not degrade cell electrochemical performance significantly. Whereas under wet hydrogen and potential bias conditions, nickel phosphate (such as $Ni(PO_3)_2$) is preferred to form at Ni/YSZ interface. It is proposed that the deleterious effect of PH_3 on the Ni-YSZ anode performance is due to the formation of Nickel Phosphate.

INTRODUCTION

Solid Oxide Fuel Cell (SOFC) is an electrochemical device that converts fuel chemical energy into electric energy at high temperature ($>600^\circ C$). Several types of materials have been studied in the literature for the SOFCs anode, Nickel-Yttria Stabilized Zirconia (Ni-YSZ), however, is the most widely used anode material due to its excellent performance in hydrogen and simulated syngas [1-3]. However, its electrochemical performance and materials properties are affected when exposed to phosphine (PH_3) that is commonly present in coal-derived syngas due to the formation of secondary phases [4-6]. Krishnan et al. [6] and Trembly et al. [7, 8] concluded that the phosphorus vapors react with the Ni contained in the anode material and caused the reduction of the SOFC power output. However, the detailed degradation mechanisms are still unclear. Marina [4, 5] and Trembly et al. [7, 8] attributed the loss of cell performance to a series of nickel phosphide phase formation, while Krishnan [6] and Zhi et al. [9] reported the formation of zirconium phosphate. Zhi et al. [9] also reported the formation of nickel phosphate.

In the present work a modified Sagnac interferometry, integrated with infrared thermometry, is utilized for in-situ anode surface temperature measurement as a function of applied current density under different test conditions, while the effects of PH_3 contaminant on the SOFC performance are studied under hydrogen environment with and without steam. The in-situ experimental results and post-test analyses could be used for the validation and development of SOFC electrochemical models, and understanding the anode-contaminant interaction.

EXPERIMENTAL APPROACH

Anode-supported SOFC button cells, manufactured by Materials and Systems Research Inc. (MSRI), were tested in this study. The MSRI cell fabrication procedure and material parameters are described in details by Zhao, F., et al. [10]. Each button cell was about 3.0 cm in diameter with an anode composed of a 1 mm thick Ni-8YSZ support structure and a 25 μm thick interlayer of a highly catalytic Ni-8YSZ mixture. The electrolyte (8YSZ) was 25 μm thick and cathode (2.27 cm^2 active area) was composed of a 25 μm thick $La_{0.8} Sr_{0.2}MnO_3$ (LSM)-8YSZ interlayer and a 25 μm thick current-collection layer of LSM. A gold (Au) current collection mesh was attached to the anode using nickel contact paste. Similarly, a platinum (Pt) mesh was attached to the cathode using Pt paste. Silver current cables and voltage taps were spot welded onto opposite sides of each current collection mesh. The button cell was mounted inside the NexTech Probostat^{TM} button cell test apparatus using AREMCO-516 high temperature cement. Alicat^{TM} mass flow controllers (MFCs) were used to control fuel/air flow rates, pressure and fuel compositions. A temperature-controlled humidifier was used to control the water content of fuel gas supplied to the anode side.

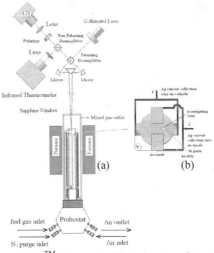

Figure 1. (a) NexTech Probostate^{TM} integrated with Sagnac interferometry and IR thermometer
(b) button cell with open anode surface at the center

A NexTech Probostat^{TM} button cell test apparatus was modified and integrated with a Sagnac interferometric optical setup [12-14] and PhotriX infrared (IR) pyrometer, as shown in Figure 1 (a). The optical setup consisted of a 20mW diode laser (wave length λ = 658 nm, laser spot size diameter = 2 mm), beam splitters, polarizer and beam directing mirrors. PhotriX infrared (IR) pyrometer was integrated to monitor anode surface temperature under operation. With a modified and narrower field of view (FOV), the surface IR emission from the same spot of the anode surface was captured by the PhotriX IR Pyrometer, albeit with only limited spectrum transmission (300 nm to 1650 nm). Concurrently, electrochemical performances were measured using Reference 300 Potentiostat/Galvanostat/ZRA (Gamry Instruments, Warminster, PA). The impedance spectra were collected with AC amplitude of 10 mV at frequencies ranging from 1 MHz to 0.1 Hz. Detailed description of the experimental setup and approach can be found in reference [12]. In order to investigate the impurities effects on the electrode, the anode surface was directly exposed to the injected fuel without any intervening current collector or metal paste as shown in Figure 1(b).

RESULT AND DISCUSSION

PH₃ Effects on Ni-YSZ Anode Materials

It has been reported that the exposure to phosphorous leads to irreversible degradation of the SOFC anode [4-9]. However, there are few studies focused on the relationship among the phosphorus compounds, steam in fuel and potential bias. In this study, the button cells were exposed to ppm level of PH_3 in dry and moist hydrogen under open circuit voltage (OCV) and loading conditions to study PH_3 poisoning effect on SOFC performance at 800°C. We have earlier reported that the poisoning effect of PH_3 was associated with the content of steam in the fuel gas [15]. The cell power density did not show degradation during exposure to PH_3 with dry H_2 until the steam was introduced into the fuel gas. The cell performance degradation was more severe under higher steam content in the PH_3-containing fuel gas.

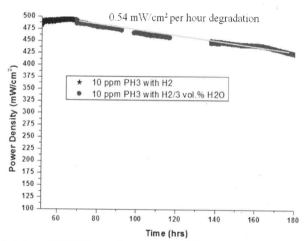

Figure 2. Variation of SOFC Power Density under different testing conditions at i = 0.6 A/cm²

To investigate these phenomena and further understand the secondary phase formation mechanisms, PH_3 exposure tests were carried out under dry and wet hydrogen conditions w/o loading potential bias. The repeated experimental results illustrated that no cell performance degradation occurs during exposure to PH_3 with dry H_2, as shown in Figure 2, while cell degrades once exposed to PH_3 with steam.

Figure 3 reveals the XRD patterns obtained from the Ni-YSZ electrode tested under different conditions. Ni_3P and YSZ were the major phases observed on the electrode after exposure to ppm level of PH_3 under either dry or wet hydrogen condition at open circuit voltage condition (OCV), as shown in the Figure 3 (a) and (b). The corresponding SEM images show that the surface corrosion and Ni agglomeration are more significant under wet hydrogen than under dry condition. Under cell operating condition (e.g. 0.3V bias), the dominant P-based secondary phase was found to be nickel phosphate such as $Ni(PO_3)_2$, at the active Ni/YSZ interface especially in the presence of steam as shown in Figure 3 (c) and (d). From the SEM micrograph shown in Figure 4, it is found that more significant Ni migration and agglomeration occur on the open anode surface which is directly exposed to the injected fuel gas as compared to under the Au mesh (current collector) with Ni paste. No YSZ phase was found on the open anode surface after exposure of PH₃ as shown in Figure 3(d). In this region, nickel reacted to form Ni_3P on the surface which is similar as the reaction between Ni paste (or Ni plate) and PH₃. However, under the Au current collector nickel phosphate formed (Figure 3(e)) instead which is proposed to be more deleterious to the SOFC performance rather than nickel phosphide according to the performance history shown in Figure 2.

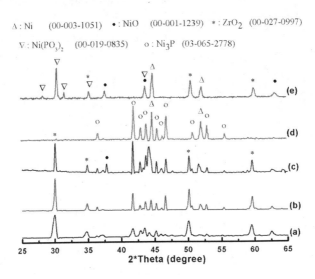

Figure 3. XRD patterns of the Ni-YSZ electrode: (a) exposure to 20 ppm PH₃ in dry H_2 for 4 days at OCV; (b) exposure to 20 ppm PH₃ in H_2/26 vol.% H_2O for 4 days at OCV; (c) exposure to 20 ppm PH₃ in dry H_2 for 4 dyas at 0.3 V of potential bias; (d) the electrode open surface after exposure to 10 ppm PH₃ in H_2/3 vol.% H_2O for 4 days at 0.3 V of potential bias and (e) the electrode under Au mesh after exposure to 10 ppm PH₃ in H_2/3 vol.% H_2O for 4 days at 0.3 V of potential bias

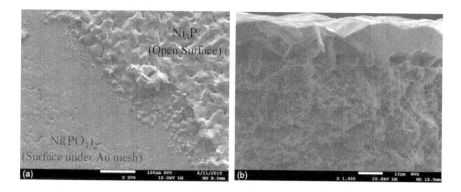

Figure 4. SEM Micrograph of the Ni-YSZ anode surface after exposure to 10 ppm PH₃ in H₂/3 vol.% H₂O for 4 days at 0.3 V of potential bias: (a) surface (b) cross section

To understand the formation of P-based secondary phases, the following reaction may be considered under different testing conditions:

$$yPH_3(g) + xNi(s) \rightarrow Ni_xP_y(s) + \frac{3}{2}yH_2(g) \qquad (1)$$

$$PH_3(g) - xH_2O(g) - HPO_x(g) - (x-1)H_2(g) \qquad (2)$$

$$2HPO_x(g) - Ni(s) - Ni(PO_x)_2(s) \qquad (3)$$

It has been reported that PH₃ can be converted to phosphoric acid or HPO$_x$ (g), in the presence of steam between 700 °C and 900 °C. It can attack the nickel electrode and YSZ electrolyte and affect on the SOFC performance [6]. The potential bias across a fuel cell might accelerate the phosphoric acid absorption on TPB and reaction with Ni/YSZ electrode. Under OCV or without steam, nickel phosphide (Ni₃P) is formed rather than nickel phosphate while the nickel phosphide shows little deleterious effect on Ni/YSZ electrodes.

IR Temperature Measurements

To investigate the polarization effect on electrode reaction, in-situ surface IR temperature measurement was carried out as a function of current densities at 800 °C under different testing conditions. Due to the optical alignment, on-line calibration is needed in order to get accurate temperature measurement. After complete reduction of anode material, IR Pyrometer calibration was carried out with 5 °C steps, from 750 °C to 800 °C. The thermal calibration data was found to be linear that could be extrapolated to transform the IR signal into temperature under cell operating conditions.

As described in reference [15], the final steady state surface temperature is independent of the intermediate current steps and only depends on the final loading current density. The changes of steady-state button cell surface temperature are plotted against loading current densities and the comparison under different cell operating conditions is illustrated in Figure 5. As shown, similar trends of button cell surface temperature variation were obtained under dry hydrogen and $H_2/3$ vol.% H_2O. The major source of temperature rise is entropic heating from fuel gas oxidation. The heat generation within a cell increases with the current due to the fuel cell internal resistance. The relation between surface temperature and current density is found to be nonlinear. It also shows that the correlation coefficient is little higher under moist hydrogen than under dry hydrogen fuel. The experimental data is also used to assess the validity of the numerical model [16].

The surface temperature variations as the fuel cell performance degrading over time due to the exposure of ppm level PH_3 in wet hydrogen environment are shown in Figure 5. Figure 6 and 7 show the corresponding EIS plots, IV-curves and power densities under different conditions. It is illustrated that the change of surface temperature increases with the polarization as cell performance degrades over time.

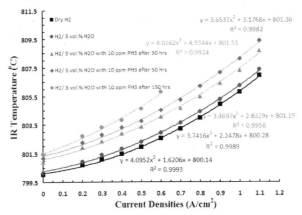

Figure 5. Surface temperature comparisons under different operating conditions

The electrochemical impedance data were obtained at the open circuit condition. Two arcs were observed in the Nyquist plots of impedance (Figure 6). The Z'-intercept of the high frequency arc yielded the ohmic polarization resistance of the cell. It exhibited a low value of about 0.2 $\Omega \cdot cm^2$ and showed little shift during the successive gas exposures. The high frequency arc is attributed to the impedance of the charge transfer processed occurring at the electrolyte/electrode interface and at the Ni/YSZ interface. The low frequency arc is associated with the bulk capacitance and the chemical processes including adsorption, surface migration, bulk migration and other mechanistic steps.

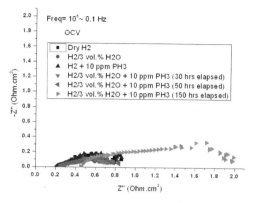

Figure 6. Impedance plots for SOFC operating under the open circuit voltage condition (OCV) at 800 °C

It is illustrated that the polarization resistance (the total arc width) increased with time during exposure to PH₃-containing moist hydrogen fuel gas. When exposed to 10 ppm PH₃ with 3 vol.% H₂O, the polarization resistance increased 158% in 150 hrs. The corresponding ohmic resistance increased 125%, which indicated the cell degradation due PH₃ effects. On the other hand, the cell resistance showed little increase under 10 ppm of PH₃ in dry conditions, as shown in Figure 6. Hence, the poisoning effect of PH₃ was sincerely associated with the content of steam in the fuel gas, which is consistent with the cell degradation indicated in the Figure 7.

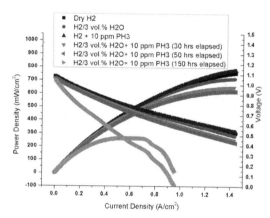

Figure 7. I-V curves of SOFC operating at 800 °C

CONCLUSIONS

PH_3 poisoning tests, conducted with the exposure to ppm level of PH_3 in hydrogen with/out steam, indicates that the Ni-cermet-based SOFCs anode is more susceptible to degradation due to PH_3 in the presence of steam than under dry conditions. The fuel cell performance degradation is attributed to the type of P-based secondary phase. Nickel phosphate, which is preferred to be formed due to the combine effects of PH_3, H_2O and potential bias, has a deleterious effect on SOFC performance. In the meantime, Ni migration and agglomeration occur on the surface directly exposed to the injected P-containing fuel gas, which also results in the formation of nickel phosphide, such as Ni_3P. Besides of the electrochemical performance measurements, a modified Sagnac interferometry method and Infrared thermometry are utilized to monitor the anode surface temperature of a button cell under operating conditions as a function of loading current densities. A non-linear surface temperature increase is observed with polarization curve under dry and moist hydrogen fuel conditions, while the temperature change increases as cell performance degrades by phosphine effect over time.

ACKNOWLEDGEMENT

This work is conducted under US DOE (Department of Energy) EPSCoR Program. It is jointly sponsored by US DOE Office of Basic Energy Sciences, NETL (National Energy Technology Laboratory), WV State EPSCoR Office and the West Virginia University under grant number DE-FG02-06ER46299. Dr. Tim Fitzsimmons is the DOE Technical Monitor. Dr. R. Bajura is the Administrative Manager and Dr. I. Celik is the Technical Manager and the Principal Investigator of this project.

REFERENCES

[1]Radovic, M., and Lara-Curzio, E., 2004, "Mechanical Properties of Tape Cast Nickel-Based Anode Materials for Solid Oxide Fuel Cells before and after Reduction in Hydrogen," Acta Materialia, 52(20), pp. 5747–5756.
[2]Selcuk, A., Merere, G., and Atkinson, A., 2001, "The Influence of Electrodes on the Strength of Planar Zirconia Solid Oxide Fuel Cells," J. Mater. Sci., 36(5), pp. 1173-1182.
[3]N. Q. Ming , T. Takahashi. Science and Technology of the Ceramic Fuel Cells. Amsterdam: Elsevier; 1995.
[4]O. Marina, C. Coyle, EC Thomsen, D. Edwards, G. Coffey, L. Pederson, Degradation mechanisms of SOFC anodes in coal gas containing phosphorus, Solid State Ionics, 181, 430-440 (2010).
[5]O. Marina, L. Pederson, C. Coyle, EC Thomsen, D. Edwards, C Nguyen, G. Coffey, Interactions of Ni/YSZ Anodes with Coal Gas Contaminants, Proceeding of the 9th Annual Solid State Energy Conversion Alliance (SECA) Workshop, Pittsburgh, PA, 2008.
[6]G. Krishnan, P. Jayaweera, J. Perez, Effect of Coal Contaminants on Solid Oxide Fuel System Performance and Service Life, SRI Technical Progress Report, 2006.
[7]J. P. Trembly, R. S. Gemmen, D. J. Bayless, The effect of IGFC warm gas cleanup system conditions on the gas-solid partitioning and form of trace species in coal syngas and their interactions with SOFC anodes, J. Power Sources 163, 986-996 (2007).
[8]J. Trembly, A. Marquez, T. Ohrn, and D. Bayless. Effects of Coal Syngas and H_2S on the Performance of Solid Oxide Fuel Cells: Single-Cell Tests, J. Power Sources., 163, 263–273 (2006).

[9]M.J. Zhi, X.Q. Chen, H. Finklea, I. Celik, N.Q. Wu, Electrochemical and microstructural analysis of nickel-yttria-statilized zirconia electrode operated in phosphorus-containing syngas, J. Power Sources., 183, 485-491(2008).

[10]F. Zhao, A. V. Virkar, Dependence of palarization in anode-supported solid oxide fuel cells on various cell parameters, J. Power Sources., 141, 79–95 (2005).

[11]C.C. Xu, J.W. Zondlo, H. Finklea, O. Demircan, M.Y. Gong, X.B. Liu, The effect of phosphine in syngas on Ni-YSZ anode-supported solid oxide fuel cells, J. Power Sources, 193, 739-746 (2009).

[12]H. Guo, G. Iqbal, and B. Kang, Development of an In-Situ Deformation and Temperature Measurement Technique for a Solid Oxide Fuel Cell Button Cell, Int. J. Appl. Ceram. Technol., 7[1], 55-62 (2010).

[13]G. Sagnac, L'ether lumineux demontre par l'effect du vent relatif dether dans un interferometre en rotation uniforme, Acadamie der Sciences, Paris Comptes Randus, 1913.

[14]B. Kang, and S.M. Anderson, Experimental Investigation of 3-D Crack-Tip Deformation using Combined Moire-Sagnac Interferometry, ASME J. of Pressure Vessel Technology, 123, (2001).

[15]H. Guo, G. Iqbal, and B. Kang, Phosphine Effects on Ni-based Anode Material and Related SOFC Button Cell Performance Investigation, Proceeding of FuelCell2010, ASME 2010 8th International Fuel Cell Science, Engineering and Technology Conference, Brooklyn, New York June 14-16, (2010).

[16]S. Pakalapati, A New Reduced Order Model for Solid Oxide Fuel Cells, Ph.D. thesis, West Virginia University, Morgantown, WV, 2006.

PEN STRUCTURE THERMAL STRESS ANALYSIS FOR PLANAR-SOFC CONFIGURATIONS UNDER PRACTICAL TEMPERATURE FIELD

Gulfam Iqbal, Suryanarayana Raju Pakalapati, Francisco Elizalde-Blancas, Huang Guo, Ismail Celik, Bruce Kang
Mechanical & Aerospace Engineering Department, West Virginia University, Morgantown WV 26506, USA

ABSTRACT

Positive-electrode/electrolyte/negative-electrode (PEN) assembly is the electrochemically active composite structure of a solid oxide fuel cell (SOFC). Thermal stresses in a PEN structure arise due to coefficient of thermal expansion mismatch between its constituent layers, spatial and temporal temperature gradients, and external physical boundary conditions. In this paper, thermal stresses in the anode-supported PEN structure are analyzed for co-, counter-, and cross-flow configurations of planar-SOFCs. The temperature fields are obtained from an in-house developed multi-physics solver: DREAM-SOFC. In the anode layer, the peak maximum principal stresses (MPS) occur at the anode/electrolyte interface in the high temperature region for all configurations. Whereas cathode and electrolyte layers experience highest MPS near the air inlet region in co- and counter-flow configurations. The stress discontinuities across the interfaces are due to the different material properties. Under similar operating conditions, cross-flow configuration results the highest MPS in the anode layer at the conjunction of fuel inlet and air outlet corner.

INTRODUCTION

Solid Oxide Fuel Cell (SOFC) is a high temperature fuel cell (>600°C) that converts fuel chemical energy into electrical energy and heat by means of electrochemical reactions. This technology presents many advantages over conventional power generation systems including high efficiency, fuel flexibility, quiet operation, and low levels of NO_x and SO_x emissions. SOFCs are proven to be technically feasible for industrial stationary and automotive auxiliary power sources. Further research, however, is required to make them reliable and economically practical against the conventional power sources.

The maximum power density that can be achieved using Ni-YSZ anode fueled with hydrogen at 800°C is less than 1W/cm² [1]. In order to obtain desired power output, multiple cells are combined to form a stack. Since SOFC entirely consists of solid-state components, a wide variety of stacking arrangements are possible including planar, tubular and monolithic designs. Planar design, however, has drawn much attention than other stacking options [2, 3] because it is simpler to fabricate. Moreover planar design offers higher power density compared to tubular design attributed to the shorter electric current path which reduces resistant losses as compared to tubular design. In addition to electrochemical performance, structural integrity of the SOFC components is essential for successful long-term operation. Thin layers of the PEN structure are inherently susceptible to mechanical failure when subjected to moderate stresses [3,4]. Thermal stresses in the PEN assembly may arise from coefficient of thermal expansion mismatch between its constituent layers, spatial and temporal temperature gradients, and external physical boundary conditions [4-6]. These stresses can cause delamination and microcraking in the layers of the PEN and degrade its electrochemical performance and structural integrity. Therefore a comprehensive thermal stress analysis of the PEN assembly is essential under

61

practical operating conditions to prevent its structural failure. However, very few studies in the literature have focued on the thermal stress analysis of the SOFC and most of studies have considered PEN as a single structure under uniform temperature distribution. Yakabe et al. [3] performed SOFC electrochemical analysis and determined stress distribution in the electrolyte and interconnect from simulated temperature profiles. Lin et al. [7] predicted the thermal stress distribution in a cross-flow planar-SOFC stake by using finite element analysis (FEA) during various stages of operation by considering the PEN as a unified structure. Anandakumar et al. [8] estimated thermal stresses in the PEN structure under spatially uniform temperature distribution. They determined the thermal stresses in the PEN structure for a uniform increase in temperature of 800°C and 600°C in a 1mm x 0.66mm x 1mm model.

The objective of this paper is to determine the thermal stress distribution in the anode-supported PEN assembly of a planar-SOFC by incorporating realistic temperature field. 10cm x 1mm x10cm Co-, counter- and cross-flow configurations are analyzed under similar operating conditions for comparison. Temperature fields are obtained from an integrated thermo-electrochemical model DREAM-SOFC [9] and thermal stresses are determined by incorporating the temperature field into FEA software ABAQUS^{TM} (ABAQUS Inc., Providence, RI, USA) [10] with the help of a user defined subroutine.

FEA MODEL AND MATERIAL PROPERTIES

A schematic view of a co-flow planar-SOFC is shown in Figure 1(a). The cell active area is 10cm x 10cm and it contains 18 channels for fuel and air flow each. Further details about the cell geometry can be obtained from ref [9]. Figure 1 (b) shows the PEN assembly and its dimension. In this model a typical anode-supported PEN assembly is considered which composed of 780 µm nickel-yttria stabilized zirconia (Ni-YSZ) anode, 20 µm thick YSZ electrolyte, and 200 µm thick Sr-doped LaMnO$_3$ (LSM).

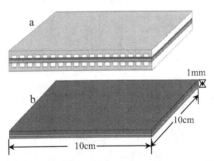

Figure 1. (a) Schematic view of a co-flow planar-SOFC with fuel channels (b) PEN assembly with anode (█████), electrolyte (█████) and cathode (█████) layers

Figure 2 represents a finite element model of one half of the co- or counter-flow PEN assembly. Due to symmetry, only one half of the co- or counter-flow cell is analyzed in the FEA, whereas in the case of cross-flow configuration complete PEN assembly is analyzed. 8-node brick elements are used to model the PEN structure as shown in Figure 2. The applied physical boundary conditions will only prevent rigid body motion of the PEN structure and that it will not bear any external mechanical load. The right side face is fixed in its perpendicular direction in order to apply the symmetry boundary conditions. The effects of sealing and PEN interaction are not considered in these simulations.

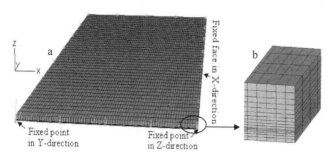

Figure 2. (a) FEA model of half of the co-flow/counter-flow PEN structure and applied physical boundary conditions (b) zoomed view of a corner for clarity

Temperature dependent material properties (Figure 3) for the PEN structure are primarily taken from the MENTAT-FC GUI [11-12] and ref [13] material database although MENTAT-FC software is not utilized in these simulations.

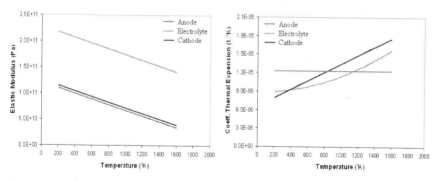

Figure 3. Temperature dependent (a) Young's modulus and (b) coefficient of thermal expansion of anode, cathode and electrolyte, drawn from [11]

Table I. Material properties of the PEN layers [11]

Material	Density (Kg/m^3)	Conductivity (W/m-^{o}K)	Sp. Heat (J/kg-^{o}K)	Poisson Ratio
Anode	4423.2	55.2	426.4	0.32
Electrolyte	6050.0	2.0	400.0	0.31
Cathode	2470.5	3.0	400.0	0.30

RESULTS AND DISCUSSION

The temperature fields are taken from DREAM-SOFC and incorporated into commercial FEA software ABAQUS^{TM} with the help of a user defined subroutine to calculate thermal stresses.

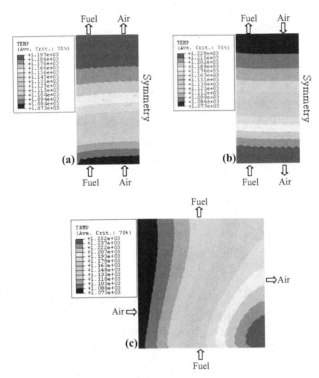

Figure 4. Temperature (^{o}K) profile on a co-flow planar SOFC anode imported in ABAQUS^{TM}

DREAM-SOFC utilized finite volume method to calculate the temperature distribution and species concentration inside the cell. Therefore the temperature data obtained from DREAM-SOFC is interpolated from the cell center to the nodal point or to the integration points of the FEA mesh used in ABAQUS™. The temperature distribution can be read in ABAQUS™ at the nodal point with the help of user subroutine DISP or at the integration point with the help of user defined subroutine, e.g. UMAT without changing the material behavior. Three flow configuration are analyzed i.e. co-, counter-, and cross-flow for comparison as shown in Figure 4. Due to symmetry in a co-flow and counter-flow temperature fields with respect to the axis of flow, only one-half of a 10cm x 10cm PEN structure is analyzed for these configurations as shown in Figure 4(a) and Figure 4(b) respectively. In the case of cross-flow configuration complete PEN assembly is analyzed as shown in Figure 4(c). The temperature variation through the PEN thickness is insignificant and therefore not considered in the model. The stress analysis is assumed to be fully elastic.

The thermal stresses are determined by applying the temperature field obtained from DREAM-SOFC, and boundary conditions shown in Figure 4, into FEA code ABAQUS™. From room to the operating temperature, the PEN assembly behaves as an elastic brittle material [7]. Likewise, F. Mora et al. [14] found that deformation behavior of an anode-supported bilayer consisting of Ni-YSZ substrate and YSZ layer is controlled by the deformation of nickel phase in the temperature range of 1000°C – 1200°C. However, they concluded that anode-supported bi-layer behaves as brittle material below 1100°C [14]. Therefore, maximum principal stress failure criterion is normally used for the PEN structure [7,8,15].

Figure 5(a)-(f) shows the resulting maximum principal stresses (MPS) through the PEN structure of co-flow configuration. The peak maximum principal stresses are found in the anode are at the anode/electrolyte interface near the fuel/air outlet as shown in Figure 5(b). Fischer et al. [5] also reported the similar stress profile in the anode material.

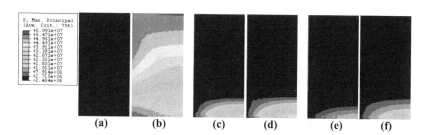

Figure 5. Principal stresses through the PEN structure in a co-flow case, anode: (a) anode exposed surface, (b) anode/electrolyte interface; electrolyte: (c) electrolyte/anode interface, (d) electrolyte/cathode interface; cathode: (e) cathode/electrolyte interface, (f) cathode exposed surface

Anode exposed surface is almost at uniform low level (<10MPa) of tensile and compressive stresses as shown in Figure 5(a). The stress profile is similar in the electrolyte at both the interfaces as shown in Figure 5(c)-(d) since its thickness is only 20μm. The highest principal stresses in the electrolyte and cathode layers occur near the fuel/air inlet as shown in Figure 5(c)-(f). The stress profile in the electrolyte layer is similar to one obtained by Yakabe et al. [3] for a co-flow case. These stress values are lower than the average strength of the PEN structure under the specified conditions [7].

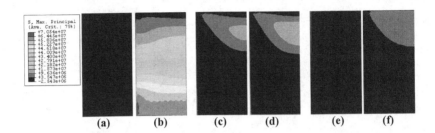

Figure 6. Principal stresses through the PEN structure in a counter-flow case, anode: (a) anode exposed surface, (b) anode/electrolyte interface; electrolyte: (c) electrolyte/anode interface, (d) electrolyte/cathode interface; cathode: (e) cathode/electrolyte interface, (f) cathode exposed surface

Similarly, the MPS distribution for counter-flow configuration is shown in Figure 6 (a)-(f) through the PEN structure. The MPS are found in the anode are at the anode/electrolyte interface near the fuel outlet as shown in Figure 6 (b). As in the case of co-flow, anode exposed surface is almost at uniform low level of MPS as shown in Figure 6 (a). The highest principal stresses in the electrolyte and cathode layers occur near the fuel inlet as shown in Figure 6 (c)-(f). The stress profile in the electrolyte layer is a bit different than the one calculated by Yakabe et al. [3] for a counter-flow case and peak MPS value is significantly smaller than predicted by them. The difference is because of the dissimilar temperature field predicted by different electrochemical models or different fuels.

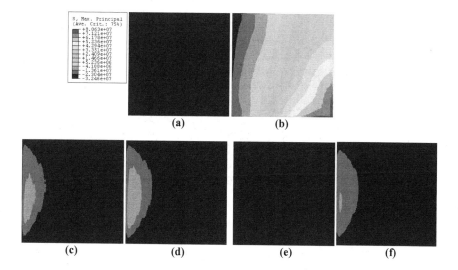

Figure 7. Principal stresses through the PEN structure in a cross-flow case, anode: (a) anode exposed surface, (b) anode/electrolyte interface; electrolyte: (c) electrolyte/anode interface, (d) electrolyte/cathode interface; cathode: (e) cathode/electrolyte interface, (f) cathode exposed surface

For cross-flow configuration, complete PEN structure is analyzed as there is no axis of symmetry. In the anode layer, peak MPS occurs at the anode/electrolyte interface in the high temperature region similar to the co- and counter-flow configuration. The corresponding MPS in the electrolyte is noted at the anode/electrolyte interface near the air and fuel inlet corner.

The location of the peak MPS can be explained by the temperature gradient and the greater coefficient of the expansion mismatch between electrolyte and anode/cathode in that temperature range as can be seen in Figure 3(b). Similar stress profile in the electrolyte is also documented by Selimovic et al. [15] although they estimated a different temperature profile but the trend is similar.

CONCLUSION

Thermal stress analyses are performed for the case of co-, counter- and cross-flow configurations of planar-SOFC PEN structure using finite element method. The analyses are conducted under realistic temperature fields obtained from DREAM-SOFC. In the anode layer, the peak maximum principal stresses (MPS) occur at the anode/electrolyte interface in the high temperature region for all configurations whereas cathode and electrolyte layers experience highest MPS near the air inlet region. In electrolyte, MPS is highest in the co-flow configuration due to the steep temperature gradient near the air/fuel inlet. The stress discontinuities across the interfaces are due to the different material properties. Under similar operating conditions, cross-flow configuration produces the highest MPS in the anode layer at the conjunction of fuel inlet and air outlet corner.

ACKNOWLEDGEMENTS

This work is conducted under US DOE (Department of Energy) EPSCoR Program. It is jointly sponsored by US DOE Office of Basic Energy Sciences, NETL (National Energy Technology Laboratory), WV State EPSCoR Office and the West Virginia University under grant number DE-FG02-06ER46299. Dr. Tim Fitzsimmons is the DOE Technical Monitor. Dr. R. Bajura is the Administrative Manager and Dr. I. Celik is the Technical Manager and the Principal Investigator of this project.

REFERENCES

[1] H. Koide, Y. Somaya, T. Yoshida, and T. Maruyama. Properties of Ni/YSZ cermet as anode for SOFC, Solid State Ionics, 132, pp. 253–260 (2000).

[2] M.A Khaleel, Z. Lin, P. Singh, W. Surdoval, and D. Collin. A finite element analysis modeling tool for solid oxide fuel cell development: coupled electrochemistry, thermal and flow analysis in MARC®, J Power Sources, 130, pp. 136-148 (2004).

[3] H. Yakabe, T. Ogiwara, M. Hishinuma, and I. Yasuda. 3-D model calculation for planar SOFC, J Power Sources, 102 pp. 144-154 (2001).

[4] W. Li, K. Hasinska, M. Seabaugh, S. Swartz, and J. Lannutti. Curvature in solid oxide fuel cells, J Power Sources, 138, pp. 145-155 (2004).

[5] W. Fischer, J. Malzbender, G. Blass, and R.W. Steinbrech. Residual stresses in planar solid oxide fuel cells, J Power Sources 150, pp. 73–77 (2005).

[6] J. Laurencin, B. Morel, Y. Bultel, and F. Lefebvere-Joud. Thermo-mechanical model of solid oxide fuel cell fed with methane. Fuel Cell, 06, pp. 64-70 (2006).

[7] C.K. Lin, T.T. Chen, Y.P. Chyou, and L.K. Chiang. Thermal stress analysis of a planar SOFC stack, J Power Sources, 164, pp. 238-251 (2007).

[8] G. Anandakumar, Na Li, A. Verma, P. Singh, and Jeong-Ho Kim. Thermal stress and probability of failure analyses of functionally graded solid oxide fuel cells. J Power Sources, 195, pp. 6659–6670 (2010).

[9] S. Pakalapati. A new reduced order model for solid oxide fuel cells. Ph.D. thesis, West Virginia University, Morgantown, WV, 2006.

[10] ABAQUS, Analysis User's manual, version 6.5, Hibbitt, Karlsson & Sorensen, Inc., 2004.

[11] K.I. Johnson, V.N. Korolev, B.J. Koeppel, K.P. Recknagle, M.A. Khaleel, D. Malcolm, and Z. Pursell. Finite element analysis of solid oxide fuel cells using SOFC-MP™ and MSC. Marc/Mentat-FC™, Pacific Northwest National Laboratory, Report No. PNNL-15154, Richland, WA. 2005

[12] K. Johnson. Finite element SOFC analysis with SOFC-MP and MSC. Marc/Mentat-FC. SECA 6th Annual Workshop, April 18-21, 2005.

[13] K. Scott Weil, John E. Deibler, John S. Hardy, Dong Sang Kim, Guan-Guang Xia, L.A. Chick, and Chris A. Coyle. Rupture testing as a tool for developing planar solid oxide fuel cell seals" J. Mater. Eng. Perform, 13, pp. 316–326 (2004).

[14] F. Gutierrez-Mora, J.M. Ralph, J.L. Routbort. High-temperature mechanical properties of anode-supported bilayers" Solid State Ionics 149 pp. 177– 184 (2002).

[15] A. Selimovic, M. Kemm, T. Torisson, M. Assadi. Steady state and transient thermal stress analysis in planar solid oxide fuel cells, J Power Sources, 145, pp. 463–469 (2005).

ELECTROLESS COATING OF NICKEL ON ELECTROSPUN 8YSZ NANOFIBERS

Luping Li[1], Peigen Zhang[2], and S.M. Guo[2]
[1] Department of Materials Science & Engineering, University of Florida, Gainesville, Florida 32601, USA
[2] Department of Mechanical Engineering, Louisiana State University, Baton Rouge 70803, USA

ABSTRACT

The coating of a thin, uniform layer of nickel on 8YSZ nanofibers using electroless plating was presented in this paper. 8YSZ nanofibers were fabricated by electrospinning a mixture of yttria and zirconia nano dispersions. The fibers were then densified by sintering to 1000 °C. To obtain the best coatings, the optimal concentrations of acidified hydrosols of $SnCl_2$ and $PdCl_2$ for pre-treatment were found. Ni-deposition behavior was studied by varying the coating variables, including the coating time and the pH values. The Ni-YSZ fibers can be used as potential anode material for solid oxide fuel cells (SOFCs).

1 INTRODUCTION

Electrospinning is a simple and versatile technique to fabricate 1-D nanofibers[1]. When combined with heat treatment, ceramic fibers of required chemical compositions with different surface morphologies and structural characteristics can be obtained. In a typical electrospinning process, a polymer-containing precursor is loaded into a syringe and a high voltage is applied between the syringe needle and a counter electrode. When the electrical field reaches beyond a critical value, the precursor will be pulled out of the needle tip and shoot towards the counter electrode plate, and will go though much whipping and bending in the air. Consequently, a non-woven mat is formed on the plate when enough time is given.

8YSZ is an extensively investigated material because of its interesting properties, such as good thermal stability, high chemical resistance, low thermal conductivity, and excellent ionic conductivity, etc[2]. These attributes have enabled it's usage on thermal barrier coatings (TBCs)[3], oxygen sensors[4], and especially solid oxide fuel cells (SOFCs)[5]. SOFCs are electrochemical devices that convert chemical energies in fuels, such as hydrogen and natural gas, into electrical energies without direct combustion of these fuels. SOFCs boast traits such as high efficiency, environmental friendliness, diversity of fuels, etc. Ni-YSZ is the state-of-the –art anode material for SOFCs. For the Ni-YSZ anode, Ni is to facilitate the electron conduction, as well as catalyzing electrochemical reaction of the fuel; on the other hand, YSZ provides matrix for Ni dispersion to prevent Ni agglomeration, and it is also the key component for realization of ionic conduction along the anode/electrolyte interface.

Aside from SOFCs, Ni coating is also useful in many other areas, such as the formation of carbon filaments[6], hydrogenation of benzene, hydrogen generation, separation, and storage[7]. In these applications, Ni is preferred to have high surface area for better performance. One effective way of achieving high surface area is by depositing Ni on structures that already had high surface areas.

Although coatings on powder particles have been studied[8,9,10], Ni coatings with nanofibers were rarely reported in literature. In this work, Ni electroless coating was conducted with YSZ nanofibers, which was first obtained by electrospinning a mixture of zirconia and yttria dispersions and subsequent heat treatment; the dependence of coated fibers' morphology on coating variables was examined and presented.

2 EXPERIMENTAL

Tin chloride dihydrate (>98% purity), palladium chloride (99% purity), nickel acetate tetrahydrate (>99.0% purity), hydrazine monohydrate (64-65 wt%, 99.0% purity), lactic acid, ethylenediaminetetraacetic acid (EDTA), and sodium hydroxide were purchased from Sigma-Aldrich. For pH measurement, a digital pH meter (model PHH-80BMS/TDS, Omega Engineering Inc. USA) was calibrated before use, which gave an accuracy of ±0.02 pH.8YSZ fibers were prepared by electrospinning, which was described in our previous work[11]. The as-spun fibers were fired to 1000 °C for 2 h to remove organics and to form pure 8YSZ fibers.

For successful coating, 8YSZ fiber surfaces need to be sensitized and activated by Sn-Pd systems. There are two ways of accomplishing it: one approach is to dip the sample into a mixture of acidified $SnCl_2$ and $PdCl_2$ solution (called 1-step procedure); the other is to treat the sample in acidified $SnCl_2$ solution first, followed by a brief wash in deionized water, and lastly dip it in acidified $PdCl_2$ solution (called 2-step procedure). The 2-step procedure was adopted in this research. Based on the reported pre-treating solutions in literature[10,12,13], we came up with 5 solution pairs and attempted to find out the best and most effective combination. These 5 systems are listed in Table 1.

Table I. Five different Sn-Pd systems for treating the fibers before Ni coating.

	$SnCl_2$ (g/l)	$PdCl_2$ (g/l)
System 1	0.00	0.00
System 2	30.00	0.05
System 3	4.30	3.30
System 4	2.30	3.30
System 5	10.00	0.38

After fibers' pre-treatment, Ni coating started either immediately or after baking out the fibers overnight (see Sec. 3.2.1 for the results). The coating was conducted in a water bath sitting on a hot plate. The water-based coating solution used in this work was based on Wen et al's formula[8] and it typically consists of nickel acetate (50 g/l), EDTA (20 g/l), lactic acid (40ml/l), hydrazine (80 ml/l), and NaOH (~35g/l to reach the target pH values).

To investigate the effect of plating variables on the morphology of the resulting coating, plating time was varied from 4 min to 7 min, and pH values were changed between 12.57 and 13.22.

Transmission electron microscopy (TEM) (Model Jeol 2010, Japan) was used to inspect the initial particle size in the colloids. Field emission scanning electron microscopy (FESEM) (Model Quanta 3D FEG, FEI Company, USA) was utilized to examine the fibers' surface morphologies. The fibers were also characterized by powder X-ray Diffraction (XRD) (Bruker/Siemens D5000 X-ray diffractometer with Rietveld software).

3 RESULTS AND DISCUSSIONS

3.1 Zirconia colloid and 8YSZ fibers

Figure 1 shows the TEM image of the as-received zirconia particles in the colloid. It can be seen that the particle sizes are around 10 nm.

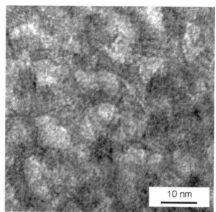

Fig. 1 TEM image of the initial zirconia particles in the colloid.

Figure 2 shows the typical FESEM images of the electrospun fibers. We can see that the as-spun fibers (see (a)) are uniform in size and the fiber surfaces are extremely smooth. The defect-free fibers have diameters of ~300 nm. After the fibers were sintered at 1000 °C in air for 2 h (see (b)), the fibers' cylindrical shapes were perfectly kept although the surfaces became much rougher. The diameters of the fibers did not shrink much during firing. The YSZ particles grew to ~ 100 nm from the initial 10 nm.

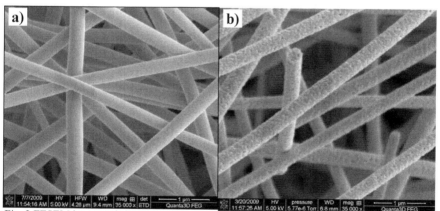

Fig. 2 FESEM images of electrospun fibers: (a) the as-spun fibers; (b) the fibers fired at 1000 °C for 2h.

Figure 3 is the XRD pattern of the YSZ fibers after firing. The XRD spectra were refined and indexed by Jade (software ver. 5.0). We found that the fibers had cubic crystalline structure and the PDF number is 30-1468.

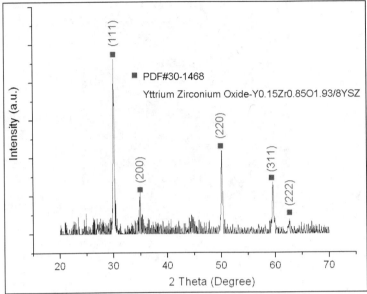

Fig. 3 Typical XRD pattern of the fired YSZ fibers.

3.2 The impact of pre-treatment by Sn-Pd systems to Ni plating

3.2.1 Importance of drying the fibers after pre-treatment

After the fibers were treated by $SnCl_2$ and $PdCl_2$ solutions sequentially, we found it is very important to dry the fibers thoroughly before conducting the Ni coating. Figure 4 shows the results of two types of coatings: for (a) the coating was done right after the fibers were taken out of the $PdCl_2$ solution, and for (b) the coating was formed when the fibers were baked out overnight in a furnace. It is apparent that (a) had sparse and very discreet Ni particles on fibers surfaces, while (b) exhibited uniform Ni coating on the fibers.

Fig. 4 Comparison of two different steps after pre-treatment: a) coating was done right after PdCl₂ activation; b) coating was done after drying the PdCl₂-activated fibers overnight.

It was reported[10] that during the pre-treating process, $SnCl_2$ and $PdCl_2$ react with each other in an acidic environment and metallic Pd will form. The chemical reaction can be expressed as:

$$SnCl_2 + PdCl_2 = SnCl_4 + Pd \qquad (1)$$

The generated metallic Pd is formed in the vicinity of the fibers and tends to be adsorbed on fiber surfaces. The adhered Pd particles will act as catalyst for the electrochemical reduction of Ni during the following coating step. If the fibers were put into the coating solution right after the pre-treatment when the fibers were still wet, the "unsecured" Pd particles on the fiber surfaces would have a tendency to travel away from the fibers, which would render the Ni coating unsuccessful because of the lack of catalyst.

3.2.2 Effect of varying the concentrations of the pre-treating solutions

Figure 5 shows different Ni coating behaviors when examined by FESEM after the fibers were treated by 5 different Sn-Pd systems. We found that when the fibers were treated by deionized water only (see (a)), the fibers would not be coated at all. This result proved the necessity and importance of the pre-treating procedure. When comparing the coated fibers that were treated by System 2-5 (see Fig. 5 (b) - (e)), it can be seen that these fibers all had Ni coating on their surfaces. Additionally, for all coatings there are large Ni agglomerations adhering on fiber surfaces except the one treated by System 5. We believe the reason for obtaining agglomeration-free coating in Fig. (e) is that System 5 had the optimal ratio of Sn to Pd in the treating solutions, which according to equation (1) uniform Pd particles were formed on fiber surfaces. On the other hand, in Systems 2 to 4, local high concentrations of Pd particles were created on the surfaces of the fibers, which would cause excess Ni precipitation during the following coating.

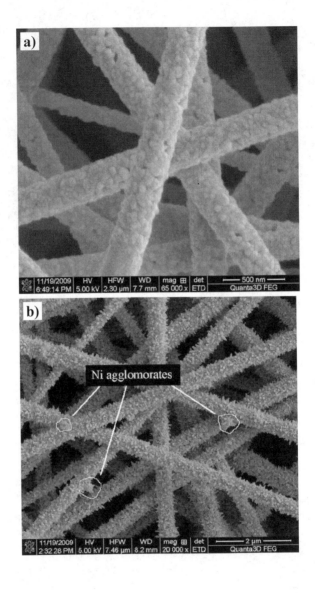

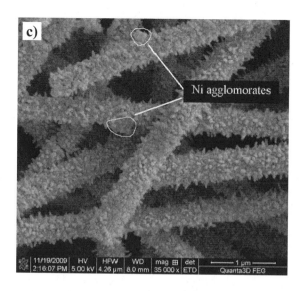

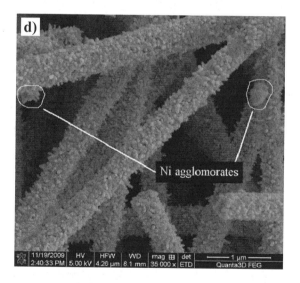

Fig. 5 Different Ni coating behaviors after the fibers were treated by different Sn-Pd systems: (a) fibers were treated by System 1 (acidified deionizer water); (b) fibers were treated by System 2; (c) fibers were treated by System 3; (d) fibers were treated by System 4; (e) fibers were treated by System 5.

3.3 Ni plating- Impact of coating time

To observe the evolvement of Ni coating with respect to time, the fibers were taken out of the plating solution at different times. Figure 6 shows the results when the coating was conducted for 3 min (Fig. 6 (a)) and 7 min (Fig. 6 (b)). It can be seen that the fibers' surfaces in (b) is much rougher than (a), which meant more Ni precipitation on fibers took place over longer times.

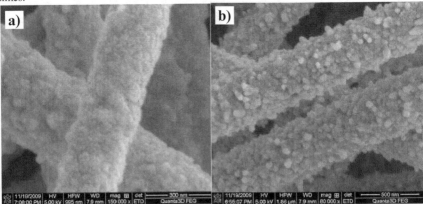

Fig. 6 Time evolvement of Ni coating: (a) coating time was 3 min; (b) coating time was 7 min.

3.4 Ni plating- Impact of pH values

Figure 7 shows the FESEM images of the fibers coated at 3 different pH values: 12.57 (shown in (a)), 12.98 (b), and 13.22 (c). We can see that as the pH values increased, the coated Ni particles became larger.

The dependence of the particle size of the coated Ni on the pH values is attributed to the kinetics of Ni precipitation during the coating process. The chemical reactions involved during the coating of fibers can be summarized as[8]

$$N_2H_4 + OH^- + Ni^{2+} \longrightarrow Ni + N_2 + H_2O \qquad (2)$$

Therefore, when pH value increases, a larger amount of OH$^-$ ions will be available for the reaction, which will result in a shift of the above reaction to the right-hand side and thus faster sedimentation of Ni in a shorter period of time. However, since all fibers were sensitized and activated by the same pre-treating solutions, the number of catalytic Pd particles was the same, which meant the nucleation sites were constant for all 3 coating processes. Consequently there would be more Ni precipitation on each nucleation core at higher pH values. The strong effect of pH values may also be related to the gas release kinetics during the reaction[8]. In equation (2) above N$_2$ was formed as a gas and would tend to bubble out, which would impact and disturb the Ni precipitation and growth. Because higher pH value would lead to faster gas release, the uniform nucleation of Ni on fiber surfaces might be interrupted, resulting in larger Ni particles and porous coating structure.

| 6/3/2010 4:27:43 PM | HV 5.00 kV | pressure 9.68e-6 Torr | WD 9.5 mm | mag □ 65 000 x | 1 µm Quanta3D FEG |

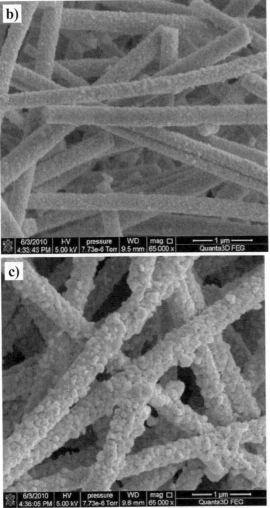

Fig. 7 Ni coating results at different pH values: (a) pH=12.57; (b) pH=12.98; (c) pH= 13.22.

3.5 Ni-coated-YSZ fibers as the anode material for SOFCs

The Ni-coated-YSZ fibers have been demonstrated in our previous work to be a superior material as the anode for SOFCs against conventional powder-derived anode structures[11]. It was shown that the power density of a single cell with the fibrous anode structure was twice as large as the cell with the traditional powder-derived anode. For the application as the anode material on SOFCs, the formation of contiguous, agglomeration-free, and fine-sized Ni particles on fiber surfaces is preferred.

4 CONCLUSIONS

8YSZ fibers were successfully fabricated by electrospinning. By using an electroless plating technique, the fiber surfaces were coated with a layer of Ni. It was found that in order to obtain the uniform Ni coating, the fibers need to be dried after pre-treatment. Agglomeration-free coating has been obtained when the fibers were consecutively treated by 10 g/l of $SnCl_2$ solution and 0.38 g/l of $PdCl_2$ solution. More Ni deposition on fiber surfaces were seen when more coating time was allowed. Higher pH values would result in larger Ni particles, possibly due to faster Ni precipitation on constant nucleation sites. Alternatively, it may be caused by gas release kinetics as described in the article.

ACKNOWLEDGEMENTS

This work was supported by Louisiana Board of Regents under contract LEQSF (2007-10)-RD-A-08.

REFERENCES

[1] Luping Li, Peigen Zhang, Jiandong Liang and S.M. Guo, Phase Transformation and Morphological Evolution of Electrospun Zirconia Nanofibers during Thermal Annealing, Ceram. Int., 36, 589-594 (2010).

[2] G. Laukaitis, J. Dudonis, D. Milcius, YSZ Thin Films Deposited by E-beam Technique, Thin Solid Films, 515, 678 (2006).

[3] Mohsen Saremia, Abbas Afrasiabia, and Akira Kobayashib, Microstructural Analysis of YSZ and YSZ/Al_2O_3 Plasma Sprayed Thermal Barrier Coatings after High Temperature Oxidation, Surf. Coat. Technol., 202, 3233 (2008).

[4] Teiichi Kimura, Takashi Goto, Ir-YSZ Nano-composite Electrodes for Oxygen Sensors, Surf. Coat. Technol., 198, 36 (2005).

[5] James R. Wilson, Worawarit Kobsiriphat, Roberto Mendoza, Hsun-Yi Chen, Jon M. Hiller, Dean J. Miller, Katsuyo Thornton, Peter W. Voorhees, Stuart B. Adler, and Scott A. Barnett, Three Dimensional Reconstruction of a Solid Oxide Fuel Cell Anode, Nat. Mater., 5, 541-544 (2006).

[6] G. G. Kuvshinov, Yu. I. Mogilnykh, D. G. Kuvshinov, D. Yu. Yermakov, M. A. Yermakova, A. N. Salanov and N. A. Rudina, Mechanism of Porous Filamentous Carbon Granule Formation on Catalytic Hydrocarbon Decomposition, Carbon, 37, 1239 (1999).

[7] G. H. Jonker, J. W. Veldsink, and A. A. C. M. Beenackers, Intraparticle Diffusion Limitations in the Hydrogenation of Monounsaturated Edible Oils and Their Fatty Acid Methyl Esters, Ind. Eng. Chem. Res., 37, 4646 (1998).

[8] G. Wen, Z.X. Guo, C.K.L. Davies, Microstructural Characterization of Electroless-nickel Coatings on Zirconia Power, Scr. Mater., 43, 307 (2000).

[9] A.K. Grag, L.C. De Jonghe, Metal-coated Colloidal Particles, J. Mater. Sci., 28, 3427 (1993).

EFFECT OF SURFACE CONDITION ON SPALLATION BEHAVIOR OF OXIDE SCALE ON SS 441 SUBSTRATE USED IN SOFC

Wenning Liu, Xin Sun, Elizabeth Stephens, Moe Khaleel
Pacific Northwest National Laboratory
P.O Box 999, 906 Battelle Blvd, Richland, WA 99352

ABSTRACT

As operating temperature of SOFC decreases, ferritic stainless steel has attracted a great deal of attention for its use as an interconnect in SOFCs because of its gas-tightness, low electrical resistivity, ease of fabrication, and cost-effectiveness. However, oxidation reaction of the metallic interconnects in a typical SOFC working environment is unavoidable. The growth stresses in the oxide scale and on the scale/substrate interface combined with the thermal stresses induced by thermal expansion coefficient mismatch between the oxide scale and the substrate may lead to scale delamination/buckling and eventual spallation during stack cooling, which can lead to serious cell performance degradation. Therefore, the interfacial adhesion strength between the oxide scale and substrate is crucial to the reliability and durability of the metallic interconnect in SOFC operating environments. In this paper, we investigated the effect of the surface conditions on the interfacial strength of oxide scale and SS441 substrate experimentally. Contrary to the conventional sense, it was found that rough surface of SS441 substrate will decrease the interfacial adhesive strength of the oxide scale and SS441 substrate.

INTRODUCTION

The operating temperature of solid-oxide fuel cells (SOFCs) has decreased from 1100°C to 800°C. This has led to a great deal of attention being focused on ferritic stainless steel for use as an interconnect (IC) in SOFCs because of its gas-tightness, low electrical resistivity, ease of fabrication, and cost-effectiveness [1]. In addition to providing cell-to-cell electrical connections, ICs in SOFCs also act as separator plates in separating the anode side fuel flow from the cathode side airflow for each cell. Compared to chromium-based alloys, iron-based alloys have advantages in terms of high ductility, good workability, and low cost. Iron-based alloys, especially Cr-Fe based alloys, e.g., Crofer 22 APU and SS441, are by far the most attractive metallic IC materials for SOFCs [2, 3]. However, under a typical SOFC working environment, an oxide scale will grow on the metallic ICs in an oxidation environment.

An oxide layer inevitably forms on the surface of these ferritic stainless steels in the atmospheres representative of SOFCs. Even though various coating techniques have shown great potential in slowing down the oxide growth kinetics and protecting the ferritic stainless steel IC under SOFC operating conditions [4-6], the sub-coating oxide scale growth is still inevitable. The appearance and growth of the oxide scale will cause growth stress in the oxide scale [7-11]. In addition, the CTE mismatch between the oxide and the substrate creates thermal stresses in the scale and on the scale/substrate interface during cooling [12-14]. The growth stresses in the oxide scale and on the scale/substrate interface combined with the thermal stresses induced by the CTE mismatch between the oxide scale and the substrate may lead to scale delamination/buckling and eventual spallation during stack cooling. This can lead to serious degradation of cell performance [14, 15]. Therefore, it is necessary to investigate the scale delamination/spallation of the ferritic ICs at the operating environments of SOFCs.

In this paper, we investigated the effect of the surface conditions on the interfacial strength of oxide scale and SS441 substrate experimentally. Contrary to the conventional sense, it was found that rough surface of SS441 substrate will decrease the interfacial adhesive strength of the oxide scale and SS441 substrate.

EXPERIMENTAL

SS 441 is one of the most powerful contenders for ferritic ICs used in SOFCs and is used in the present paper. The chemical composition is listed in Table 1 [16].

Table 1. Chemical composition of SS 441 (wt%)

Element	C	Mn	P	S	Si	Cr	Ni	Ti	Nb	N	Al	Fe
SS 441	0.009	0.35	0.023	0.002	0.34	18.0	0.3	0.22	0.5	0.014	0.05	rest

A total of two batches of SS441 sheet were used, called B1 and B2, respectively. The surface roughness of the SS 441 sheets was quantified by a profilometer. The typical measurement of the surface roughness is illustrated in Table 2 for the as-received B1 sheets, where Ra is the mean of absolute values of the profile deviations from the mean line, Rz is the sum of the mean height of the five highest peaks and the mean depth of the five deepest valleys, and Rq represents the square root of mean of the squares of profile deviations from the mean line. The average Ra roughness is 0.68 μm.

Table 2. Measured surface roughness of as-received SS 441

Profile#	Ra	Rz	Rq
1	0.85	5.48	1.21
2	0.67	4.60	0.84
3	0.66	4.77	0.82
4	0.62	4.56	0.79
5	0.63	4.83	0.80
6	0.64	4.39	0.70
7	0.69	6.54	0.97
8	0.64	4.85	0.87
9	0.69	6.06	0.99
Mean	**0.68**	**5.12**	**0.89**
Std Dev	0.07	0.74	0.15

First, the as-received SS 441 sheet was sheared into the 1-in. × 2-in. test coupons with the edges filed. To investigate the effect of the surface roughness on the interfacial strength, surface treatment was applied to some coupons. The specimens without any surface finishing are called as-received (AR), and the specimens with surface modification are called surface-modified (SM). The roughness of the AR and SM specimens was measured for both batches as listed in Table 3. For the second batch of SS441 material, the specimen coupon was fabricated where half of each sample was taped off to allow fine polishing of the neighboring surface while leaving the taped surface in the as-received condition. A one micron diamond paste was used in the final polishing step of the polished surface for the second batch. These specimens with different surface roughness were used to reveal the effect of the surface condition on the interfacial strength. These specimens were oxidized for 600 h to 900 h at 850°C and then furnace cooled to room temperature. Metallography was performed to determine the oxide scale thickness for each specimen.

Table 3 Measurement of surface roughness

SS441 samples	B1		B2	
	AR	SM	AR	SM
Roughness (μm)	0.7	0.25	0.4	0.02

RESULTS AND DISCUSSIONS

After oxidation of 600 h to 900 h at 850°C, the coupons were cooled down to room temperature. Figures 1 to 3 show the typical images of the B1 and B2 coupons with AR and SM surfaces, respectively, when they were moved out of the furnace. It may be seen that spallation of the oxide scale on the SS 441 substrate was observed for some test coupons. For the as-received B1 coupon, the spallation of the oxide scale occured almost everywhere as in Figure 1. With surface modification, the B1 coupon avoided the spallation except at the edges in Figure 2. Image of the B2 coupons in Figure 3 shows that spallation happened in the half with AR surface and SM surface almost eliminated the spallation except some edges.

Figure 1 Typical image of B1 coupon with as-received surface

Figure 2 Typical image of B1 coupon with surface modification

Figure 3 Typical image of B2 coupon with AR and SM surface

The experimental observation suggests a strong relationship of the surface roughness with the spallation behavior of the oxide scale on the ferritic substrate. With decrease of the surface roughness, the spallation of the oxide scale reduced. There is threshold value of the surface roughness. When the surface roughness is higher than this value, the smoothness of the surface could effectively alleviate spallation. When the surface roughness reaches or is lower than this value, the further smoothness of the surface could not enhance the behavior of the oxide scale and avoid spallation.

CONCLUSIONS

In the paper, we investigated the effect of the surface conditions on the interfacial strength of oxide scale and SS441 substrate experimentally. It was observed that the surface roughness has a relatively large effect on the spallation behavior of the oxide scale on the ferritic substrate. The surface modification/finish mitigates the spallation of the oxide scale. There is threshold value of the surface roughness. When the surface roughness is higher than this value, the smoothness of the surface could effectively alleviate spallation. When the surface roughness reaches or is lower than this value, the further smoothness of the surface could not enhance the behavior of the oxide scale and avoid spallation.

ACKNOWLEDGEMENTS

The Pacific Northwest National Laboratory is operated by Battelle Memorial Institute for the United States Department of Energy under Contract DE-AC06-76RL01830. The work was funded as part of the Solid-State Energy Conversion Alliance (SECA) Core Technology Program by the U.S. Department of Energy's National Energy Technology Laboratory (NETL).

REFERENCES

[1]A. Boudghene Stambouli and E. Traversa, Solid Oxide Fuel Cells (SOFCs): A Review of an Environmentally Clean and Efficient Source of Energy, *Renew. Sust. Energ. Rev.*, **6**, 433–55 (2002).J. W. Fergus, Metallic Interconnects for Solid Oxide Fuel Cells, *Materials Science & amp; Engineering A (Structural Materials: Properties, Microstructure and Processing)* 397(1-2):271-83 (2005).

[2]Y. Li, J. Wu, C. Johnson, R. Gemmen, X. M. Scott and X. Liu, Oxidation behavior of metallic interconnects for SOFC in coal syngas, *International Journal of Hydrogen Energy*, Vol.34, No.3, 1489-1496 (2009).

[3]W.N. Liu, X. Sun, E. Stephens, M. Khaleel, Interfacial Shear Strength of Oxide Scale and SS 441 Substrate, *Materials Science & amp; Engineering A (Structural Materials: Properties, Microstructure and Processing)*, (2010) (in press)

[4]M.J. Garcia-Vargas, M. Zahid, F. Tietz, and A. Aslanides: "Use of SOFC metallic interconnect coated with spinel protective layers using the APS technology," *ECS Transactions*, 2007, vol. 7, no. 1, Part 2, pp. 2399-2405.

[5]D.G. Ivey, W. Qu, J. Li, and J.M. Hill: "Electrical and microstructural characterization of spinel phases as potential coatings for SOFC metallic interconnects," *Journal of Power Sources*, 2006, vol. 153, no. 1, pp. 114-24.

[6]M. Bertoldi, T. Zandonella, D. Montinaro, V.M. Sglavo, A. Fossati, A. Lavacchi, C. Giolli, and U. Bardi: "Protective coatings of metallic interconnects for IT-SOFC application," *Journal of Fuel Cell Science and Technology*, 2008, vol. 5, no. 1, pp. 011001-1-5.

[7]M. Schulte and M. Schutze: "The role of scale stresses in the sulfidation of steels," *Oxidation of Metals*, 1999, vol. 51, no. 1-2, pp. 55-77.

[8]S.R. Pillai, N.S. Barasi, H.S. Khatak, and J.B. Gnanamoorthy: "Effect of external stress on the behavior of oxide scales on 9Cr-1Mo steel," *Oxidation of Metals*, 1998, vol. 49, no. 5-6, pp. 509-30.

[9]V.K. Tolpygo, J.R. Dryden, and D.R. Clarke: "Determination of the growth stress and strain in α-Al_2O_3 scales during the oxidation of Fe-22Cr-4.8Al-0.3 Y alloy," *Acta Materialia*, 1998, vol. 46, no. 3, pp. 927-37.

10W. Przybilla and M. Schutze: "Growth stresses in the oxide scales on TiAl alloys at 800 and 900°C," *Oxidation of Metals*, 2002, vol. 58, no. 3-4, pp. 337-59.

[11]J. Mougin, A. Galerie, M. Dupeux, N. Rosman, G. Lucazeau, A.M. Huntz, and L. Antoni: "In-situ determination of growth and thermal stresses in chromia scales formed on a ferritic stainless steel," *Materials and Corrosion*, 2002, vol. 53, no. 7, pp. 486-90.

[12]X. Sun, W.N. Liu, P. Singh and M.A. Khaleel: "Effects of Oxide Thickness on Scale and Interface Stresses under Isothermal Cooling and Micro-Indentation," *Technical Report PNNL-15794*, Pacific Northwest National Laboratory, Richland, WA 2007.

[13]Y. M. Xu and H. M. Wang: "Oxidation behavior of γ/Mo$_2$Ni$_3$Si ternary metal silicide alloy," *Journal of Alloys and Compounds*, 2008, vol. 457, no. 1-2, pp. 239-43.

[14]P.Y. Hou, A.P. Paulikas, B.W. Veal, and J.L. Smialek, "Thermally grown Al$_2$O$_3$ on a H$_2$-annealed Fe$_3$Al alloy: stress evolution and film adhesion," *Acta Materialia*, 2007, vol. 55, no. 16, pp. 5601-13.

[15]J.Y. Kim, V.L. Sprenkle, N.L. Canfield, K.D. Meinhardt, and L.A. Chick: "Effects of chrome contamination on the performance of La$_{0.6}$Sr$_{0.4}$Co$_{0.2}$Fe$_{0.8}$O$_3$ cathode used in solid oxide fuel cells," *J. Electrochem. Soc.* 2006, vol. 153, no. 5, pp. A880–A886.

[16]Technical Data Blue Sheet, Stainless Steel AL 441HP Alloy, ATI Allegheny Ludlum: Available at: http://www.alleghenyludlum.com/ludlum/documents/441.pdf. Accessed 10-01-2010.

EFFECT OF FUEL IMPURITY ON STRUCTURAL INTEGRITY OF Ni-YSZ ANODE OF SOFCS

Wenning Liu, Xin Sun, Olga Marina, Larry Pederson, Moe Khaleel
Pacific Northwest National Laboratory
P.O Box 999, 902 Battelle Blvd, Richland, WA 99354

ABSTRACT

Electricity production through the integration of coal gasification with solid oxide fuel cells (SOFCs) may potentially be an efficient technique for clean energy generation. However, multiple minor and trace components are naturally present in coals. These impurities in coal gas not only degrade the electrochemical performance of Ni-YSZ anode used in SOFCs, but also severely endanger the structural integrity of the Ni-YSZ anode. In this paper, the effect of trace impurity in coal syngas on the mechanical degradation of Ni-YSZ anode was studied by using an integrated experimental/modeling approach. Phosphorus is taken as an example of impurity. Anode-support button cell was used to experimentally explore the migration of phosphorous impurity in the Ni-YSZ anode of SOFCs. X-ray mapping was used to show elemental distributions and new phase formation. The subsequent finite element stress analyses were conducted using the actual microstructure of the anode to illustrate the degradation mechanism. It was found that volume expansion induced by the Ni phase change produces high stress level such that local failure of the Ni-YSZ anode is possible under the operating conditions.

INTRODUCTION

Fuel cells are high-efficiency energy conversion devices and are environmental friendly with little or no toxic emissions. The solid oxide fuel cell continues to show great promise as a future power source, with potential applications in both stationary power generation and auxiliary power units (APUs) [1]. As a high temperature fuel cell ($600^{\circ}C$-$1000^{\circ}C$), solid oxide fuel cell (SOFC) has many advantages over other types of fuel cells such as high kinetic activity, no need of precious catalysts, fuel flexibility and possibility of internal fuel reforming [2, 3]. Despite the fact that SOFC technology has been proved, its high costs have limited implementation [4]. A primary advantage of SOFCs over other non-combustion energy sources, i.e., the ability of SOFCs to operate using complex fuels such as hydrocarbons or alcohols, may help to reduce the cost of SOFCs [5].

With coal being the single largest domestic fuel source for the foreseeable future, the integration of coal gasification with SOFCs can potentially be a viable and efficient technology for clean power generation in the US. SOFCs have high, direct fuel-to-electricity conversion efficiencies of 45–50% (with the total efficiency of more than 80% when the cogenerated heat is also employed), environmental compatibility (low NOx production) and modularity, and no need for the high cost precious metal use [6, 7]. Among various SOFC designs, anode-supported planar cells seem to provide the best performance at a reasonable cost [8]. Currently, porous Ni–YSZ (yttria stabilized zirconia) cermet is the most widely used anode material for SOFC application because of its high catalytic activity, low cost, and high chemical and mechanical stability in reducing atmosphere at high temperature [9,10]. However, multiple minor and trace components are naturally present in coals [7, 11, 12]. Impurities exit in the fuel gas stream either as a vapor phase or in the form of fine particulate matter even after the gas stream cleanup process. Cayan *et al.* presented a literature review in [14] to summarize the studies on the identification of impurities in coal syngas and their effects on the performance of Ni-yttria stabilized zirconia (Ni-YSZ) anode of solid oxide fuel cells (SOFCs). Coal syngas typically contains major species, CO, H_2, CO_2, H_2O, CH_4, N_2, and H_2S as well as trace impurities. While many of the fuel gas impurities, such as Hg, Si, Zn and NH3, do not significantly affect the SOFC anode performance [13,14], the Ni-YSZ anodes are usually vulnerable to degradation due to the presence of other impurities in a coal-derived fuel gas stream such as P, S and As. Different impurities in the fuel stream can lead to different types of performance degradation depending on the impurity's thermo-chemical nature, concentration level, and mechanism of poisoning. They can block the reactive surface sites of the nickel particles in the anode deteriorating the catalytic ability (surface effect); react with the YSZ network impeding the transportation of the oxygen ions (bulk effect); and change the morphology of the nickel particle–YSZ network bringing down the electrical conductivity [6]. For example, impurity species such as Cl, Sb, As, and P cause severe cell voltage degradation due to attack on the Ni-YSZ anode. Sb, As and P have the potential to react with Ni to form secondary phases in the Ni-YSZ anode, which deteriorate the catalytic activity of the anode.

Demircan et al. [15] reported in situ measurements of sheet resistance, R_S, and resistivity of the anode of a commercial SOFC during prolonged exposure to wet hydrogen, then clean syngas, and then syngas with 10ppm phosphine. The effect of the prolonged exposure of the anode to phosphine with current flow was investigated using extensive post-mortem analyses. These in situ VdP measurements reveal that resistivity values under different fuel compositions, even with 10ppm PH_3 impurity in syngas, are constant with an uncertainty of ~ 5%. Electrochemical impedance spectroscopy (EIS) measurements indicate that an increase in both the ohmic resistance, R_{ohmic} and polarization resistance, R_p is caused by PH_3 exposure. Marina et al. [16, 17] evaluated the performance of SOFC with nickel/zirconia anodes on synthetic coal gas in the presence of low levels of phosphorus, arsenic, selenium, sulfur, hydrogen chloride, and antimony impurities. The presence of phosphorus and arsenic in low, 1-2 ppm, concentrations led to the slow and irreversible SOFC degradation due to the formation of the secondary phases with nickel in the upper part of the nickel-based anode close to the gas inlet.

Though coal syngas cleanup techniques such as absorption, diffusion through a membrane, chemical conversion and filter have been used to bring down the impurity concentration to lower level, these processes are not cost effective. Moreover, since they are carried out at ambient temperature, a significant portion of thermal energy is also lost [18]. Trembly et al. pointed out in [18] that the operational temperature and pressure of the gas cleanup system was found to have a small influence on the partitioning and phase equilibrium behavior of phosphorous. At a gas cleanup temperature of 200°C and pressures of 5, 10, and 15 atm the condensed phase $(NH_4)H_2PO_4$ (s) was found to account for 98.1, 100, and 100% of the total phosphorus in the system, respectively. The dominant vapor phase form of phosphorous found in the system was $(P_2O_3)_2$ (g), however trace amounts of PH_3 (g) were also found to form at varying amounts based upon the temperatures and pressures. It is important to establish the tolerance of SOFCs for fuel impurities to facilitate proper design of the fuel feed system that would not catastrophically degrade the SOFCs electrochemical performance or structural integrity during their expected lifetime.

Currently, most of the research efforts concerning effect of the impurities of syngas on SOFCs anodes are focused on the electrochemical degradation of SOFC anode [19-21]. It should be noted that, in anode-supported cells, the porous anode provides the main structural support for the thin ($\sim 10 \mu m$) yttria-stabilized zirconia (YSZ) electrolyte layer which is inherently susceptible to mechanical failure when subjected to moderate stresses [22, 10]. Impact of fuel gas impurities on the microstructure and the associated structural integrity of anode material due to oxidation/reaction of Ni particles in the anodes has not been given much attention. Marina *et al.* [23] reported that anode degradation occurred in the presence of PH_3 and the degradation is irreversible due to the formation of Ni_5P_2 phase. Zhi *et al.* [24] have shown that a Ni–YSZ anode had a reduced porosity after exposure to 20ppm PH_3 at 900 °C and the phosphorous species had reacted with Ni and Zr, leading to the formation of $Ni_3(PO)_4$ and ZrP_2O_7. When the impurities of the coal syngeses react with nickel to form NiX, the phase change will bring about considerable amount of the volume expansion, leading to the localized stress concentrations in the contaminated region in the anode. Pilling and Bedworth Ratio (PBR), which is defined as the ratio of the oxide molar volume to that of the metal which gives the oxide [25, 26], is used to measure the volume change induced by oxidation. Huntz [27] emphasized that PBR is an important origin of stresses in scale growth by preferential anionic diffusion. PBR also yields local stress concentration because of impurity oxidation of Ni in Ni-YSZ anodes at the operating temperature of SOFCs. Such stress may be high enough to result in initiation and propagation of local crack and to severely degrade the structural integrity of the anode, leading to ultimate SOFC failure due to the loss of the anode structural integrity.

In this paper, we study the effects of the trace impurity in coal-derived syngases on the mechanical degradation of Ni-YSZ anode using an integrated experimental/modeling approach. Phosphorus is considered as an example of impurity. Anode-supported button cell was used to experimentally explore the migration of phosphorous impurity in the Ni-YSZ anode of SOFCs. X-ray mapping was used to show elemental distributions and new phase formation. Scanning electron microscopy (SEM) was used to capture the microstructure of Ni-YSZ anodes and distribution of Ni and YSZ elements was identified by electron backscatter diffraction (EBSD). The subsequent finite element model is constructed based on the actual microstructure obtained by SEM. Stresses in the Ni-YSZ anode induced by the phase conversion of Ni to Ni_xP_y phase was investigated. Also effect of contaminated area on the stress was obtained with different diffusion depth of the impurity into the anode. It was found that phase conversion induced volume change by impurities of fuel results in very high stress level so that the localized failure of the Ni-YSZ anode may be possible under the operating conditions of SOFC. With time, the propagation of the local failure will crack the anode and lead to failure of the anode structure. In the mean time, the high stress induced by the Ni phase conversion to a series of binary nickel phosphide phase is localized within the contaminated region, and will not propagate outside the contaminated region.

EXPERIMENTAL

Button cells were used to investigate the interaction of phosphorus in synthetic coal gas with the nickel-based anode of solid oxide fuel cells. The fabrication of the anode-supported button cells with an active area of 2 cm^2 was described previously [16]. The anode support was comprised of a Ni/YSZ bulk layer with a 40/60 vol.% solids ratio and approximately 30 vol.% porosity, and was approximately 0.9 mm in thickness. The active anode was approximately 10 μm in thickness and consisted of Ni/YSZ in a 50/50 vol.% solids ratio and 3 vol.% porosity. The electrolyte was 8YSZ (8 mol% yttria-stabilized zirconia), approximately 8 μm in thickness. A 4 μm thick samaria-doped ceria (SDC) barrier layer was applied to the anode/electrolyte structure (pre-sintered at 1375 °C) by screen printing followed by sintering at 1200 °C. A 30–50 μm thick (La0.80Sr0.20)0.98MnO3 (LSM-20) cathode was applied in the same manner and sintered at 1100 °C for 2 h. The cathode current collector was silver mesh, embedded in LSM-20. The anode current collector, comprised of nickel mesh, was embedded in a grid of NiO paste, and was co-fired with the ceria. Cells were sealed to alumina test fixtures with a barium aluminosilicate glass by heating to 850°C in air. Eight cells were configured in a single box furnace, each of which had independent control of anode gases. The Ni/YSZ anode was reduced in situ at 800 °C by moist hydrogen saturated with water at room temperature (ca. 3% H_2O). Thick nickel or alumel wires were utilized as electrical leads. To decrease possible interactions between the leads and phosphine, wires were enclosed into a 2-bore alumina tube, with only an exposed wire loop touching the current collector. The exposed wire was partially coated with a barium aluminosilicate glass to limit the unwanted interactions with the gas environment. Electrolyte-supported button cells were purchased from Fuel Cell Materials. These cells were comprised of an 8YSZ electrolyte of 165±8 μm in thickness, with a Ni/8YSZ anode approximately 30 μm thickness. The cathode consisted of LSM-20, with a ~2 μm gadolinia-doped cerium oxide barrier layer between the cathode and electrolyte. Electrolyte-supported button cells were 25 mm in diameter, with 1.26 cm^2 active anode and cathode areas. A nickel oxide current-collecting grid, approximately 5 μm in thickness, was screen printed onto the cells to facilitate electrical contact. Cells were sealed to alumina test fixtures with a barium aluminosilicate glass, similar to anode-supported cells.

The fuel gas consisted of synthetic coal gas, created by equilibrating 55 parts by volume hydrogen with 45 parts by volume carbon dioxide over a nickel catalyst, yielding a nominal composition of $H_2/CO/CO_2/H_2O=30\%/23\%/21\%/26\%$ at 800°C. Phosphine from cylinder was added to the synthetic coal gas downstream of the catalyst bed using calibrated electronic flow controllers. The operating temperature of the test cell was ranged 700 to 800°C. All tests were performed at a constant current (0.05 to 1 A/cm^2) to ensure that the operating conditions on the cathode were constant throughout the test duration. The readers are hereby referred to Ref. [28] for details of the cell test setup. Following termination of electrochemical tests, surfaces and cross sections of individual cells were analyzed using scanning electron microscopy (SEM) and energy dispersive spectroscopy (EDS), providing information on whether alteration phases were formed, the depth of interaction, and the composition of the reacted anode. X-ray mapping was used to show elemental distributions and new phase formation. Selected samples were additionally analyzed using electron backscatter diffraction (EBSD) to identify specific crystal structures of alteration products. This approach is particularly useful in distinguishing among multiple possible alteration phases that are similar in composition.

During the button cell test, binary nickel phosphide phases Ni_3P, Ni_5P_2, $Ni_{12}P_5$, and Ni_2P all have been identified by EBSD to have been formed for phosphorus concentrations ranging from 0.5 to 10 ppm. Phosphorus was essentially completely captured by the anode, with a sharp boundary formed between reacted and un-reacted portions of the anode. Figure 1 depicts the depth and extent of the nickel–phosphorus interactions during exposure to 1 ppm of PH_3 at 800 °C. The binary phase Ni_3P appeared first, and was predominantly adjacent to the un-reacted nickel. Progression to phases with higher phosphorus content was favored by high phosphine concentration, low temperature, and long exposure time. No phase with phosphorus content higher than that in Ni_2P was observed in this study. Overall, the experimental studies have shown a very strong nickel–phosphorus interaction through the capture of a high fraction of phosphorus in coal gas, and the formation of a sharply defined boundary between converted and unconverted portions of the anode [28].

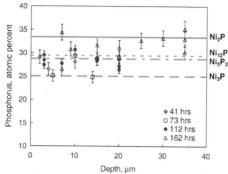

Figure 1 Depth and extent of the nickel–phosphorus interactions during Ni/YSZ anode of the electrolyte-supported cell exposure to 1 ppm of PH_3 at 800 °C. Expected nickel phosphide phases are marked [28]

With the conversion of Ni phase to Ni_xP_y phase, the volume of the Ni phase particle will change. Ni density is well known as 8.902 g/cm³ at 25°C, e.g., [29]. Ni_3P density is 7.86 g/cm³ from [30], and Ni_5P_2 density is 7.71 g/cm³ from [30]. Calculation of the volume change induced by the Ni phase conversion to Ni_xP_y phase is listed in Table 1.

Table 1. Volume change of Ni conversion to Nickel phosphide phase Ni_xP_y

	Formula Weight (FW)	Density (g/cm^3)	d/FW	Ni in FW	Number of Ni in d/FW (Ni*d/FW)	Volume change vs Ni	Length ratio change
Ni	58.7	8.902	0.1516525	1	0.15165247	1	1
Ni_2P	148.37	7.33	0.0494035	2	0.098807036	1.5338014	1.153
Ni_3P	207.07	7.86	0.0379582	3	0.113874535	1.331750509	1.10
Ni_5P_2	355.45	7.71	0.0216908	5	0.108454072	1.39831052	1.118

CONSTRUCTION OF MICROSTRUCTURE-BASED STRESS ANALYSIS MODEL

As illustrated in the experimental study in the previous section, the phosphorus impurity in the fuel stream will react with nickel in the Ni-YSZ anode, leading to a considerable amount of volume expansion. In this section, we study the stress induced by the volume expansion from nickel to nickel phosphide using finite element (FE) stress analysis.

The microstructure of Ni-YSZ anode was obtained by SEM. Figure 2 shows an X-ray map with a view of 42 μm x 32 μm with phases identified by the Gatan Images. The X-ray maps were used to create a mixed elemental map image showing nickel as red and zirconium as blue. The two elements are assumed to be partitioned into pure nickel and yttria stabilized zirconia (YSZ) phases. The mixed map images were then analyzed using ImageJ software to determine the area fractions of nickel, zirconium and pores within the field of view.

The microstructure shown in Figure 2 is taken as the representative volume elements (RVE) and used in the subsequent finite element analysis. The finite element discretization for the microstructure shown in Figure 2 is generated as shown in Figure 3. In this model, only two initial phases are considered to be present in addition to the voids and perfect bonding is assumed between the two phases. The RVE here is taken to represent the cross-section of Ni-YSZ anode. 2D plane strain element is used since the in-plane dimension of the cell is much larger than the thickness of Ni-YSZ anode. In this model, 32,360 plane strain linear triangular elements were used with 18127 nodes. Both left and right sides of the microstructure-based model shown in Figure 3 are constrained along the horizontal direction, and the vertical displacement is allowed. The bottom side is constrained of vertical displacement while allowing horizontal displacement. The top side of the model is treated as free surface. The volume expansion induced by the Ni phase conversion to the nickel phosphide phase is simulated by an isotropic straining model. This model is implemented in the commercial finite element software MSC.MARC by user subroutine.

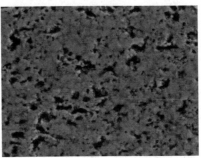

Figure 2 Microstructure and distribution of two phases identified by Gatan Images.
Red: nickel; Blue: YSZ

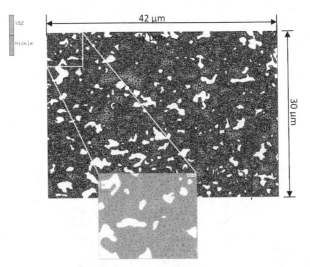

Figure 3 Finite element model used in analysis

The mechanical properties of pure Ni phase and YSZ phase are temperature dependent. Their Young's moduli are plotted in Figure 4 as functions of temperature [33, 34]. In this analysis, the operating temperature is constant, therefore, CTE does not play any role without temperature change. As listed in Table 1, the volume change induced by the Ni phase change to nickel phosphide is different for the series of binary nickel phosphide, from Ni_2P, Ni_3P, Ni_5P_2, to $Ni_{12}P_5$. The conversion from Ni phase to Ni_2P phase causes the largest volume expansion, see Table 1. For analysis purpose here, all the nickel phosphide phase is considered to be Ni_2P. The modulus of Ni_2P is taken as 130 GPa [35].

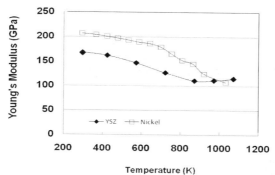

Figure 4 Temperature dependent Young's moduli of YSZ and Ni

STRESSES IN NI-YSZ ANODE

In general, the depth of the nickel phosphide phase penetration into the Ni-YSZ support was non-uniform due to the strong Ni – P interactions and a mix of different Ni-P phases were found to coexist [28]. In this section, all the Ni crystallites in the RVE were assumed to be fully converted to Ni_2P. No partial conversion cases were considered.

Figure 5 shows the distribution of equivalent von Mises stress in the YSZ phase, Ni-YSZ combined and Ni phase only, respectively. Figure 5(a) depicts the distribution of equivalent von Mises stress only in the YSZ: many 'hot' spots with high localized stress were predicted. The highest stress level in the YSZ was predicted to be near 5 GPa, much higher than the strength of the YSZ tested at the operating temperature [33, 36]. Therefore, localized failure in the form of localized micro-cracks is likely to appear first in the YSZ phase, and the cracks would subsequently propagate through the rest of the support, leading to the macro-cracks in the Ni-YSZ, see Figure 6(b). White arrows in Figure 5(b) indicate the directions of potential propagation of the micro-cracks to form macro-cracks. These findings confirm that Ni volume expansion due to the transition from Ni to nickel phosphide is responsible for stress striation in the Ni-YSZ support. These high stress zones may result in transverse crack propagation.

Figure 5(c) illustrates how many 'hot' spots (high stress level) were developed in the Ni-P phase, also with a very high magnitude, ~5 GPa. This is due to the fact that Ni_2P was treated as elastic material. The melting point of Ni_2P is near 870-890°C [37], and because all of the tests were performed at 700-800°C, Ni-P phase was at temperatures very close to melting. Therefore, considerable degree of softening and creep is expected to occur for Ni_2P at operating temperatures, which should result in extensive grain growth and coalescence under a small amount of stress. These effects were not considered in the present study. Also, no stress assisted Ni-P conglomeration due to creep deformation is considered in the current study, and Ni coalescence and agglomeration is considered to be solely due to the high mobility of Ni-P phase at the operating temperatures. Any consideration of a stress assisted creep deformation at temperature will only exacerbate the above described anode structural degradation.

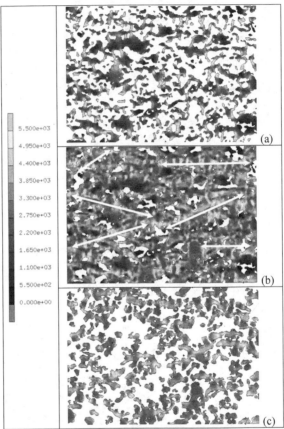

Figure 5 Distribution of von Mises stress (in MPa) of the Ni-YSZ support induced by volume change due to Ni conversion to Ni_2P: (a) in YSZ phase; (b) Ni-YSZ support; (c) in Ni phase.

Figure 6 shows the SEM image of the cross-section of Ni-YSZ anode after 650 h test with 1 ppm phosphorous impurity in the fuel stream. Horizontal stripe appears after the diffusion of phosphorus into the depth of the Ni-YSZ anode. These horizontal stripes are origins of the potential cracks. At this moment, the Ni phase has not yet converted into nickel phosphide phase completely. Note that stress assisted/induced Ni diffusion/clustering/conglomeration as shown in Figure 9 is not included in the current study. It will only cause higher localized stress and exacerbate the anode structural degradation. The Ni striation shown in Figure 6 is a clear demonstration of this effect.

Figure 6 SEM image of cross-section of Ni-YSZ anode after 650h test with 1ppm phosphorous impurity of fuel

PREDICTED STRESSES IN Ni-YSZ WITH VARIED FRACTION OF Ni-P PHASE
    In this section, the transformation of Ni to Ni-P phase was assumed to occur within a certain distance from the anode surface toward the electrolyte at a given percentage. As shown in Figure 7, 33%, 66%, and 100 % of the total Ni-YSZ thickness, respectively, were considered to be affected by phosphorous, i.e., Ni particles within 1/3, 2/3 from the surface, or the whole Ni-YSZ support including the active anode were fully converted to Ni-P. Again, Ni_2P was assumed to be the only nickel phosphide phase.

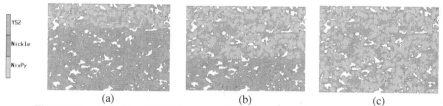

(a)   (b)   (c)

Figure 7 Fraction of the Ni-YSZ support thickness where Ni was converted to Ni-P:
(a) 33%; (b) 66%, and (c) 100%

    Figure 8 shows the equivalent von Mises stress in the Ni-YSZ support with varied phosphorus penetration depths. It can be seen that independent of the phosphorus penetration, the regions of very high stress were always predicted within the transformed zone. The nickel phosphide appears to act as a localized stress riser due to volume expansion. It is obvious that the stratified microstructure with horizontal alternating layers of Ni-P and YSZ was a consequence of stress. The horizontal stripes were origins of the potential cracks. Compared to the reported strength of YSZ at operating temperatures [33, 36], the stress induced by phosphorous was high enough to initiate the micro-cracks in the YSZ. The horizontal propagation of the micro-cracks would then form near-horizontal crack stripes in the anode support. Again, several main high stress stripes were predicted within the contaminated region, and no driving force was predicted for the cracks to spread out of the contaminated region. Such a localized degradation phenomenon indicates that only the structural integrity of the affected region will be compromised by the presence of phosphorus. Unaltered Ni-YSZ should not show any signs of structural degradation.

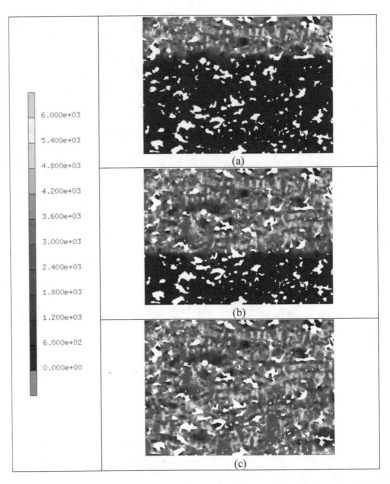

Figure 8 Effect of impurity diffusion on stress distribution. (a) 1/3 depth of anode; (b) 2/3 depth of anode; full depth of anode.

Figure 9 and 10 show SEM images of cross-section of the tested Ni-YSZ anode from the same Ni-YSZ anode-supported cell. They display the areas near to fuel flow surface and the interface of anode and electrolyte, respectively. This Ni-YSZ anode-supported cell was tested for 1660 h at 750 °C in synthetic coal gas with 5 ppm PH_3. Ni was converted to Ni_5P_2 at the active interface and to Ni_2P in the upper part, respectively. In general, the conversion ratio of Ni phase to nickel phosphide in the vicinity of the interface of anode and electrolyte is less than that in the area of fuel flow surface. Also, the volume change induced by the Ni phase conversion to Ni_5P_2 is less than that from Ni phase conversion to Ni_2P as listed in Table 1. Therefore, the stress level in the area near the interface of the anode and electrolyte is less than the stress in the area of fuel flow surface. The high stress in the Ni_2P contaminated area causes initiation and propagation of cracks and forms several failure stripes. It is also observed that the cracks are localized within the contaminated area, and does not spread to uncontaminated area. In Figure 10, the low volume change induced by Ni phase conversion to Ni_5P_2 in the vicinity of the anode and electrolyte interface has not created high enough stresses in the contaminated region to initiate any micro-cracks.

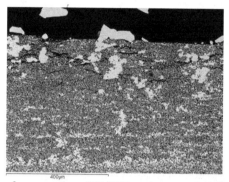

Figure 9 SEM images of cross-section of Ni-YSZ anode near to fuel flow surface after test of 1660 h at 750 °C in synthetic coal gas with 5 ppm PH_3

Figure 10 SEM images of cross-section of Ni-YSZ anode near to active interface of anode and electrolyte after test of 1660 h at 750 °C in synthetic coal gas with 5 ppm PH_3

CONCLUSIONS

In the paper, effect of the trace impurity of the coal syngas on the mechanical degradation of Ni-YSZ anode was studied using a combined experimental/modeling approach. Phosphorus is taken account as an example of impurity, and the conclusions can be summarized as: (1) Phase conversion from Ni to Ni phosphide due to phosphorus in fuel stream results in large localized volume increase in the anode. The volume change is dependent on the phase state of nickel phosphide. The phase conversion from Ni phase to Ni_2P causes the highest volume change; (2) Volume change induced by the phase conversion of Ni phase to nickel phosphide creates high level localized stress in the anode. The stress level is closely related to the phase state of Ni phase to nickel phosphide; (3) The high stress induced by the Ni phase conversion to a series of binary nickel phosphide phase is localized within the contaminated region, and will not create enough driving force for the cracks to propagate outside of the contaminated region.

ACKNOWLEDGEMENTS

The Pacific Northwest National Laboratory is operated by Battelle Memorial Institute for the United States Department of Energy under Contract DE-AC06-76RL01830. The work was funded as part of the Solid-State Energy Conversion Alliance (SECA) Core Technology Program by the U.S. Department of Energy's National Energy Technology Laboratory (NETL).

REFERENCES

[1]A. Boudghene Stambouli and E. Traversa, Solid Oxide Fuel Cells (SOFCs): A Review of an Environmentally Clean and Efficient Source of Energy, *Renew. Sust. Energ. Rev.*, **6**, 433–55 (2002).

[2]H. Yakabe, T. Ogiwara, M. Hishinuma, and I. Yasuda, 3-D model calculation for planar SOFC, *J. Power Sources.*, **102**, 144-154 (2001).

[3]M.A. Khaleel, Z. Lin, P. Singh, W. Surdoval, and D. Collin, A finite element analysis modeling tool for solid oxide fuel cell development: coupled electrochemistry, thermal and flow analysis in MARC®, *J. Power Sources.*, **130**, 136-148 (2004).

[4]M. C. Williams, J. P. Strakey, and W. A. Surdoval, U.S. Department of Energy's Solid Oxide Fuel Cells: Technical Advances, *Int. J. Appl. Ceram. Technol.*, **2** (4), 295–300 (2005).

[5]Michael B. Pomfret, Daniel A. Steinhurst, David A. Kidwell, Jeffrey C. Owrutsky, Thermal imaging of solid oxide fuel cell anode processes, *Journal of Power Sources*, **195**, 257–262 (2010).

[6]JianEr Bao, Gopala N. Krishnan, Palitha Jayaweera, Jordi Perez-Mariano, Angel Sanjurjo, Effect of various coal contaminants on the performance of solid oxide fuel cells: Part I. Accelerated testing, *Journal of Power Sources*, **193**, 607–616 (2009).

[7]U.S. Dept. of Energy Office of Fossil Energy, *FutureGen: Integrated hydrogen, electric power production and carbon sequestration research initiative*, March 2004.

[8]Teagan WP, Thijssen JHJS, Carlson EJ, Read CJ. Current and future cost structures of fuel cell technology alternatives. in: McEvoy AJ, Editor. *Proc. 4th European Solid Oxide Fuel Cell Forum.* Lucerne, Switzerland, 2000. p. 969-980.

[9]W.Z. Zhu, S.C. Deevi, A review on the status of anode materials for solid oxide fuel cells, *Mater. Sci. Eng. A* **362**, 228–239 (2003).

[10]W. Li, K. Hasinska, M. Seabaugh, S. Swartz and J. Lannutti, Curvature in solid oxide fuel cells, *J. Power Sources.*, **138**, 145-155, (2004).

[11]M. Diaz-Somoano, M.R. Martinez-Tarazona, Trace element evaporation during coal gasification based on a thermodynamic equilibrium calculation approach, *Fuel*, **82**, 137-145 (2003).

[12]Díaz-Somoano, M., Kylander, M.E., López-Antón, M.A., Suárez - Ruiz, I., Martínez-Tarazona, M.R., Ferrat, M., Kober, B., Weiss, D.J., Stable lead isotope compositions in selected coals from around the world and implications for present day aerosol source tracing, *Environmental Science and Technology*, **43**(4), 1078-1085 (2009).

[13]G. Krishnan, P. Jayaweera, and J. Perez, Effect of Coal Contaminants on Solid Oxide Fuel System Performance and Service Lif, *SRI Technical Progress Report*, (2006).

[14]F. N. Cayan, M. Zhi, S. R. Pakalapati, I. Celik, N. Wu, and R. Gemmen, Effects of coal syngas impurities on anodes of solid oxide fuel cells, *J. Power Sources*, **185**, 595-602 (2008)

[15]Oktay Demircan, Chunchuan Xu, John Zondlo, Harry O. Finklea, *In situ* Van der Pauw measurements of the Ni/YSZ anode during exposure to syngas with phosphine contaminant, *Journal of Power Sources*, **194**, 214–219 (2009).

[16]O. A. Marina, L. R. Pederson, D. J. Edwards, C. A. Coyle, J. W. Templeton, M. H. Engelhard, and Z. Zhu, Effect of Coal Gas Contaminants on Solid Oxide Fuel Cell Operation, *ECS Transactions*, **11** (33), 63-70 (2008).

[17]O. A. Marina, L. R. Pederson, C. A. Coyle, E. C. Thomsen, and G. W. Coffey, Ni/YSZ Anode Interactions with Impurities in Coal Gas, *ECS Transactions*, **25** (2), 2125-2130 (2009).

[18]J. P. Trembly, R. S. Gemmen, and D. J. Bayless, The effect of IGFC warm gas cleanup system conditions on the gas-solid partitioning and form of trace species in coal syngas and their interactions with SOFC anodes, *J. Power Sources*, **163**, 986-996 (2007).

[19]Natasha M. Galea, Eugene S. Kadantsev, and Tom Ziegler, Studying Reduction in Solid Oxide Fuel Cell Activity with Density Functional Theory-Effects of Hydrogen Sulfide Adsorption on Nickel Anode Surface, *J. Phys. Chem. C*, **111**, 14457-14468 (2007).

[20]Natasha M. Galea, Eugene S. Kadantsev, and Tom Ziegler, Modeling Hydrogen Sulfide Adsorption on Mo-Edge MoS2 Surfaces under Solid Oxide Fuel Cell Conditions, *J. Phys. Chem. C*, **113**, 193–203 (2009).

[21]J. Mougin, S. Ravel, E. de Vito, M. Petitjean, Effect on Performance and Degradation Mechanisms, *ECS Transactions*, **7** (1) 459-468 (2007).

[22]Liu, W.N., Sun, X., Khaleel, M.A., Qu, J.M., Global failure criteria for positive/electrolyte/negative structure of planar solid oxide fuel cell, *Journal of Power Sources*, **192** (2), 486-493 (2009).

[23]O.A.Marina, L.R. Pederson, D.J. Edwards, C.W. Coyle, J. Templeton, M. Engelhard, Z. Zhu, *Proceedings of the 8th Annual SECA Workshop*, San Antonio, TX, 2007.

[24]M. Zhi, X. Chen, H. Finklea, I. Celik, N.Q. Wu, *J. Power Sources*, **183**, 485–490 (2008)

[25]Kofstad, P., Fundamental aspects of corrosion by hot gases. *Materials Science & Engineering A: Structural Materials: Properties, Microstructure and Processing*, **A120-1**(1-2 pt 1), 25-29 (1989).

[26]Huntz, A.M. Scale growth and stress development, *Materials Science and Technology*, **4**(12) 1079-1088 (1988)

[27]Huntz, A.M., Stresses in NiO, Cr2O3 and Al2O3 oxide scales. *Materials Science & Engineering A: Structural Materials: Properties, Microstructure and Processing*, **A201**(1-2), 211-228 (1995).

[28]O.A. Marina, C.A. Coyle, E.C. Thomsen, D.J. Edwards, G.W. Coffey, L.R. Pederson, Degradation mechanisms of SOFC anodes in coal gas containing phosphorus, *Solid State Ionics*, **181** (8-10), 430-440 (2010).

[29]*CRC Handbook of Chemistry and Physics*, 80th Edition, 1999-2000.

[30]G.S. Cargill III, Structural Investigation of nano-crystalline nickel phosphorus alloys, *Journal of Applied Physics*, **41** (1), 12-29 (1970).

[31]S. Oryshchyn, V. Babizhetskyy, S. Chykhriy, L. Aksel'rud, S. Stoyko, I. Bauer, R. Guerin, Y. Kuz'ma, Crystal structure of Ni5P2, *Inorganic materials*, **40** (4), 380-385 (2004).

[32]MARC, 2007. Version 2007r1. http://www.mscsoftware.com/products/marc.cfm?Q=131&Z=396&Y=400

[33]http://www.seca.doe.gov/~secamat

[34]Nickel 201, Alloy Digest, published by Engineering Alloys Digest, Inc, Upper Montelair, New Jersey, 1969.

[35]X.D. Li, B. Bhushan, K. Takashima, C.K. Baek,Y.K. Kim, Mechanical characterization of micro/nanoscale structures for MEMS/NEMS applications using nanoindentation techniques, *Ultramicroscopy*, **97**, 481–494 (2003).

[36]M. Mori, Y. Hiei, H. Itoh, G.A. Tompsett, N.M. Sammes, Evaluation of Ni and Ti-doped Y2O3 stabilized ZrO2 cermet as an anode in high-temperature solid oxide fuel cells, *Solid State Ionics*, **160**, 1-14 (2003).

[37]C. Schmetterer, J. Vizdal, H. Ipser, A new investigation of the system Ni-P, *Intermetallics*, **17**, 826–834 (2009).

STRATEGIES TO IMPROVE THE RELIABILITY OF ANODE-SUPPORTED SOLID OXIDE FUEL CELLS WITH RESPECT TO ANODE REOXIDATION

Manuel Ettler, Norbert H. Menzler, Georg Mauer, Frank Tietz, Hans Peter Buchkremer, Detlev Stöver
Forschungszentrum Jülich GmbH, Institute of Energy Research, IEK-1
Jülich, GERMANY

ABSTRACT

Solid oxide fuel cells (SOFCs) are highly efficient devices for converting the chemical energy of a fuel into electrical energy featuring high fuel flexibility. Forschungszentrum Jülich has been developing the concept of a planar cell design based on a nickel/yttria-stabilized zirconia anode substrate for approximately twenty years. This development work covers the full spectrum ranging from fundamental research in materials science, processing and engineering issues, such as cell and stack design and construction aiming at stationary and mobile applications, up to cell and stack testing and system modeling. The main focus at present is the improvement of long-term stability and reliability for cells and stacks. This contribution will present the concept developed in Jülich and the challenges and strategies involved in achieving better reliability here especially by the example of the problem of nickel reoxidation in the anode. Operating conditions causing nickel reoxidation may result in irreversible microstructural changes and macroscopic expansion of the anode, potentially leading to catastrophic cell failure.

INTRODUCTION

The state-of-the-art substrate and anode material for anode-supported SOFCs is a Ni/YSZ-cermet offering both excellent catalytic properties for fuel oxidation and effective current conduction. Typically, the substrate is manufactured by tape casting followed by a pre-sintering step. The substrate is then coated with the anode, electrolyte, cathode and current collector by screen printing. The raw materials for both substrate and anode are YSZ and NiO powders. Particle size distribution, composition and the resulting microstructure are different, as the supporting substrate usually has a coarser structure and has to ensure good mechanical strength, gas permeability, thermal expansion compatibility and electronic conductivity. The electrochemically active anode needs to have a fine microstructure with the largest possible three-phase boundary and high percolation in both YSZ and Ni phases. The NiO of the substrate and anode is reduced to Ni during system start-up[1]. The anode-supported SOFC developed at Forschungszentrum Jülich has very favorable prospects because of its remarkably good electrical performance and reduced operating temperature[2].

However, with respect to commercialization of SOFCs not only is high performance of great importance but also long-term stability and reliability of cells, stacks and systems. One strategic focus of our current work is the improvement of the reliability of SOFCs by addressing issues such as thermal cycling, carbon deposition and reoxidation stability. Our strategy is based on the application, characterization and adaptation of alternative processing routes and techniques as well as the development, characterization and application of alternative materials. As an example, we present our approaches and results in enhancing the reoxidation stability of our SOFCs.

REOXIDATION STABILITY – ONE MAJOR ISSUE

As long as fuel (e.g. hydrogen, reformate or a methane/water vapor mixture appropriate for internal reforming) is supplied to the anode side of the SOFC during operation, the Ni particles in the substrate and anode remain in the reduced state. However, if the fuel supply is interrupted, oxygen will diffuse from the cathode side through the electrolyte to the anode side or will enter via imperfect seals from outside the system or the BoP (balance of plant) components (e.g. reformer, afterburner). Interruption of the fuel supply may occur accidentally as a result of an error in the system control (unintentional) or deliberately upon system shutdown (intentional). The presence of oxygen on the anode side of the cell, especially at typical operation temperatures for SOFCs between 700°C and 800°C, will lead to reoxidation of Ni. Operation with undesirably high fuel utilization can also cause the oxygen activity to rise above the equilibrium condition between Ni and NiO and form NiO[3]. The reoxidized Ni can be re-reduced, but various investigations have demonstrated that the structure of the substrate and anode cannot be restored[4-10]. The structural changes in the substrate and anode upon reoxidation lead to dimensional changes that generate stresses in the substrate, anode and other cell components, potentially promoting damage in all layers of the cell and therefore degrading cell performance or even causing complete failure of the cell[3, 5, 7, 8, 11, 12].

Micrographs taken from exactly the same location of an anode-supported cell from the co-fired state to the status after the second reoxidation give further insights into the reoxidation process revealing the mechanism leading to cell damage[13-15].

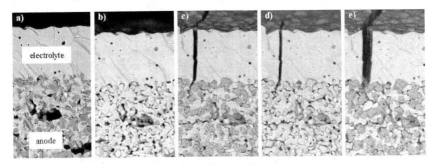

Figure 1. SEM study on microstructural changes in anode and substrate due to reduction/reoxidation (redox) cycles. a) co-fired state, b) reduced, c) reoxidized, d) re-reduced and e) re-reoxidized.

The initially dense NiO particles (Fig. 1a) shrink upon reduction (Fig. 1b). The microstructure after reoxidation differs from the initial as-prepared microstructure. The particles are fragmented, have a spongy microstructure (Fig. 1c), a greater porosity (~ 30%) and hence occupy a larger volume than in the as-prepared state. Microcracks in the anode and electrolyte are visible (Fig. 1c). Reducing the NiO again results in dense particles (Fig. 1d), which appear to occupy a larger volume than in the initial reduced state (Fig. 1b) and display a coarser structure. NiO particles after the second reoxidation appear to be even more fragmented, again showing the spongy microstructure (Fig. 1e). The electrolyte crack is seen to have opened even further. Macroscopically, the crack density after the second reoxidation is larger than after the first. This could be the result of the coarser structure of the Ni in the re-reduced state. Additional TEM investigations have also revealed the increased porosity in the reoxidized NiO particles[13].

An extensive study of the behavior of state-of-the-art anode-supported SOFCs manufactured at Forschungszentrum Jülich upon reoxidation has shown that cells can be reoxidized to a certain degree without being damaged catastrophically[15-17]. The limit for the degree of oxidation that does not cause complete failure as well as the degree of oxidation of the cell after reoxidation is influenced not only by the substrate characteristics, but also by the temperature and time of oxidation and the air flow rate applied. In Fig. 2, the effect of the temperature of oxidation is shown as one example.

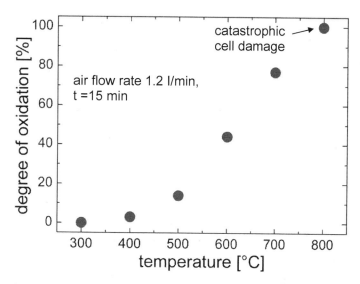

Figure 2. Temperature dependence of the degree of oxidation of anode-supported SOFCs based on a 500-μm-thick substrate reoxidized for 15 min with an air flow rate of 1.2 l/min.

The free-standing samples were reoxidized for 15 min with a continuous air flow of 1.2 l/min. At 300°C the substrate shows no mass gain and therefore is not reoxidized at all. With increasing temperature, the degree of oxidation rises reaching a value of nearly 80% at 700°C. However, this reoxidation does not lead to catastrophic cell damage. At 800°C the substrate and anode are completely reoxidized showing numerous cracks in the electrolyte. This reveals that the cells have a limited reoxidation stability. However, complete failure of the cell upon reoxidation can only be avoided in the case of intentional redox cycling, when temperature time and air flow can be controlled. One major task in enhancing the long-term stability and reliability of the cell is therefore to improve its reoxidation stability or even make it completely reoxidation tolerant. This contribution presents the approaches taken at Forschungszentrum Jülich to improve the reoxidation stability of anode-supported SOFCs.

EXPERIMENTAL AND RESULTS

In a first approach, the microstructure of the substrate was varied aiming at an optimization of the microstructure with respect to reoxidation stability. It turned out that the reoxidation stability of the cells could be improved by applying substrates with lower porosity and higher mechanical stability.

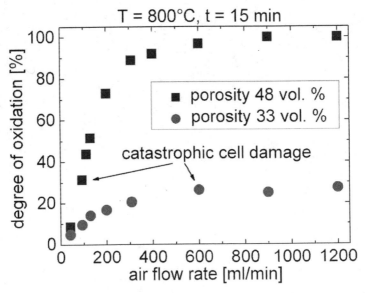

Figure 3. Comparison of the degree of oxidation of anode-supported SOFCs based on two types of 500-μm-thick substrates with different porosity reoxidized for 15 min at 800°C depending on air flow rate.

While the degree of oxidation of cells based on a substrate with 48 vol. % porosity strongly increased with increasing air flow rate, reaching 100% at a value of about 0.7 l/min, oxidation is limited to about 27% for the cells based on a substrate with about 33 vol. % porosity even at higher air flow rates. Although the damage threshold for both types of cells for reoxidation at 800°C is found at a degree of oxidation of about 25%, first cracks in the electrolyte of the cells can be observed at an air flow rate of about 0.1 l/min for cells based on substrates with higher porosity and at about 0.6 l/min for cells based on the denser substrate. This indicates that due to diffusion limitations the reaction rate of the reoxidation is much lower for the cells based on the denser substrate. Therefore the cells based on a substrate with lower porosity can tolerate a higher amount of oxygen without catastrophic failure although the damage threshold in terms of degree of oxidation may not be changed (see Fig. 3). In electrochemical measurements it was demonstrated that cells based on the substrate with a porosity of 30 vol. % show no losses compared to the cells based on more porous substrates (37 vol. %) regarding electrical performance (see Fig. 4). By optimizing the substrate microstructure it was thus possible to significantly improve the reoxidation stability. However further improvement is needed to ensure that anode-supported SOFCs are also capable of surviving unintentional and uncontrolled reoxidation.

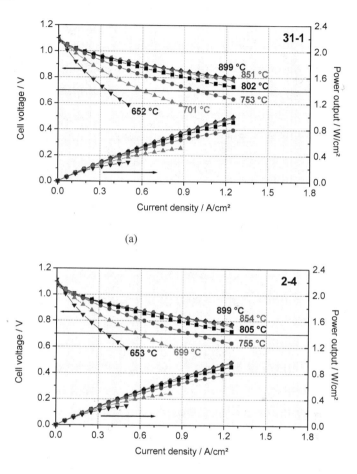

(a)

(b)

Figure 4. Current-voltage curves for 5 x 5 cm^2 anode-supported single cells with LSM cathode as a function of the temperature (fuel gas: H$_2$ (3% H$_2$O) = 1000 ml/min, oxidant: air = 1000 ml/min), anode and electrolyte here applied by vacuum slip casting, (a) cell based on a substrate with 37 vol. % porosity, (b) cell based on a substrate with 30 vol. % porosity.

To further improve the reoxidation stability of the state-of-the-art anode-supported SOFC substrate, its porous structure was coated with an Al$_2$O$_3$ layer. The aim was to coat the outer as well as the inner surface of the structure. Therefore the ALD-CVD technology was applied. Coating parameters were successfully optimized to achieve a continuous layer on the outer and inner surface. SEM pictures and also EDX analysis of the fracture surfaces revealed that the entire inner and outer substrate surface had been successfully coated with a 50- to 100-nm-thick Al_2O_3 layer (see Fig. 5 and 6).

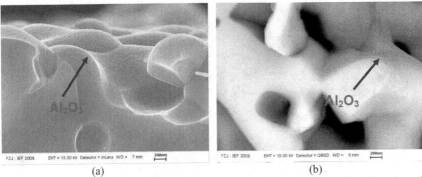

(a)   (b).

Figure 5. SEM pictures of the fracture surface of a substrate coated with Al_2O_3, (a) coating of the outer surface, (b) coating of the inner surface.

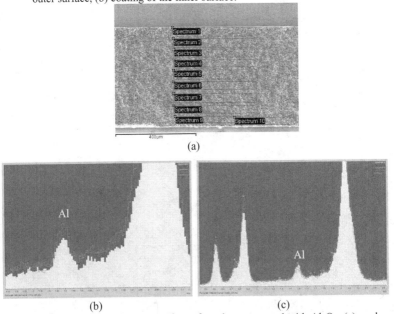

Figure 6. EDX spectra on a fracture surface of a substrate coated with Al_2O_3, (a) analyzed areas, (b) spectra of areas 1 to 5, (c) spectra of areas 6 to 10.

In Fig. 5, two examples are presented of various SEM pictures taken from the fracture surface of an Al_2O_3-coated substrate. Fig. 5 (a) shows the coating on the outer surface, Fig. 5 (b) the coating of the inner structure. In Fig. 6, the picture (a) shows the fracture surface of the coated sample. The areas analyzed via EDX are indicated and labeled "Spectrum 1" to "Spectrum 10". All corresponding EDX spectra (Fig. 6 (b) and (c)) indicate the presence of Al in the analyzed region. This reveals that the inner surface of the substrate is coated with an Al_2O_3 layer. As the peak intensities of all 10 spectra show only small variations, it can be assumed that the coating of the inner surface is relatively uniform.

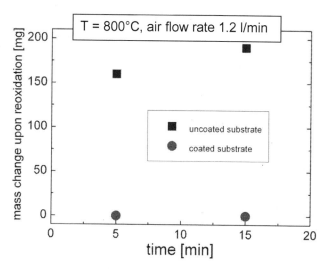

Figure 7. Mass change of coated and uncoated samples upon air exposure at 800°C.

In first reoxidation experiments with coated substrates, the potential of the Al_2O_3 layer as a protective coating was demonstrated. Coated and uncoated samples were exposed in air (flow rate 1.2 l/min) at 800°C for 5 and 15 min, respectively. While the uncoated samples showed a significant weight gain of between 150 and 200 mg, and were therefore massively reoxidized, the coated samples did not show any mass change at all and hence could not be reoxidized (see Fig. 7).

In a second approach, Y-substituted $SrTiO_3$ substrates were developed as an alternative to the state-of-the-art Ni/YSZ-cermet substrates. So far the potential of cells based on ceramic anodes, especially with respect to reoxidation stability, has only been demonstrated on small pellet-sized cells. However, the electrochemical performance was barely satisfactory. Recently, a scale-up was achieved from the small pellet-sized cells to planar cells $(50 \times 50 \ mm^2)$[18]. The electrical performance of the SYT-anode-supported cells was characterized by current-voltage (I/U) and impedance measurements. The value for the open circuit voltage (OCV) at 800°C was close to the theoretical value and reached 1.09 V. At this temperature, an average current density of 1.22 Acm^{-2} at 0.7 V was achieved. The actual data for all the tested cells varied between 1.0 and 1.5

Acm^{-2} at 0.7 V and 800°C and therefore demonstrated good electrical performance. The cells survived redox cycling at 750°C (200 cycles) and 800°C (50 cycles) without mechanical damage (see Fig. 8). The 200 redox cycles at 750°C caused a decrease in the OCV of only 1.3%. After 200 redox cycles at 750°C with 10 min in H$_2$ and 10 min in air the cell performance decreased by 35%. This is probably due to the reduced electrical conductivity of the substrate, which can be restored by extended reduction periods, as shown by the 50 redox cycles at 800°C with 2 h in H$_2$ and 10 min in air.

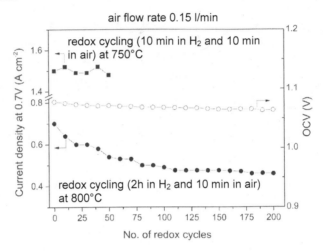

Figure 8. OCV (open circles) and current density at 0.7 V vs. the number of redox cycles with 10 min in H$_2$ and 10 min in air at 750°C (closed circles) and 2 h in H$_2$ and 10 min in air at 800°C (closed squares)[18].

CONCLUSION AND SUMMARY

The key to the next step towards commercialization of anode-supported SOFCs is long-term stability and reliability. In this respect, reoxidation stability is one major issue. At Forschungszentrum Jülich this issue has been addressed not only by an extensive study carried out to characterize the behavior of today's state-of-the-art anode-supported cells upon reoxidation, but also by developing new strategies to improve the reoxidation stability of the cells. These strategies imply the application, characterization and adaptation of alternative processing routes and techniques as well as the development, characterization and application of alternative materials. The reoxidation stability of anode-supported SOFCs developed at Forschungszentrum Jülich has been improved by optimizing the substrate microstructure. This optimization was based on reducing the porosity of the substrate. It was demonstrated that the denser substrate structure leads to a reduced reaction rate for reoxidation without negatively influencing the electrical performance of the cells. However, it turned out that further improvement was needed to enable the cells to tolerate not only intentional and controlled reoxidation to a certain degree, but also to withstand unintentional, accidental reoxidation. Therefore two different approaches were taken with promising results: On the one hand, a protective Al_2O_3 coating of the outer and inner surface of the state-of-the-art Ni/YSZ-cermet substrate was applied by the ALD-CVD technology. It was demonstrated that the complete substrate surface was successfully coated. The resulting layer was between 50 and 100 nm thick. First reoxidation tests revealed that this coating has the desired protective effect. While the mass of uncoated substrates increased significantly upon exposure in air at 800°C, the coated substrate showed absolutely no mass change.

On the other hand, cells based on Y-substituted $SrTiO_3$ substrates were developed and manufactured showing good electrical performance and reoxidation stability. Therefore, Y-substituted $SrTiO_3$ has the potential to be the future state-of-the-art material for full ceramic anode substrates solving the problem of reoxidation stability and hence improving the reliability of future SOFCs.

OUTLOOK

Further work needs to be done to explore the potential of the two approaches discussed in this contribution.

In the case of the protective coating of the state-of-the-art Ni/YSZ-cermet substrates, the reliability of the coating and its protective properties upon repeated redox cycling needs to be analyzed and demonstrated. The application of the ALD-CVD technology has to be extended to other coating materials for further optimization. In this way it could also become an interesting method for other applications. If further investigations and analyses reveal that the method is applicable for cell production it has to be integrated in the processing chain.

In the case of the reoxidation-stable Y-substituted $SrTiO_3$ anode substrates, further characterization and optimization is needed aiming at a scale-up to commercial cell dimensions and stack testing and finally integration in the cell processing as alternative option or even standard material.

ACKNOWLEDGEMENTS

The authors wish to express their thanks to Dr. Wolfgang Schafbauer for his work on the development of the substrate with lower porosity, to Dr. Quianli Ma for his work on the alternative substrate materials and to Frank Vondahlen for his work on ALD-CVD coatings. Dr. Doris Sebold and Dr. Vincent A. C. Haanappel are gratefully acknowledged for taking the SEM pictures and performing the EDX analyses as well as the electrochemical measurements.

REFERENCES

[1] W.A. Meulenberg, N.H. Menzler, H.-P. Buchkremer, and D. Stöver, Manufacturing routes and state-of-the-art of the planar Jülich anode supported concept for solid oxide fuel cells, *Ceramic Transactions*, **127**, 99 (2002).

[2] R. N. Basu, G. Blass, H. P. Buchkremer, D. Stöver, F. Tietz, E. Wessel, and I. C. Vinke, Simplified processing of anode-supported thin film planar solid oxide fuel cells, *J. Eur. Ceram. Soc.*, **25**, 463 (2005).

[3] D. Sarantaridis, and A. Atkinson, Redox Cycling of Ni-Based Solid Oxide Fuel Cell Anodes: A Review, *Fuel Cells*, **7**, 246 (2007).

[4] T. Klemensø, C. C. Appel, and M. Mogensen, In Situ Observations of Microstructural Changes in SOFC Anodes during Redox Cycling, *Solid-State Lett.*, **9**, A403 (2006).

[5] T. Klemensø, C. Chung, P. H. Larsen, and M. Mogensen, The Mechanism Behind Redox Instability of Anodes in High-Temperature SOFCs, *J. Electrochem. Soc.*, **152**, A2186 (2005)

[6] D. Waldbillig, A. Wood, and D. G. Ivey, Thermal Analysis of the Cyclic Reduction and Oxidation Behaviour of SOFC Anodes, *Solid State Ionics*, **176**, 847 (2005)

[7] D. Waldbillig, A. Wood, and D. G. Ivey, Electrochemical and Microstructural Characterization of the Redox Tolerance of Solid Oxide Fuel Cell Anodes, *J. Power Sources*, **145**, 206 (2005)

[8] J. Malzbender, E. Wessel, R.W. Steinbrech, and L. Singheiser, Reduction and Re-Oxidation of Anodes for Solid Oxide Fuel Cells, *Solid State Ionics*, **176**, 2201 (2005)

[9] Y. Zhang, B. Liu, B. Tu, Y. Dong, and M. Cheng, Redox Cycling of Ni–YSZ Anode Investigated by TPR Technique, *Solid State Ionics*, **176**, 2193 (2005)

[10] D. Fouquet, A. C. Müller, A. Weber, and E. Ivers-Tiffée, Kinetics of Oxidation and Reduction of Ni/YSZ Cermets, *Ionics*, **9**, 103 (2003)

[11] A. Wood, M. Pastula, D. Waldbillig, and D.G. Ivey, Initial Testing of Solutions to Redox Problems with Anode-Supported SOFC, *J. Electrochem. Soc.*, **153**, A1929 (2006)

[12] M. Pihlatie, A. Kaiser, P. H. Larsen, and M. Mogensen, **Dimensional Behavior of Ni–YSZ Composites during Redox Cycling**, *J. Electrochem. Soc.*, **156**, B322 (2009)

[13] J. Malzbender, and R.W. Steinbrech, Advanced measurement techniques to characterize thermo-mechanical aspects of solid oxide fuel cells, *J. Power Sources*, **173**, 60 (2007)

[14] E. Ivers-Tiffée, H. Timmermann, A. Leonide, N. H. Menzler, and J. Malzbender, Methane reforming kinetics, carbon deposition, and redox durability of Ni/8 yttria-stabilized zirconia (YSZ) anodes, in: W. Vielstich, H. A. Gasteiger, H. Yokokawa (Eds.), *Handbook of Fuel Cells – Fundamentals, Technology and Applications*, Volume 5: Advances in Electrocatalysis, Materials, Diagnostics and Durability, John Wiley & Sons Ltd., Chichester, England, 933 (2009)

[15] M. Ettler, H. Timmermann, J. Malzbender, A.Weber, and N.H. Menzler, Durability of Ni anodes during reoxidation cycles, *J. Power Sources*, **195**, 5452 (2010)

[16] M. Ettler, N. H. Menzler, H. P. Buchkremer, and D. Stöver, Characterization of the re-oxidation behaviour of anode-supported SOFCs, in: P. Singh, N. P. Bansal (Eds.), *Advances in Solid Oxide Fuel Cells IV: Ceramic Engineering and Science Proceedings*, **29**, 33 (2008)

[17] M. Ettler, *Schriften des Forschungszentrums Jülich, Reihe Energie & Umwelt / Energy & Environment, Band / Volume 36*, Einfluss von Reoxidationszyklen auf die Betriebsfestigkeit von anodengestützten Festoxid-Brennstoffzellen, (PhD Diss., Bochum Univ., 2008) (only available in German)

[18] Q. L. Ma, F. Tietz, A. Leonide, and E. Ivers-Tiffée, Anode-supported planar SOFC with high performance and redox stability, *Electrochem. Commun.*, 2010 (in press), http://dx.doi.org/10.1016/j.elecom.2010.07.011

MIXED COMPOSITE MEMBRANES FOR LOW TEMPERATURE FUEL CELL APPLICATIONS

Uma Thanganathan*
RCIS (Research Core Interdisciplinary Science), Okayama University, Tsushima-Naka, Kita-Ku, Okayama, 700-8530, Japan

ABSTRACT

The development of new class of $PVA/PMA/P_2O_5/SiO_2$ hybrid composite membranes via sol-gel chemistry with the goals to increase the proton conductivity at low temperature as well as to improve the thermo-mechanical properties was described and their characterization by the various experimental techniques. The fabrication and properties of a hybrid composite membrane based on phosphomolybdic acid (PMA) of heteropolyacid, phosphosilicate (P_2O_5-SiO_2) and polyvinyl alcohol (PVA) were described. The conductivity of the membranes was up to 10^{-2}-10^{-1} S cm^{-1} at room temperature and low humidity conditions.

INTRODUCTION

Proton exchange membrane (PEM) fuel cells are one of the most promising clean energy technologies under development [1-5]. The major advantages include electrical efficiencies of up to 60 %, high energy densities (relative to batteries), and low emissions. The most commonly used polymer for electrolyte membranes is the perfluorinated ionomer known by its trade name Nafion[TM]. Despite its attractively high proton conductivity and chemical stability, Nafion[TM] is costly and has high methanol permeability. The high methanol permeability allows undesirable transport of methanol from the anode side of the fuel cell, through the membrane, to the cathode side, a phenomenon known as methanol crossover. As a consequence, excessive methanol permeability of any PEM leads to an unacceptable decrease in cell performance [6-8]. To address the problem of methanol crossover, numerous membranes as alternatives to Nafion[TM] have been developed and studied. The development of PEM fuel cells is largely tracked by the history of the membranes. Because fuel cells of various types were known prior to PEM technologies, the catalysts, fuels, and oxidants used in PEM fuel cells were reasonably well-established materials.

Hybrid composite materials consisting of organic polymers and inorganic components have been attracting attention for their use in new high-performance materials. One interesting alternative to polymeric membranes in order to reach higher fuel cell operation temperature is the hybrid organic-inorganic membranes with nano-sized phases and interfaces. One of the research fields in hybrid organic-inorganic materials is based on the reaction between monomers and silicon alkoxides to form covalent bonds that permit the preparation of nano-structured interfaces. Of all the inorganic proton conductors that have been exploited for fuel cell applications the HPA may have the greatest potential as they not only have high proton conductivities but they have significant synthetic versatility [9]. Heteropoly acids (HPAs) are strong Bronsted acids as well as solid electrolytes [10]. For example, hydrated silicotungstic acid ($H_4SiW_{12}O_{40} \cdot 28H_2O$) has an ionic conductivity of 2×10^{-2} S cm^{-1} at room temperature [11-13]. However, HPAs are generally water-soluble. Taking the advantage of a property that HPAs are highly soluble in solvents, HPA-blended polymer membranes can be prepared by blending these two materials using suitable solvents. PMA ($H_3PMO_{12}O_{40}$, PMA) is a crystalline inorganic acid, but at the same time it also acts as an oxidizing agent. It is highly soluble in polar solvents such as water and also several oxygenated solvents but it is insoluble in non-polar solvents. PMA has shown higher proton conductivity at room temperature [14]. There are several reports are published on poly vinyl alcohol (PVA) and PMA membranes for fuel cell applications [15, 16]. The goal of the present study was to investigate a high proton conducting new membranes by introducing fabrication of hybrid composites. The development of new composite membranes was characterized with various techniques and their results were presented in this paper.

EXPERIMENTAL SECTION

To prepare 5 % (w/v) PVA solution, PVA was dissolved in distilled water at 80 °C with continuous stirring on a magnetic stirrer. Then the solution was cooled down to room temperature. PMA was dissolved directly into distilled water at room temperature. To 20 g of 5 wt % PVA solution, different amount of PMA (5, 10, 15, and 20 % w/v) was added and the resulting mixture was stirred until a homogeneous solution was obtained. To this solution various amount of TEOS (20, 15, 10 and 5 wt %) was added and the mixed solution was stirred for 2 h. After that constant amount of (5 v %) H_3SO_4 was added to the final solution further few hours stirred to obtained homogeneous solution. The resultant solution was poured into Petri dish and allowed to dry at room temperature for few days. After the heat treatment at 60 °C, the membranes were peeled off and were characterized. Distilled water was used throughout the study.

The non-crystallinity of the membranes was investigated with a Rigaku, multiflex X-ray diffractometer (CuK 40 kV, 20 mA, 5°min^{-1}). The chemical structure of the composite membranes was investigated using a JASCO, FTIR-460 spectrometer. The thermal stability of the membranes was studied using thermal gravimetric analysis with a DTG-50, Shimadzu instrument, and simultaneous DTA-TG instrument under an air atmosphere at a heating rate of 10 °C min^{-1}. All the membrane samples were stored in a vacuum oven at 100 °C before analysis. The proton conductivity of the membranes was measured in a two-point-probe conductivity cell using the AC impedance spectroscopy method as described [17]. A frequency response analyzer (Solartron 1260) over a frequency range of 1 Hz to 1 MHz with an oscillating voltage of 10 mV was used to measure the membrane impedance. Scanning electron microscopy (SEM) images were collected with a variable pressure Scanning Electron Microscope Hitachi S-3500N using an accelerating voltage of 20 kV.

RESULTS AND DISCUSSION

Figure 1 (a-d) shows the X-ray diffraction of the PVA/PMA/P_2O_5/SiO_2 hybrid composite membranes were showed the broad peak at 20° for all the samples. The amorphous phases in the membrane, which is favorable for proton transport as a result of an enhanced segmental motion of polymer chains.

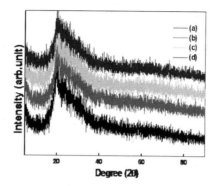

Figure 1 XRD graph for the PVA/PMA/P$_2$O$_5$/SiO$_2$ (70/5/5/20 wt %), (b) PVA/PMA/P$_2$O$_5$/SiO$_2$ (70/10/5/15 wt %), (c) PVA/PMA/P$_2$O$_5$/SiO$_2$ (70/15/5/10 wt %) and (d) PVA/PMA/P$_2$O$_5$/SiO$_2$ (70/20/5/5 wt %) hybrid composite membranes.

Figure 2 (a-d) shows the FT-IR spectra of the PVA/PMA/P$_2$O$_5$/SiO$_2$ hybrid composite membranes. The peaks of composite membrane near 1105 and 1083 cm^{-1} [(Si–O–C) and (Si–O–Si)], 482 cm^{-1} [(Si–O)] indicate the development of an inorganic/organic hybrid network during the gel formation. Identification of water, H$_3$O$^+$ and H$_5$O$_2^+$ vibrations and other hydrogen bond vibrations for the origin of the high proton conductivity of heteropolyacid reveal the presence of the most characteristic vibrational bands of the oxonium ions at 1700 cm^{-1} [18]. The peak of Si–O–Si is seen at wave number 846 cm^{-1} while the W–O–W stretching vibrations in the composite membrane are observed at wave numbers 770–798 cm^{-1}. The observable peak at 998 cm^{-1} represents the vibration moiety of the W$=$O functional group. Therefore, it is evident from these data that the SiO$_2$ and PWA are indeed present in the composite membranes.

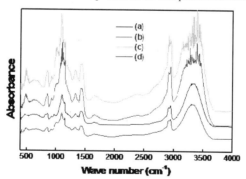

Figure 2 FTIR spectra of (a) PVA/PMA/P$_2$O$_5$/SiO$_2$ (70/5/5/20 wt %), (b) PVA/PMA/P$_2$O$_5$/SiO$_2$ (70/10/5/15 wt %), (c) PVA/PMA/P$_2$O$_5$/SiO$_2$ (70/15/5/10 wt %) and (d) PVA/PMA/P$_2$O$_5$/SiO$_2$ (70/20/5/5 wt %) hybrid composite membranes.

The thermal stability of PVA/PMA/P₂O₅/SiO₂ hybrid composite membranes was investigated by thermogravimetric (TGA) measurement from room temperature to 600 °C in air. The TGA curves are shown in Figure 3 (a-d). Four main weight loss stages were clearly observed for the PVA/PMA/P₂O₅/SiO₂ composite membranes. Decomposition of the materials is similar for all samples, showing multistep decomposition events at temperatures near 185, 235, 381 and 467 °C. The initial weight loss of around 9 %, the first broad decomposition stage ranging from 40 to 185 °C was attributed to the gradually degradation of side-groups. The second and third sharp weight loss stages of PVA/PMA/P₂O₅/SiO₂ membrane starting at 235 and 380 °C, respectively, with a weight loss of 27.9 and 36.9 %, were mainly ascribed to the degradation of phosphosilicate matrix and PVA side-groups. The fourth weight loss stage starting at 465 °C was mainly due to the degradation of PVA chains. This composites membrane was thermally stable up to 250 °C.

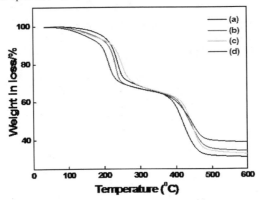

Figure 3 TG curves of PVA/PMA/P₂O₅/SiO₂ (70/5/5/20 wt %), (b) PVA/PMA/P₂O₅/SiO₂ (70/10/5/15 wt %), (c) PVA/PMA/P₂O₅/SiO₂ (70/15/5/10 wt %) and (d) PVA/PMA/P₂O₅/SiO₂ (70/20/5/5 wt %) hybrid composite membranes.

Figure 4 (a-d) shows the cole-cole plots of PVA/PMA/P₂O₅/SiO₂ composite membranes of the impedance study at room temperature. The trend of the impedance obtained exhibits almost semicircle indicating that the conduction mechanism [19] was involved in the composite system. The proton conductivity of composite membranes measured at room temperature was in the range from 10^{-2}-10^{-1} S cm⁻¹ while that for PVA-heteropolyacids membranes mixed with phosphosilicate in various ratios. Two mechanisms for proton transfer were commonly adopted: Grotthus or 'hopping' mechanism and vehicle mechanism [20]. 'Hopping' mechanism refers to the proton transport through hydrogen-bonded network of water molecules and the ion-exchange sites. Vehicle mechanism refers to the coupled proton-water transport, which is strongly associated with the water content in the membrane. High proton conductivities of phosphosilcate mixed PVA-heteropolyacid membrane were attributed to the high density of acid and metal alhoxide groups, which provided enough sites for proton jumping from one to another. Grotthus or 'hopping' mechanism was considered to play an important role in proton transfer. These composite membranes were reasonably expected to be suitable for low/high operating temperatures in terms of proton conductivity.

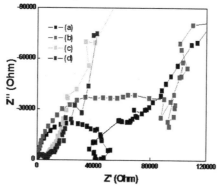

Figure 4 Impedance plots of PVA/PMA/P$_2$O$_5$/SiO$_2$ (70/5/5/20 wt %), (b) PVA/PMA/P$_2$O$_5$/SiO$_2$ (70/10/5/15 wt %), (c) PVA/PMA/P$_2$O$_5$/SiO$_2$ (70/15/5/10 wt %) and (d) PVA/PMA/P$_2$O$_5$/SiO$_2$ (70/20/5/5 wt %) hybrid composite membranes measured at room temperature.

Figure 5 (a, b) illustrates SEM image of composite membranes obtained by the distribution of inorganic particles is relatively uniform in the organic matrix. Even composition changes, the membranes are compact with no phase separation suggesting that the synthesized films were homogeneous in nature and hence formed dense membrane.

Figure 5 Scanning electron micrographs (SEM) for the PVA/PMA/P$_2$O$_5$/SiO$_2$ hybrid composite membranes: (a) (70/5/5/20 wt %) and (b) (70/20/5/5 wt %).

CONCLUSION

The present research work was thus focused on the development of nanocomposite membranes for proton exchange membranes which are chemically and mechanically more stable at low/high temperatures. These are composite materials with inorganic metal alkoxide nanoparticles incorporated within a polymer PVA and heteropolyacid. FTIR analysis shows that there are strong interaction between silica particles of TEOS and heteropolyacids. The protonic conductivity of the hybrid membranes was in the range of 10^{-2}-10^{-1} S cm^{-1} at room temperature. The characterization studies with XRD, FTIR, TGA, SEM and conductivity confirmed the PVA/PMA/P$_2$O$_5$/SiO$_2$ composite membranes are most promising candidate for the PEMFC applications.

ACKNOWLEDGEMENT

The author gratefully acknowledges financial support from the Ministry of Education, Sport, Culture, Science and Technology (MEXT) and the Special Coordination Funds for Promoting Sciences and Technology of Japan. Special thanks go to Mr. A. Suresh Kumar for his help with the XRD, FTIR and TG/DTA experiments.

REFERENCES

[1] K. D. Kreuer, In Handbook of Fuel Cell Fundamentals, Technology, and Applications; Vielstich et al., Eds.; John Wiley: Chichester, UK, Part 3, Chapter 33, Vol. 3, 420 (2003).
[2] F. Babir and T. Gomez, Efficiency and economics of proton exchange membrane fuel cells, Int. J. Hyd. Energy, 21, 891-901 (1996).
[3] M. Rikukawa and K. Sanui, Proton-conducting polymer electrolyte membranes based on hydrocarbon polymers, Prog. Polym. Sci. 25, 1463-1502 (2000).
[4] T.E. Springer, T.A. Zawodzinski and S. Gottesfeld, Polymer electrolyte fuel cell model, J. Electrochem. Soc. 138, 2334-2344 (1991).
[5] S. Gottesfeld and T. Zawodzinski, Polymer elelyte fuel cell, Adv. Electrochem. Sci. Eng. 5, 195-381 (1997).
[6] X. M. Ren, T. A. Zawadzinski, F. Uribe, H. Dai and S. Gottesfeld. Methanol cross-over in direct methanol fuel cells. Electrochem. Soc. Proc. 95, 284-285 (1995).
[7] B. Guraua, E. S. Smotkin, Methanol crossover in direct methanol fuel cells: a link between power and energy density. J. Power Sour. 112, 339-352 (2002).
[8] V. M. Barragan, C. Ruiz-Bauza, J. P. G. Villaluenga and B. Seoane, Transport of methanol and water through Nafion membranes. J. Power Sour. 130, 22-29 (2004).
[9] Z. Jiang, X. Zheng, H. Wu, J. Wang and Y. Wang, Proton conducting CS/P(AA-AMPS) membrane with reduced methanol permeability for DMFCs, J. Power Sources, 180, 143-153 (2008).
[10] A.M. Herring, Inorganic-Polymer Composite Membranes for Proton Exchange Membrane Fuel Cells, Polymer Reviews, 46, 245-296 (2006).
[11] M. Misono, Heterogeneous catalysis by heteropoly compounds of molybdenum and tungsten, Catal. Rev. Sci. Eng. 29, 269-321 (1989).
[12] O. Nakamura, T. Kodama, I. Ogino and Y. Miyake, Homogeneous permanganate oxidation in non-aqueous organic solution selective oxidations of olefins in to 1, 2- diols or aldehydes, Chem. Lett. 1, 17-18 (1979).
[13] O. Nakamura, I. Ogino and T.Kodama, Temperature and humidity ranges of some hydrates of high-proton-conductive dodecamolybdophosphoric acid and dodecatungstophosphoric acid crystals under an atmosphere of hydrogen or either oxygen or air, Solid State Ionics, 3-4, 347-351 (1981).
[14] K.-D. Kreuer, Fast proton transport in solids, J. Mol. Struct. 177, 265-276 (1988).

[15]J. Sauk, J. Byun and H. Kim, Composite Nafion/polyphenylene oxide (PPO) membranes with phosphomolybdic acid (PMA) for direct methanol fuel cells, *J. Power Sources*, **143**, 136-141 (2005).

[16]A. Anis and A. K. Banthia, *Int J Plas*, **9**, 539-545 (2006).

[17]A. Anis, A. K. Banthia, S. Mondal and A. K. Thakur, Synthesis and characterization of hybrid proton conducting membranes of poly(vinyl alcohol) and phosphomolybdic acid, *Chin J Polym Sci.*, **24**, 449-456 (2006).

[18]X. Weilin, L. Changpeng, X. Xinzhong, S. Yi, L. Yanzhuo, X. Wei and L. Tianhong, New proton exchange membranes based on poly (vinyl alcohol) for DMFCs, **171**, 121-127 (2004).

[19]E. Schmidbauer and P. Schmid-Beurmann, Electrical conductivity and thermopower of Fe-phosphate compounds with the lazulite-type structure, *J. Solid State. Chem.*, **177**, 207-215 (2004).

[20]B. S. Pivovar, Y. Wang and E. L. Cussler, Pervaporation membranes in direct methanol fuel cells, *J. Membr. Sci.*, **154**, 155-162 (1999).

CARBONATE FUEL CELL MATERIALS AND ENDURANCE RESULTS

C. Yuh, A. Hilmi, G. Xu, L. Chen, A. Franco, and M. Farooque
FuelCell Energy, Inc.
Danbury, CT, USA

ABSTRACT

The high-temperature carbonate fuel cell is an ultra-clean and high-efficiency power generator. Its intermediate operating temperature, ~550-650°C, is considered optimum to facilitate fast fuel cell reaction kinetics, utilize waste heat efficiently in a combined heat and power or bottoming power cycle, and at same time allow use of commercial commodity materials for stack and balance-of-plant (BOP) construction. MW size power plants manufactured by FCE are being operated at customer sites throughout the world. The stack and BOP materials selections are founded on many years of focused research. Long-term stack endurance as well as field operation results to date show that the baseline construction materials meet the endurance goals. Material durability is well understood and solutions are available to further extend the durability. The paper will review materials durability experience and development approaches that would allow to enhance performance, reduce cost and to further extend life.

INTRODUCTION

The Direct Fuel Cell[®] (DFC[®]) power plant of FuelCell Energy (FCE) is a high efficiency, ultra-clean electric power generation system utilizing a variety of hydrocarbon fuels, including natural gas, methanol, diesel, biogas, coal gas, coal mine methane and propane. It is based on high-temperature carbonate fuel cell technology, which uses alkali carbonate mixtures as electrolyte immobilized in a porous ceramic matrix and operates at approximately 550-650°C.

Details on carbonate fuel cell chemistry and overall material/product status are available in the literature[1-5]. The basic electrochemistry of carbonate fuel cells involves the following reactions:

Cathode Reaction: $CO_2 + \frac{1}{2} O_2 + 2e^- \rightarrow CO_3^=$

Anode Reaction: $H_2 + CO_3^= \rightarrow H_2O + CO_2 + 2e^-$

Overall: $H_2 + \frac{1}{2} O_2 \rightarrow H_2O$

DFC adopts an internal reforming approach, with its key characteristics the ability to generate hydrogen and subsequently electricity directly from hydrocarbon fuels. As such, the thermal and chemical features of the fuel cell and reforming reactions are uniquely complementary for efficient integration. Two-thirds of the heat produced by the fuel cell reaction is used up in the internal reforming reaction. This "one-step" internal-reforming-fuel-cell process results in a simpler, more efficient and cost-effective energy conversion system compared with external reforming fuel cells. A simple DFC system also consists of single-pass air flow, single-pass fuel flow, and atmospheric pressure operation. Such a cost-effective DFC system has the minimum balance-of-plant equipment among all types of fuel cell power plants.

119

In order to achieve commercial viability, performance, life and cost goals need to be reached. The life targets are at least 5 years for the fuel cell stack and >20 years for the module/BOP (balance-of-plant) piping/equipment. FCE has accumulated extensive operating and material experiences through long-term (>4 years) field operations. Concerted efforts have resulted in design promising >5-year material durability, plant power output increase and substantial components cost reduction. The current-generation stacks (350kW net AC rated 5-year life) were introduced in 2009. The paper will review materials durability experience and development approaches that would allow to enhance performance, reduce cost and to further extend life.

DFC PRODUCT STATUS

Global demand for electric power is growing rapidly, projected to grow by ~60% from 2006 to 2025, with associated increase of electricity cost. It is also projected that hydrocarbons will still remain the dominant fuel source. However, large-scale greenhouse-gas emitting hydrocarbon-burning central generation plants are costly and difficult to site. Therefore, the ability to reduce electricity cost and greenhouse gas emissions is dependent on increasing energy efficiency. Distributed clean energy generation such as FCE's DFC meets the need for efficient, economical clean power when and where needed. The key characteristics of DFC are:

- Higher electrical efficiency than competing technologies
- Near-zero NOx, SOx and significantly lower CO_2 emissions allows siting in polluted urban areas
- Reliable, secure, 24/7 uninterrupted power
- Competitive economics
- Cleaner & quieter operation
- Distributed generation puts power where needed enables smart grid
- Connects to existing electric and fuel infrastructure
- More control over power costs, emissions, and reliability

The DFC products are designed to meet the power requirements of a wide range of commercial and industrial customers including wastewater treatment plants, telecommunications/data centers, manufacturing facilities, office buildings, hospitals, and universities, mail processing facilities, hotels and government facilities, as well as in grid support applications for utility customers. These customers demand ultra-clean, efficient and reliable distributed power generation and the DFC products are responding well to these demands.

FCE adopted a building block approach, where identical cell packages are stacked together and installed in single- or 4-stack modules, rated at 300kW or 1.4MW, respectively. As shown in Figure 1, FCE's current products are rated in capacity at 300kW, 1.4MW, and 2.8MW and are scalable for various distributed applications up to 50 MW.

As of September 2010, the DFC power plants have attained 99 MW capacity installed/backlog. In addition, FCE products have been included in 43.5 MW of projects awarded (not yet contracted) under the State of Connecticut Project 150 program. DFC fuel cells are generating power at over 50 locations worldwide. To date, DFC power plants have generated over 550 million kWh of electricity. The DFC plants achieved an electric simple-cycle efficiency of 45 to 49%, the highest of any distributed generation technology in a comparable size range.

Using the high value heat byproduct for cogeneration applications, an overall energy efficiency of 65 to 95% can be achieved based on the application. These field plants also confirmed significantly lower emissions of greenhouse gases and particulate matter than conventional combustion-based power plants.

DFC commercialization effort is now focusing at larger, more cost-competitive >1-3 MW-size units. The world largest single-site fuel cell installation is being constructed at Daegu City, South Korea with 4×2.8MW DFC plants (11.2MW total capacity). The larger plants can better realize the high-efficiency benefit (58-65% electric efficiency) of hybrid systems such as DFC/T® (DFC-Turbine) and DFC/ERG™ (DFC-Energy Recovery Generation), capable of an electrical efficiency of approximately 60% (LHV).

The first generation DFC stacks delivered 250kW net AC power and 3-year life. Through comprehensive technology advancement, the current generation stacks delivers 350kW rated power and 5-year life. In summary, FCE is well on its way to realize commercialization of the high-efficiency DFC systems. The performance and endurance of the materials on which the DFC is based is discussed next.

MATERIAL PERFORMANCE AND ENDURANCE RESULTS

The DFC fuel cell package construction is illustrated in Figure 2. The major characteristics of the DFC fuel cells are:

1. Large cell area (~1m^2)
2. High cell area utilization for power production, greater than 90% of the geometric area is active area,
3. Ease of fabrication, and,
4. Lower cost.

Each cell package is composed of active components (anode, cathode, electrolyte matrix) and cell hardware (bipolar plate, current collectors, and seal area). The bipolar plate and the corrugated current collectors are made of stainless steel. The electrodes are made from porous nickel-based materials. The matrix is ceramic lithium aluminate, and holds an electrolyte mixture of lithium and potassium/sodium carbonate salts, which melt between 450 and 510°C. In addition to these active repeating cell packages, a fuel cell stack also consists of non-repeating stack hardware: manifold, end plate, compression system, etc., for providing compressive sealing pressure, gas distribution and power takeoff. A fuel cell module is an insulated vessel enclosing fuel cell stacks and associated gas distribution system. Finally a fuel cell power plant consists of fuel cell modules and balance-of-plant (BOP): piping, expansion joints, valves, and various equipments (heat exchangers, recuperator, fuel/water cleanup systems, anode exhaust gas oxidizer, blowers, power-conditioning unit, etc.) FCE has gathered extensive endurance experiences on all DFC power plant materials.

Material development status for carbonate fuel cell has been reviewed extensively[6-18]. Material properties such as creep, sintering, compaction, oxidation, hot corrosion, carburization, etc. will impact durability. In addition, stable, long-term anode and cathode electrochemical activity is also necessary. The selection of DFC materials is based on intensive materials research carried out during the past three decades, focusing on performance and endurance improvements, and cost reduction, confirmed by the extensive DFC operating and material experiences accumulated through long-term field operations.

Electrodes and Electrolyte

Based on understanding of rate-controlling processes and decay mechanisms, electrode and electrolyte designs have been advanced to improve cell power output and to achieve the stack useful life of >5 years. As with most other fuel cell systems, the cathode is the major source of cell polarization. The polarization loss at the anode is considered to be relatively small and insensitive to temperature due to the fast hydrogen oxidation kinetics (Figure 3). However, unlike the anode polarization, the cathode polarization has strong temperature sensitivity mainly due to a much larger charge-transfer resistance. Since a substantial portion of a DFC stack is operated below 600°C, the larger cathode polarization controls overall stack power output.

The anode materials are ODS (oxide dispersion strengthened) Ni-based alloys (Ni-Cr, Ni-Al or Ni-Al-Cr). The baseline cathode is *in-situ* oxidized and lithiated NiO with a dual-porosity microstructure. The anode has been demonstrated to be sufficiently mechanically and chemically durable. The current anode development effort is mainly focused on reducing the manufacturing cost and optimizing the pore structure and surface wettability (for improving electrolyte distribution and cell performance). FCE had developed an improved anode structure with tailored pores for enhanced mass transfer and electrolyte retention. This anode showed ~5-10 mV lower polarization than baseline and demonstrated excellent stability (Figure 4). Since the lab-scale cell is operated at a different condition (higher temperature and lower fuel utilizations) than typical stacks, the 5-10mV voltage improvement here can be projected to >25mV in stacks, translating to about 10% stack power increase.

Cathode polarization and dissolution, strongly influenced by cathode material/morphology, electrolyte compositions and operating conditions, are the two very important performance and life controlling factors. The cathode polarization loss is mainly controlled by effective surface area and oxygen solubility in the melt. Post-test analysis of the NiO cathode after long-term field operation showed that the desired agglomerate morphology and surface area remained sufficiently stable[8]. Therefore, to achieve cathode performance and endurance goals, a desired electrolyte distribution needs to be maintained. Extensive efforts had been carried out to develop alternate cathode material and advanced electrolyte compositions to enhance cell performance and life. FCE has developed an advanced cathode that showed ~10-15mV improvement over the baseline NiO[16].

In terms of cathode dissolution, NiO dissolves slowly into the electrolyte as Ni^{+2} ions according to an acidic dissolution mechanism ($NiO \leftrightarrow Ni^{+2} + O^=$), which are then reduced to metallic nickel by the hydrogen diffused from the anode side. The deposition of Ni into the matrix may cause electric shorting, decreasing stack life. In order to improve the stability of the NiO cathode, several approaches have been investigated. The first approach is to select operation conditions to increase melt basicity ($[O^=]$). The second one is to develop a new cathode material that is more stable than NiO. The third is the selection of an alternative, more basic electrolyte composition than the standard $LiKCO_3$ eutectic electrolyte.

It has been demonstrated by FCE that for atmospheric-pressure DFC systems with a proper operational control (e.g., a lower CO_2 partial pressure) the nickel shorting is controlled to achieve desired useful life. Also, Ni-shorting life extension by increasing matrix thickness or using more basic electrolytes has been verified at FCE in numerous long-term single cells and stacks. Figure 5 shows Ni deposition reduction in the matrix at FCE by as much as ~60%.

Proper electrolyte management is essential to stack performance and endurance. Electrolyte composition and distribution affects corrosion, ohmic loss, evaporation loss, NiO dissolution and cathode polarization. Proper electrolyte management scheme needs to be developed to assure desired electrolyte composition and distribution within cell components during endurance operation. Analyzing electrolyte inventory in detail in long-term operated DFC stacks, it can be confidently concluded that sufficient electrolyte inventory for >5-year service can be retained with selected cell and material designs. Potential concern for electrolyte migration has been addressed by developing a new gasket design that has reduced the rate by 80%[8]. Electrolyte composition can play a major role on both cathode polarization and cell stability. FCE has developed an advanced electrolyte composition that showed ~25 mV improvement at 650°C and more than 40 mV at low temperatures (<600°C) compared to baseline[16].

Matrix

The electrolyte matrix, sandwiched between the electrodes, is a key cell component, isolating the fuel from the oxidant, as well as storing electrolyte and facilitating ionic transport. It has a microporous ceramic structure consisting of ultra-fine $LiAlO_2$ powders. The electrolyte is immobilized within the micropores by capillary force. Stable matrix fine pore size is necessary for long-term operation.

FCE has demonstrated in long-term field operation virtually no phase change of the $LiAlO_2$ phase[8]. Current effort at FCE is focused on optimizing powder morphology and tape-casting process to deliver a well-packed pore structure and consistent and scalable process. FCE has identified controlling variables including raw material particle size distribution, purity level and slurry processing parameters that can improve the slurry and matrix tape consistencies and yield. The manufacturing process and the tape pore structure have been consistently improved. An advanced matrix and electrolyte design has demonstrated significantly reduced cell resistance increase rate (Figure 6) in endurance bench-scale cells. The resistance stability improvement can be attributed to better matrix electrolyte retention due to improved particle packing.

Cell Hardware Materials

Metallic heat-resistant alloys, particularly austenitic stainless steels, are extensively used. They are exposed to oxidation and hot corrosion conditions during operation. Alloy selection needs to balance corrosion, oxide adhesion, contact ohmic resistances, electrolyte loss, and cost.

Upon reaction with the thin creeping film of the alkali carbonate electrolyte, the alloy materials form a multi-layered corrosion scale. The less protective outer layer is Fe-rich while the underlying Cr-rich layer is relatively dense to provide corrosion protection. Metal loss rate, oxide ohmic resistance and mechanical integrity are the primary endurance considerations. Contrary to SOFC (Solid Oxide Fuel cell), Cr vapor from the hardware materials does not poison the cathode. Figure 7 summarizes the corrosion rate data of cell hardware materials from long-term field-operated stacks. It has been determined that the cell hardware metal loss rate is acceptable for a minimum of 5-year field service, even for the cathode current collector that experiences the fast corrosion rate among all cell hardware components. In addition to corrosion metal loss, oxide scale formed between the cathode and cathode current collector may contribute to cell resistance increase and performance decay. The oxide resistance of the cathode current collector after >4-year service was found still well within allowable limit (only ~15 mV voltage loss per cell after ~5 years). The current collector material became brittle during service (further discussed later). However, the mechanical fracture strength of the current collector after >4-year service has been determined to be well above the minimal requirement[15]. FCE has also identified an approach to modify alloy structure to enhance cathode current collector corrosion resistance by ~30%, verified in ~2-year operated stacks. This modification has the potential to increase the cathode current collector life to beyond 7 years.

During long-term high-temperature service, gradual precipitation of secondary phases such as carbide and σ phases occurs in stainless steels. The formation of these precipitates embrittles the alloys[19]. The source of carbon for the carbide precipitation could be from the inherent carbon impurity in the steels (sensitization) or from high carbon-activity hydrocarbon fuels (carburization). The low-carbon activity oxidant gas is non-carburizing. Sensitization/carburization depletes Cr near the carbide precipitates (mainly at grain boundaries), accelerating grain-boundary corrosion during plant transients when aqueous condensate forms. Alloy materials and operating conditions need to be carefully selected to mitigate the effect of these precipitates.

The σ-phase precipitates have been reported to reduce alloy creep strength[19]. The reduced creep strength may result in cell dimensional change under operational thermo-mechanical stress. Furthermore, the carbon intake (carburization) may also increase the volume of the alloys, resulting in further dimensional change. Alternate alloys to reduce the formation of the σ phase to improve creep strength are desired. FCE has identified an advanced alloy that showed significantly reduced σ-phase formation and better creep strength than the baseline alloy (Figure 8).

The wet seal area simultaneously experiences reducing and oxidizing environments and posed the most significant challenge to the durability. Only alumina-forming alloys have sufficient corrosion resistance. A low-cost aluminized coating developed by FCE has demonstrated excellent protection for the substrate stainless steels during a >4-year field operation[15]. The results project that the stability of the selected coating is adequate for well beyond 5-year use.

Stack Module and BOP Hardware Materials

The module and BOP hardware experiences temperature from low (<200°C) to as high as 900°C (e.g., for recuperator in a hybrid system), and can experience various thermal and gas-atmosphere transients. The life goals for the stack external hardware and BOP materials are >20 years. These materials are not in the path of current conduction and not in direct contact with the liquid electrolyte; therefore, a conductive oxide scale is not required. Only oxidation resistance and scale adhesion are important. Another important consideration is the cost. High cost materials such as superalloys are only used sparingly, for applications experiencing very high stress. Significant long-term stability data of high-temperature alloys has been gathered from simulated out-of-cell (up to 1100°C) and field operations, allowing a proper selection of alloys with balanced oxidation resistance and cost. Common austenitic stainless steels such as 304 are generally adequate for the thick-wall piping/equipment application. For lower temperature service (<600°C), lower-cost standard ferritic stainless steels such as 430 may even be acceptable. However, for more stringent service conditions such as under dual-atmosphere condition (e.g., for manifold body) or at higher temperatures (>700°C), higher-Cr stainless steels such as 309/310 are required. For applications that require more oxidation or creep resistance, alloys such as alumina-forming, Ni-based or superalloys should be used.

MANUFACTURING AND COST REDUCTION

FCE has ongoing efforts to lower DFC product costs by improving performance, life, design and manufacturing, developing new vendors, competitive bidding, and global sourcing. The DFC employs well-known manufacturing processes such as metal forming, welding, tape casting, powder doctoring, sintering, etc. The cells and stack are manufactured at FCE's Torrington, CT manufacturing facility, which can produce up to 70 MW of DFC units per year. This facility is the largest capacity fuel cell manufacturing facility in the world. Plans are also in place to increase production capacity as demand increases.

As shown in Figure 9, FCE has been successful in achieving product cost reductions by almost 80% in the last fifteen years through continued value engineering, technology improvements, and increased volume production. With continued cost reduction efforts and stack power density increase, the product cost is expected to decrease further, making the DFC products a profit-generating business.

CONCLUSION

The ultra clean DFC plants are suitable for high efficiency electricity or combined heat and power for stationary applications. Significant progress has been made in DFC technology development, manufacturing, product engineering, and field operation. With a vast experience gathered from >4-years field operations, material considerations are well understood and materials designs have been validated for the current stack life, performance and cost targets. As the technology matures through further material advancement and cost reduction efforts, the product are expected to capture broader market acceptance paving the way for larger multi-megawatt systems.

ACKNOWLEDGEMENT

The authors would like to acknowledge FCE's technical, engineering and manufacturing staffs on the combined contribution to progress reported in this paper. The financial support by US DOE, EPRI, GRI, and others in the early stage of development are greatly appreciated.

REFERENCES
[1] J. R. Selman and H. Maru, Physical Chemistry and Electrochemistry of Alkali Carbonate Melts, *Advances in Molten Salt Chemistry Vol. 4*, G. Mamantov and J. Braunstein, Editors, Plenum Press, New York, 159-390 (1981).

[2]M. Bischoff, Large Stationary Fuel Cell Systems: Status and Dynamic Requirements, *J. Power Sources*, **154**, 461-466 (2006).

[3]M. Bischoff, Molten Carbonate Fuel Cells: A High Temperature Fuel Cell on the edge to Commercialization, *J. Power Sources*, **160**, 842-845 (2006).

[4]M. Farooque and H. Maru, Carbonate Fuel Cells: Milliwatts to Megawatts, *J. Power Sources*, **160**, 827-834 (2006).

[5]J.R. Selman, Molten-Salt Fuel Cells: Technical and Economic Challenges, *J. Power Sources*, **160**, 852-857 (2006).

[6]C. Yuh, R. Johnsen, M. Farooque and H. Maru, Status of Carbonate Fuel Cell Materials, *Journal of Power Sources*, **56**, 1-10 (1995).

[7]J. Hoffmann, C. Yuh and A. Godula Jepek, Electrolyte and Material Challenge, *Handbook of Fuel Cells - Fundamentals, technology and Applications, Vol. 4*, W. Vielstich, A. Lamm, and H.A. Gasteiger, Editors, 921-941, John Wiley and Sons, Hoboken, NJ (2003).

[8]C. Yuh, A. Hilmi, M. Farooque, T. Leo, and G. Xu, Direct Fuel Cell Materials Experience, *ECS Transactions*, **17(1)**, 637-654 (2009).

[9]M. Farooque, C. Yuh and H. Maru, Carbonate Fuel Cell Technology and Materials, *MRS Bulletin*, **30**, 602-606 (2005).

[10]D. Brdar and M. Farooque, Materials Shape up for MCFC Success, *The Fuel Cell Review*, October/November, 15-20 (2006).

[11]C. Yuh and M. Farooque, Carbonate Fuel Cell Materials, *Advanced Materials and Progress*, **160**, 31-34 (2002).

[12]Y. Fujita, Durability, *Handbook of Fuel Cells - Fundamentals, technology and Applications Vol. 4*, W. Vielstich, A. Lamm, and H.A. Gasteiger, Editors, 969-982, John Wiley and Sons, Hoboken, NJ (2003).

[13]C. Yuh, J. Colpetzer, K. Dickson, M. Farooque and G. Xu, Carbonate Fuel cell Materials, *J. Materials Engineering and Performance*, **15**, 457-462 (2006).

[14]C. Yuh, A. Hilmi, G. Xu, J. Colpetzer and M. Farooque, Carbonate fuel cell Materials and Life Updates, *ACS Fuel Chemistry Division Preprint* **52(2)**, 644-645 (2007).

[15]C. Yuh, L. Chen, A. Franco, and M. Farooque, Review of High-Temperature Fuel cell Hardware Materials, *Proc. ASME 2010 Eighth International Fuel cell Science, Engineering and Technology Conference*, FuelCell2010-33163 (2010)

[16]A. Hilmi, C. Yuh, and M. Farooque, Performance and Stability of Carbonate Fuel Cell Electrodes, *Proc. ASME 2010 Eighth International Fuel cell Science, Engineering and Technology Conference*, FuelCell2010-33109 (2010).

[17]C. Yuh and M. Farooque, Materials and Life Considerations, *in Fuel Cells-Molten Carbonate Fuel Cells, Encyclopedia of Electrochemical Power Sources*, 497-507 (2009).

[18]A. Hilmi, C. Yuh, and M. Farooque, Anodes, *in Fuel Cells-Molten Carbonate Fuel Cells, Encyclopedia of Electrochemical Power Sources*, 454-461 (2009).

[19]T. Sourmail, Precipitation in Creep Resistant Austenitic Stainless Steels, *Materials Science and Technology*, 17, 1-13 (2001).

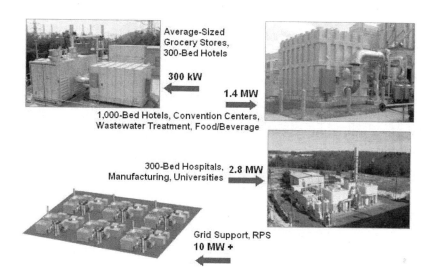

Figure 1 DFC products and typical applications: Building block approach allows easy packaging and scale-up.

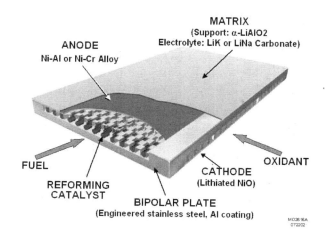

Figure 2. The DFC cell construction employs stainless steel sheet metal and electrodes are nickel-based: Commercial commodity materials are used.

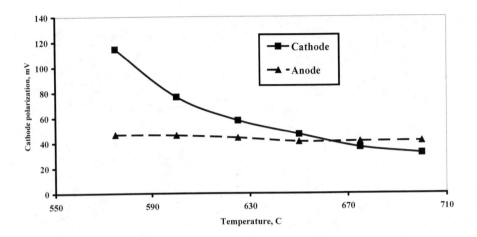

Figure 3. Anode and cathode polarization in a lab-scale cell at 160mA/cm²: Anode reaction exhibits low sensitivity to cell temperature; majority of performance loss at T<600°C is attributed to cathode.

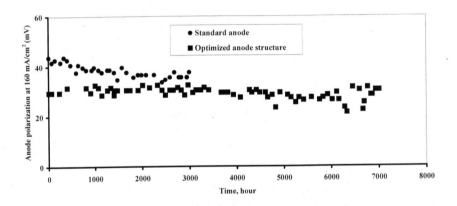

Figure 4. Performance of standard anode vs. advanced anode: An advanced anode showed less polarization and excellent performance stability in lab-scale cells at 160mA/cm².

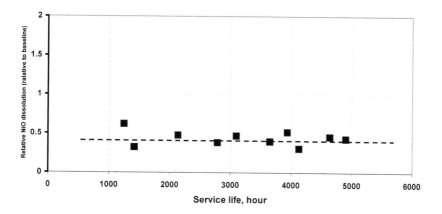

Figure 5. Improvement of cathode stability: Advanced electrolyte design reduced Ni deposition in matrix by ~60% in endurance cells.

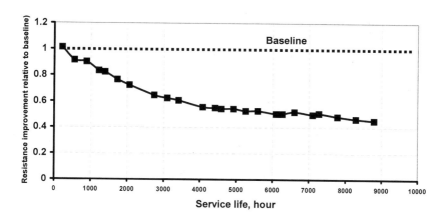

Figure 6. Cell internal resistance baseline vs. advanced cell designs: The advanced design showed >50% resistance stability improvement.

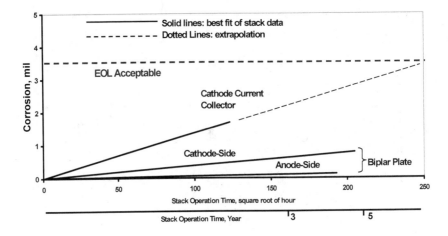

Figure 7. Cell hardware corrosion: Cell hardware materials have no concerns for meeting the 5-year life goal.

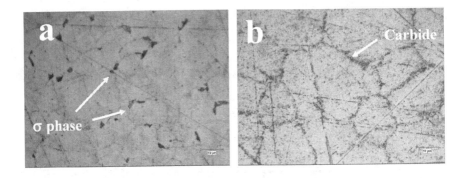

Figure 8. Comparison of (a) baseline cell hardware alloy with (b) advanced alloy (15,000h operation): Advanced alloy with less Cr, more Ni, and with additive showed less σ-phase formation, fine carbide precipitates at grain boundary and inside grains, and better creep strength.

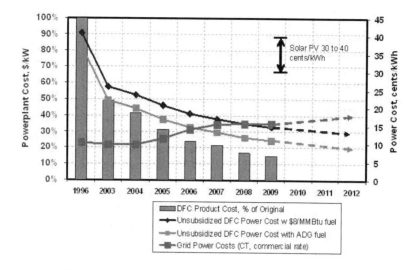

Figure 9.Product cost reduction with time: Significant cost reduction has been achieved during the past 15 years through value engineering, technology advancement and volume increase.

Materials Solutions for the Nuclear Renaissance

CHARACTERIZATION OF CORE SAMPLE COLLECTED FROM THE SALTSTONE DISPOSAL FACILITY

A.D. Cozzi and A.J. Duncan
Savannah River National Laboratory
Aiken, SC, USA

ABSTRACT

During the month of September 2008, grout core samples were collected from the Saltstone Disposal Facility, Vault 4, cell E. This grout was placed during processing campaigns in December 2007. Core samples were retrieved to initiate the historical database of properties of emplaced Saltstone and to demonstrate the correlation between field collected and laboratory prepared samples.

The density and porosity of the Vault 4 core sample, 1.90 g/cm^3 and 59.90% respectively, were comparable to values achieved for laboratory prepared samples. X-ray diffraction analysis identified phases consistent with the expectations for hydrated Saltstone. Microscopic analysis revealed morphology features characteristic of cementitious materials with fly ash and calcium silicate hydrate gel.

When taken together, the results of the density, porosity, x-ray diffraction analysis and microscopic analysis support the conclusion that the Vault 4, Cell E core sample is representative of the expected waste form.

INTRODUCTION

During the month of September 2008, grout core samples were collected from the Saltstone Disposal Facility (SDF), Vault 4, cell E.[1] This grout was placed during processing campaigns in December 2007 from Deliquification, Dissolution and Adjustment (DDA) Batch 2 salt solution. The 4QCY07 Waste Acceptance Criteria (WAC) sample collected on 11/16/07 represents the salt solution in the core samples.[2]

A total of 9 samples were obtained from three different locations. Three samples were collected at each location. A two-inch diameter concrete coring bit was used to collect the samples. Coring could be performed either dry or with water (to assist in coring and removal of sample). The cored samples were transferred from the core bit to stainless steel tubes and evacuated.

EXPERIMENTAL

Simulant

Salt Solution Preparation

Table I is the composition of Tank 50 as reported in the Waste Characterization System (WCS) at the time the Saltstone Production Facility (SPF) was operating. This composition and operational window correspond to the field samples collected during coring activities[1]. The table omits radionuclides or organics. The composition of the simulant salt solution used for the laboratory prepared samples was based on the composition in Table I using the components with concentrations greater than 0.001M with the exception of chloride. The omission of chloride was based on work that indicates no effect of chloride on grout properties of interest.[3] Sulfate and phosphate were included as Reference [3] demonstrated that, even at low concentrations, the presence of either sulfate or phosphate anions affected grout properties.

135

Table I. Composition of Tank 50 as of 11/30/07 as Reported by WCS.

Component	Concentration	
	mg/L	M
NH_4^+	83.4	0.00
CO_3^{2-}	3194.3	0.05
Cl^-	229.7	0.01
F^-	16.2	0.00
OH^-	21973.5	1.29
NO_3^-	200835.9	3.24
NO_2^-	1466.6	0.03
$C_2O_4^-$	665.5	0.01
PO_4^-	225.1	0.002
SO_4^-	422.9	0.004
As	0.1	0.00
Ba	1.8	0.00
Cd	0.3	0.00
Cr	12.5	0.00
Pb	1.3	0.00
Hg	57.5	0.00
Se	0.1	0.00
Ag	0.6	0.00
Al	6689.7	0.25
Na	--	4.5

Table II is the composition of the salt solution used for the simulant to produce all of the Saltstone samples used in this study. To prepare the simulant salt solution from Table I, reagents were selected to meet the desired concentrations of the individual analytes and maintain charge balance in the solution. To meet these criteria, the source of aluminum was divided between sodium aluminate and aluminum nitrate to obtain the appropriate aluminum concentration and nitrate concentration. The sodium value was the variable used to achieve charge balance. The initial simulant was prepared in a volumetric flask. The components were added to water and mixed until no visible solids remained. Water was added to attain the desired volume. Subsequent simulant were prepared in 15-L carboys using water additions scaled from the water needs determined in the initial simulant. The measured density and the weight percent solids of the simulant correspond well with density and the weight percent solids of the salt solution— 1.222 g/mL and 26.20% respectively— sampled from Tank 50 in November 2006.[4] Density and weight percent solids measurements were performed on each simulant batch to verify that the batches were properly prepared.

Table II. Composition of Simulant Salt Solution

Component	Concentration	
	g/L	M
NaOH (w/w 50.5%)	103.3	1.29
NaNO$_3$	212.3	2.50
NaNO$_2$	2.19	0.03
Na$_2$CO$_3$	5.63	0.05
Na$_2$C$_2$O$_4$	1.01	0.01
Na$_2$SO$_4$	0.62	0.004
Na$_2$Al$_2$O$_4$ •2H$_2$O	12.71	0.06 (0.127 Al)
Al(NO$_3$)$_3$ •9H$_2$O	45.15	0.12
Na$_3$PO$_4$ •12H$_2$O	0.90	0.002
Total Na	—	4.47
Properties		
Density	1.222	g/mL
Wt % total solids	28.52	%

Grout Preparation

Premix components cement, granulated blast furnace slag and Class F fly ash were obtained from the Z-Area Saltstone facility. Premix components were blended in the ratio of 10/45/45 cement/slag/fly ash and mixed with the appropriate amount of salt solution to make the Saltstone grout run during 4Q07 SPF operations.[5] Table III is the mix used for all of the Saltstone simulants prepared in this study.

Table III. Saltstone Mix Proportions for Simulant Grout.

Component	Premix	Saltstone
	%	%
Salt Solution	0	46.8
Water (71.48% of salt solution)	0	(33.5 as salt solution)
Water to Premix Ratio	--	0.63
Cement	10	5.3
Slag	45	23.9
Fly Ash	45	23.9

Sample Preparation

Simulant grout slurries were cast into either 2-inch PVC cylinders or 5-gallon pails, Figure 1. The cast cylinders were used to develop the sample cutting technique and to provide a laboratory-cast sample for permeability to compare to the permeability of a cored sample of the same sample composition. The 5-gallon pail was cored using the same method developed for collecting the core samples from Vault 4, Cell E, Figure 2.[6] Cores were used to demonstrate sample sectioning technique and provide a permeability link between lab prepared and field collected (cored) samples.

Figure 1. Simulated grout samples cast for method development.

Figure 2. Cored 5-gallon pail and core.

Vault 4, Cell E Core Samples
Visual Inspection
    The core samples were removed from the evacuated sample container, inspected, and transferred to PVC containers and backfilled with nitrogen for continued storage.[7] Samples from location 3 (furthest from the wall) were the most intact cylinder shaped cored samples. The shade of the core samples darkened as the depth of coring increased. Based on the visual inspection performed, sample 3-3 shown in Figure 3 was selected for all subsequent analysis.

Figure 3. Sample 3-3. 5.5" long cylinder and associated fragments.

Sample Sectioning

A miter box saw modified to accommodate smaller work pieces and for operation in a radiohood was used to prepare samples for hydraulic conductivity measurements. A one-half inch section was trimmed off the end of sample 3-3 to provide a newly exposed surface.

Density and Porosity

During sectioning activities, three fragments of the core were stored for density and porosity measurements using ASTM C 642 "Standard Test Method for Density, Absorption, and Voids in Hardened Concrete." In the procedure, the test specimen is specified to be no less than 350 cm^3 (~850 g normal weight concrete). This criterion was not achievable as the entire core was ~122 cm^3. The fragments used were similar in size and the fragments each weighed approximately 10–14 g. Following the procedure, samples were dried at 110 °C and weighed intermittently until the difference between any two successive values is less than 0.5 % of the lowest value obtained, W_D. The dried, cooled samples were then immersed in liquid. The samples were weighed intermittently until two successive values of mass of the surface-dried sample at intervals of 24 h show an increase in mass of less than 0.5 % of the larger value, W_I. The ASTM procedure instructs immersion of the samples in water to determine the saturated mass. Given that the core samples were produced using salt solution, the immersion fluid used for these measurements was the salt simulant in Table II. To obtain the saturated mass after boiling, the samples were boiled in the salt solution for five hours. Additional salt solution was added as needed during boiling to keep the samples submerged. The sample fragments were allowed to cool and the surface-dried mass was measured, W_{IB}. The samples were submerged and the suspended mass was measured, W_{IBS}. The bulk density of the immersed sample was calculated by the equation

$$BD_I = \frac{W_I}{W_{IB} - W_{IBS}} \times \rho \tag{1}$$

where; BD_I is the bulk density of the immersed sample as defined by ASTM C 642, W_I is the mass of the sample immersed in salt solution and surface dried, W_{IB} is the mass of the sample immersed and boiled in salt solution and surface dried, W_{IBS} is the mass of the sample boiled in salt solution, cooled, and then suspended, and ρ is the density of the salt solution after boiling.

The porosity of the samples was calculated by the equation

$$Porosity\ \% = \frac{(W_{IB} - W_D)}{(W_{IB} - W_{IBS})} \times 100 \tag{2}$$

where; W_D is the mass of the sample immersed in salt solution and surface dried, W_{IB} is the mass of the sample immersed and boiled in salt solution and surface dried, and W_{IBS} is the mass of the sample boiled in salt solution, cooled, and then suspended.

Microstructure

By design, the coring in cell position 3, traversed three SPF operating days that resulted in three pouring lifts.[1] X-ray Diffraction Analysis (XRD) analysis was performed on fragments from cores from each of the three cores collected from cell position 3. The core sampling XRD was performed on the as-collected core samples 3-1, 3-2, and 3-3. Approximately 0.25 g of each of the cores was submitted for analysis.

Fragments of the trimmed core sample 3-3 were analyzed using various microscopic techniques. Optical microscopy and scanning electron microscopy (SEM) were used to evaluate fresh fractured surfaces and polished surfaces. In addition to standard SEM imaging using secondary electrons (SEI), images were obtained using the detector for backscattered electrons (BSE). Imaging with the BSE detector provides information about the distribution of different elements in the sample. On select images, Energy Dispersive X-ray Spectroscopy (EDS) was used to determine the relative elemental composition of the sample.

Results and Discussion
Sample Sectioning
    To demonstrate the soundness of the approach, the sectioning methodology was used on the lab prepared samples, Figure 1, and the lab cored samples, Figure 2. The first cut of the lab cored sample removed the end of the core. The sample was advanced two inches and the second cut completed the sample. The samples cut under the weight of the saw. The as-cut surface was sufficiently smooth to eliminate the need for additional surface preparation for permeability measurements.

    The Vault 4, Cell E cored sample 3-3 was removed from the storage container as described in Reference 7. As with the lab cored sample, the first cut removed the end of the core. Two 2-inch samples were cut and stored in the simulated salt solution in Table II. The samples were cut using the weight of the saw. When cutting the Vault 4 core sample, more material was removed with each stroke than when cutting the lab cored sample.

Density and Porosity
    Table IV is the mass used for the core fragments after each of the procedure steps in the modified ASTM C 642. Using Equations (1) and (2), the bulk density after immersion and boiling, BD_I, and the porosity was calculated, Table V.

Table IV. Mass of Core Fragments used to Calculate Density and Porosity.

Replicate	W_D	W_I	W_{IB}	W_{IBS}
1	7.441	11.595	12.772	3.524
2	8.672	13.890	15.56	4.132
3	9.539	15.809	17.694	4.508

Table V. Bulk Density and Porosity of Core Sample 3-3 Using ASTM C 642.

Replicate	BD_I	Porosity (%)
1	1.92	57.64
2	1.89	60.27
3	1.86	61.85
Average	1.90	59.90

    The density and porosity of Saltstone prepared with simulated salt solutions representing projected salt compositions for out year processing.[8] It was demonstrated that the salt solution, water-to-premix ratio, and premix blend effect the bulk density and porosity of the final product. The values measured for the Vault 4 core sample in Table V correlate well with the bulk density and porosity values reported for the Saltstone prepared with in Reference 8—1.78-1.81 g.cm³ and 59.1-62.4%, respectively.

Microstructure
X-ray Diffraction Analysis
Figure 3 is the XRD patterns of the three core samples collected from cell position 3.

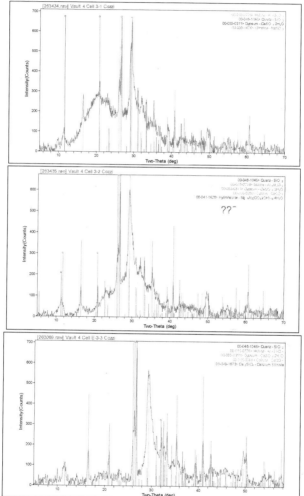

Figure 3. X-ray diffraction analysis of the three core samples collected from cell position 3.

Phases common to each of the samples are mullite, quartz, and gypsum. Mullite and quartz are associated with the crystalline portion of fly ash. Although less common, gypsum can also be attributed to the fly ash. The occurrence of gypsum in fly ash is indicative of wet treatment or storage.[9] In addition to these phases being common to all of the samples, they are also the most predominant crystalline phases. This corresponds well with the abundance of fly ash in the mix and the limited reactivity of the fly ash.[10] Hydrotalcite, a hydration phase associated with slag, is only identified in the 3-2 XRD pattern. Hydrotalcite may be present in all of the samples, however, the peaks associated with this phase can be obscured by the peaks associated with gypsum.[11] Sodium nitrate was identified in the XRD pattern for sample 3-1. The sodium nitrate is a component of the salt solution and could be expected in all of the samples. However, since the samples were prepared and analyzed with minimal handling, there may be sufficient pore water to preclude the sodium nitrate from crystallizing to an extent sufficient enough to be positively identified by XRD. The amorphous hump near 30° 2θ in the spectra is indicative of the presence of calcium silicate hydrate (CSH) gel.

Optical Microscopy
The end piece of core sample 3-3, Figure 3, was shattered to provide fragments for microscopic analysis, Figure 5. The rind of light material is due to oxidation that occurred in the ten days of storage in a plastic bag between sectioning and fracture. Fragments were collected to maximize exposure surfaces of the interior for microscopy.

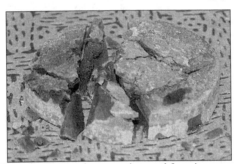

Figure 4. Shattered core section used for microscopy.

A fragment of the sample with the interior exposed was mounted for optical microscopy. Overall, the sample shows a uniform presence of porosity and inclusions. The larger round openings are where fly ash has been. Fly ash used at the time of processing had a mean particle size of ~ 52 μm and 95% of the particles > 163 μm.[12] The red coloration can be attributed to iron associated with the fly ash. Reference 12 identifies a substantial quantity of magnetic iron compounds associated with fly ash. Photographs were taken using magnifications from 5-150x with no evidence of cracking at any magnification.

After polishing, the sample was photographed at 100x magnification, Figure 6. In this photo, the fly ash particles denoted by the two upper arrows are still present, whereas the particle identified by the lowest arrow has been pulled out during sample preparation.

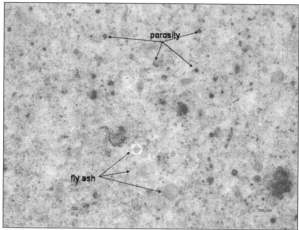

Figure 6. Polished Vault 4 core sample noting fly ash and porosity sites, 100x.

Electron Microscopy

A five millimeter fragment of fresh fractured core sample was mounted for SEM/EDS analysis. Figure 7 shows the fracture surface (SEI) and corresponding backscatter electron image (BSE) for the as mounted fragment. The gaps in the top image are approximately 80 μm wide and between 600-800 μm long. These gaps are inherent on the sample and do not appear to be an artifact of sample handling. The BSE image has spherical particles dispersed throughout the sample. Also evident is a crack crossing the sample. The crack may be attributed to sample preparation since this analysis requires placing the sample in a vacuum, which can dry the Saltstone sample and introduce cracking.

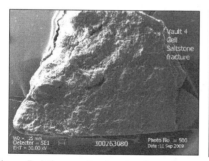

Figure 7. Micrographs showing the as-fractured surface of a fragment of the Vault 4 core sample and the corresponding BSE image, 2.5x.

Figure 8 is the area in the lower micrograph in Figure 7. These micrographs taken at a magnification of 40x show electron micrographs of the morphology and elemental distribution in the Vault 4 core sample.

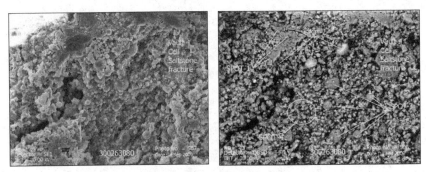

Figure 8. Representative micrographs of the fractured core sample. Left—Secondary electron image showing the morphology of the Vault 4 core sample. Right—Corresponding backscatter electron image annotated with the EDS spectra positions in Figure 9.

In the left micrograph (photo No. 507) in Figure 8, the spherical nature of the particles seen in Figure 7 is evident. In the micrograph on the right, the annotated spots were analyzed using EDS, Figure 9. The difference in contract between the spheres in spot-2 and spot-3 indicate that the spheres are of different compositions.

At the magnification of 40x, the spots are an aggregate composition of the area being evaluated. The areas in spot-1 are represented by the EDS spectrum in Figure 9. Spot-1 is predominantly calcium and silicon with smaller amounts of aluminum, magnesium and sodium. Considering the composition of the salt solution, the XRD analysis and the micrographs, it can be inferred that these points in spot-1 represent calcium silicate hydrate (CSH) gel, tricalcium aluminate (C3A), hydrotalcite, and salt from the salt solution. The spheres in spot-2 are predominantly iron with silicon, aluminum, and calcium. Recalling the abundance of magnetic iron compounds in fly ash discussed in Reference 12 and the higher atomic number of the components of these spheres as identified in the BSE micrograph in Figure 8, these spheres are the iron oxide phases from fly ash. The EDS spectrum for the spheres in spot-3 coupled with the XRD analysis would indicate that the spheres are the unreacted quartz (SiO_2) and mullite ($3Al_2O_3 \cdot 2SiO_2$) in the fly ash. Spot-4 is an overall EDS spectrum of an area and reflects the overall composition of the Saltstone (premix components and salt solution).

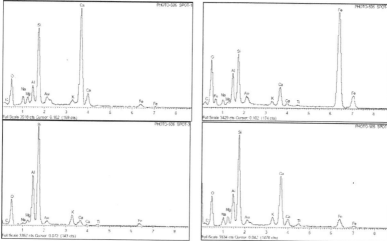

Figure 9. EDS spectra for the four locations annotated in Figure 8.

SEM/EDS analysis was performed on a polished sample of the Vault 4 core sample. Figure 10 is representative micrographs of the polished sample. The cavities in the micrographs are remnants of fly ash particles (either the glassy phase or the ferrite spinel phase) that pulled out during sample preparation. In the smaller cavities on the left of the micrograph, smaller spheres line the inside of the cavity. These spheres could be the unreacted mullite and quartz phases of fly ash. The more intact spheres that are more pronounced in the BSE micrograph are fly ash particles that remained embedded in the sample. The large cavity on the right is a glassy cenosphere, also attributed to the fly ash.[13]

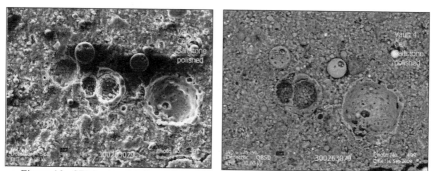

Figure 10. SEM micrographs of the polished Vault 4 core sample, 200x. Top—SEI and Bottom—BSE.

CONCLUSIONS

Core samples were collected from the Saltstone Disposal Facility Vault 4, Cell E. Visual inspection of the cores showed that the dry cored samples were darker in tone than the wet cored samples. Core 3-3, a dry cored sample, was used for all of the subsequent analyses. A sectioning methodology was demonstrated with simulants using a hand miter saw to prepare samples for future permeability measurements. This technique was applied to the Vault 4 core sample. Sectioning of the Vault 4 core sample was marginally easier than sectioning of the lab cored simulant.

The density and porosity of the Vault 4 core sample, 1.90 g/cm^3 and 59.90% respectively, were comparable to values achieved for laboratory prepared samples using water-to-cementitious materials ratios, premix blends, and simulated salt solutions.

X-ray diffraction analysis of the Vault 4 core samples from position 3 identify phases consistent with the presence of fly ash, sodium nitrate salt solution, and calcium silicate hydrates—a product of the hydration of cement and slag. Microscopic analysis revealed morphology features characteristic of cementitious materials with fly ash and CSH gel.

When taken together, the results of the density, porosity, x-ray diffraction analysis and microscopic analysis support the conclusion that the Vault 4, Cell E core sample is representative of the expected waste form.

REFERENCES
1. Smith, F.M., "Saltstone Sampling Summary for September 2008," LWO-RIP-2008-00006 Revision 0.
2. Zeigler, K.E., Bibler, N.E., "Results for the Fourth Quarter 2007 Tank 50H WAC Slurry Sample: Chemical Contaminant Results."
3. Harbour, J.R., Edwards, T.B., Hansen, E.K., and Williams, V.J., "Variability Study for Saltstone," WSRC-TR-2005-00447, Revision 0, (October 2005).
4. Cozzi, A.D., and Hansen, E.K., "Saltstone Formulation Testing for DDA Batch 3," SRNL-PSE-2008-00019 Revision 0, (January 2008).
5. Staub, A.V., "Saltstone Processing History for Third and Fourth Quarter 2007 Operation," LWO-WSE-2007-00216 revision 0, (December 2007).
6. Cheng, W.Y., Marzolf, A.D., and Milling, R.B., "Recommendations for Improving the Drilled-Core Saltstone Grout Sampling Process," SRNS-STI-2009-00481 Revision 0, (August 2009).
7. Hansen, E.K., and Crawford, C.C., "Pictures of Vault 4 Core Samples – Transfer of Samples at SRNL – April 6, 2009," SRNL-L3100-2009-00087 Revision 0, (April 2009).
8. Harbour, J.R., Williams, V.J., Edwards, T.B., Eibling, R.E., Schumacher, R.F., "Saltstone Variability Study – Measurement of Porosity," WSRC-STI-2007-00352 Revision 0.
9. Walenta, G. and Füllmann, T., "Advances in Quantitative XRD Analysis for Clinker, Cements, and Cementitious Additions," *Advances in X-ray Analysis,* **47** 287-96 (2004).
10. Harbour, J.R., Williams, V.J., Edwards, T.B., "Heat of Hydration of Saltstone Mixes-Measurement by Isothermal Calorimetry," WSRC-STI-2007-00263.
11. Gruskovnjak, A., Lothenbach, B., Winnefeld, F., Figi, R., Ko, S.-C., Adler, M., and Mäder, U., "Hydration Mechanisms of Super Sulphated Slag Cement," *Cement and Concrete Research,* **38** 993-2 (2008).
12. Hansen, E.K., and Harbour, J.R., "Comparison of Boral with Cross Station Fly Ash," SRNL-PSE-2006-00095 Revision 0, (2006).

13. Kutchko, G.G. and Kim, A.G., "Fly Ash Characterization by SEM–EDS," *Fuel,* **85** 2537-44 (2006).

14. Hansen, E.K., and Cozzi, A.D., "Vault 4, Cell E, Grout Sample Analyses," SRNL-TR-2009-00005 Revision 0.

15. Harbour, J.R., Williams, V.J., "Saltstone Performance Indicator - Dynamic Young's Modulus," SRNL-STI-2008-00488 Revision 0.

INCORPORATION OF MONO SODIUM TITANATE AND CRYSTALLINE SILICOTITANATE FEEDS IN HIGH LEVEL NUCLEAR WASTE GLASS

K.M. Fox, F.C. Johnson, and T.B. Edwards
Savannah River National Laboratory
Aiken, SC 29808 USA

ABSTRACT

Four series of glass compositions were selected, fabricated, and characterized as part of a study to determine the impacts of the addition of Crystalline Silicotitanate (CST) and Monosodium Titanate (MST) from the Small Column Ion Exchange (SCIX) process on the Defense Waste Processing Facility (DWPF) glass waste form and the applicability of the DWPF process control models. All of the glasses studied were considerably more durable than the benchmark Environmental Assessment (EA) glass. The measured Product Consistency Test (PCT) responses were compared with the predicted values from the current DWPF durability model. One of the KT01-series and two of the KT03-series glasses had measured PCT responses that were outside the lower bound of the durability model. All of the KT04 glasses had durabilities that were predictable regardless of heat treatment or compositional view. In general, the measured viscosity values of the KT01, KT03, and KT04-series glasses are well predicted by the current DWPF viscosity model. The results of liquidus temperature (T_L) measurements for the KT01-series glasses were mixed with regard to the predictability of the T_L for each glass. All of the measured T_L values were higher than the model predicted values, although most fell within the 95% confidence intervals. Overall, the results of this study show a reasonable ability to incorporate the anticipated SCIX streams into DWPF-type glass compositions with TiO_2 concentrations of 4-5 wt % in glass.

INTRODUCTION

The Savannah River Site (SRS) Tank Farm will begin a process referred to as Small Column Ion Exchange (SCIX) to disposition salt solution in fiscal year 2014. In the first step of the process, salt solution retrieved from various waste tanks will be struck with Monosodium Titanate (MST) to remove key actinides. The salt solution will then be processed using Rotary Micro Filtration (RMF) to remove the MST and any insoluble solids. The MST and insoluble solids will accumulate on the bottom of Tank 41. The filtrate from RMF will be fed to ion exchange columns, also in Tank 41, to remove the ^{137}Cs using Crystalline Silicotitanate (CST) resin. The decontaminated salt solution will be sent to the Saltstone Facility for immobilization in grout. The ^{137}Cs-laden CST resin will be sluiced and ground for particle size reduction, then sent to the Defense Waste Processing Facility (DWPF) for immobilization in glass. These processes mirror the current disposition paths for streams associated with the Salt Waste Processing Facility (SWPF), which is under construction.

The MST and insoluble solids from Tank 41 will periodically be transferred to a sludge batch preparation tank (e.g., Tank 42 or Tank 51) as part of the High Level Waste (HLW) sludge batch preparation process for DWPF. The ground, ^{137}Cs-laden CST material (hereafter referred to simply as CST) from SCIX will be periodically transferred to Tank 40 prior to being processed at DWPF. Periodic additions of CST to Tank 40 would result in a changing composition of each sludge batch as it is processed since Tank 40 serves as the feed tank for the DWPF. Work is currently in progress to determine the feasibility of dropping the ground CST into Tank 41. If ground CST can be dropped into Tank 41 (depending on heat loading issues, among others), the CST would be sent to Tank 42 or Tank 51 using an existing transfer line. Therefore, the studies of SCIX impacts on DWPF glass formulation will encompass scenarios where the CST is sent to either Tank 40 or a sludge batch preparation tank.

The MST and CST from the SCIX process will significantly increase the concentrations of Nb_2O_5, TiO_2, and ZrO_2 in the DWPF feed. Other constituents of MST and CST – Na_2O and SiO_2 – are already present in high concentrations in DWPF glass; thus their influences are well understood. The increased concentrations of Nb_2O_5, TiO_2, and ZrO_2 will likely have some impact on the properties and performance of the DWPF glass product. Properties such as the liquidus temperature, viscosity, and rate of melting of the glass may be impacted. The performance of the glass, particularly its chemical durability as it pertains to repository acceptance requirements, may also be impacted. The DWPF uses a set of semi-empirical and first-principles models referred to as the Product Composition Control System (PCCS)[1] to predict the properties and performance of a glass based on its composition since it is not possible to measure these attributes during processing. The objective of this study is to evaluate the impacts of the SCIX streams on the properties and performance of the DWPF glass product and on the applicability of the current process control models.

Previous studies have provided data on the potential impacts of MST and CST specifically on DWPF-type glasses. Edwards and Harbour completed a series of studies on the coupling of CST and MST with Purex and HM sludges for DWPF processing in 1999.[2-6] All of the glasses studied had very good measured durability values, although there were minor issues with the predictions of the durability model and the majority of the glasses failed the homogeneity constraint.[6] The glasses generally had estimated liquidus temperatures that would be acceptable for DWPF processing. The viscosities of the glasses produced with CST and the Purex sludge were acceptable for processing but not well predicted. The viscosities of the glasses fabricated with CST and HM sludge were too high for processing but were better predicted.

Fox and Edwards recently performed a paper study evaluation using updated projections for sludge batch compositions and SCIX CST and MST addition rates.[7] This study found that, as a result of the updated composition projections, several viable options were predicted to be available for incorporation of the SCIX streams into either Tank 40 or a sludge batch preparation tank. Transfer of the CST to a sludge batch preparation tank was the preferred option since it allowed more compositional flexibility for frit development while maintaining sufficient projected operating windows. The report again identified several assumptions and limitations associated with the current PCCS models, and recommended that these be further evaluated.

This study focuses on addressing the assumptions made during the paper studies, and will provide data for further evaluation of model performance.

EXPERIMENTAL PROCEDURE

Selection of Glass Compositions

Two series of glass compositions, identified as the KT01-series and the KT02-series, were developed for the first portion of the study. Composition projections developed during the paper study evaluation of the addition of the SCIX streams were used as the basis for the two series,[7] as shown in Table 1. The KT01-series compositions were derived with the intent of investigating the impact of individual glass components on the retention of TiO_2. The concentrations of Al_2O_3, B_2O_3, Fe_2O_3, K_2O, Li_2O, Na_2O, and SiO_2 were adjusted from the baseline composition to determine whether these changes would affect the ability of the glass to accommodate increased TiO_2 concentrations. The U_3O_8 and ThO_2 components were removed so that the glasses could be handled in a non-radioactive laboratory. The concentration of TiO_2 was forced to an elevated value of 8 wt % for each composition. The remaining components were then normalized to give a total of 100 wt %. These steps resulted in 13 glass compositions for the KT01-series, and their target compositions are available elsewhere.[8]

Table 1. Projected Average Concentrations (wt %) of Glass Components in Out-Year Processing at DWPF with SCIX Streams.

Oxide	Average	Oxide	Average
Al_2O_3	5.69	MnO	1.30
B_2O_3	4.80	Na_2O	15.03
BaO	0.08	Nb_2O_5	0.43
CaO	1.01	NiO	0.36
Ce_2O_3	0.21	PbO	0.12
Cr_2O_3	0.11	SO_4^{2-}	0.08
CuO	0.03	SiO_2	47.52
Fe_2O_3	12.05	ThO_2	0.23
K_2O	0.07	TiO_2	2.74
La_2O_3	0.08	U_3O_8	2.53
Li_2O	4.80	ZnO	0.05
MgO	0.15	ZrO_2	0.52

A similar process was used to develop compositions for the KT02-series glasses. The intent was to again identify any impact of individual glass components on the retention of TiO_2. The only difference from the KT01-series was the increase in TiO_2 concentrations from 8 wt % to 12 wt % for each of the glass compositions, with a subsequent normalization of the other components to 100 wt %. The KT02-series glasses were not characterized as thoroughly as the KT01-series glasses since their TiO_2 concentrations were quite high. As will be described below, X-ray Diffraction (XRD) was used to identify any obvious changes in the crystallization behavior of these glasses after the CCC when TiO_2 concentrations were high, but chemical properties were not determined.

The KT03-series of compositions was developed to further investigate the potential impacts of the addition of the SCIX streams on glass properties. The average glass composition given in Table 1 was again used, with adjusted values (minimums and maximums) for the concentrations of Al_2O_3, B_2O_3, Fe_2O_3, K_2O, Li_2O, Na_2O, and SiO_2 to identify any impacts of these individual components on the behavior of TiO_2 in the glass. The TiO_2 concentrations were fixed at an elevated value of 8 wt %. The other major constituents of CST, Nb_2O_5 and ZrO_2, were fixed at elevated concentrations of 3 wt % and 2.5 wt %, respectively, to identify any interactive effects with TiO_2. This resulted in 13 target compositions for the KT03-series glasses, which are available elsewhere.[8]

The basis for the KT04-series compositions was changed from the average glass composition used previously (Table 1) to projections of individual sludge batches incorporating the SCIX streams. These projections were again very similar to those provided in the paper study report,[7] and resulted in ten glass compositions for study. The U_3O_8 and ThO_2 components were removed from the target compositions to support non-radioactive experiments. The target compositions of the KT04-series glasses are available in detail elsewhere.[8]

Glass Fabrication

Each of the study glasses was prepared from the proper proportions of reagent-grade metal oxides, carbonates, and boric acid in 200 g batches. The raw materials were thoroughly mixed and placed into platinum/gold, 250 ml crucibles. The batch was placed into a high-temperature furnace at the melt temperature of 1150 °C. The crucible was removed from the furnace after an isothermal hold for 1 hour. The glass was poured onto a clean, stainless steel plate and allowed to air cool (quench). Approximately 25 g of each glass was heat-treated to simulate cooling along the centerline of a

DWPF-type canister[9] to gauge the effects of thermal history on the product performance. This cooling schedule is referred to as the CCC heat treatment.

Characterization

Two dissolution techniques, sodium peroxide fusion (PF) and lithium-metaborate fusion (LM), were used to prepare the glass samples, in duplicate, for chemical composition analysis. Each of the samples was analyzed, twice for each element of interest, by Inductively Coupled Plasma – Atomic Emission Spectroscopy (ICP-AES). Glass standards were also intermittently measured to assess the performance of the ICP-AES instrument over the course of these analyses. Representative samples of each quenched and CCC glass were ground for powder XRD analysis.

The Product Consistency Test (PCT) Method-A[10] was performed in triplicate on each KT01, KT03, and KT04-series quenched and CCC glass to assess chemical durability. Also included in the experimental test matrix was the Environmental Assessment (EA) benchmark glass,[11] the Approved Reference Material (ARM) glass,[12] and blanks from the sample cleaning batch. Samples were ground, washed, and prepared according to the standard procedure.[10] Fifteen milliliters of Type-I ASTM water were added to 1.5 g of glass in stainless steel vessels. The vessels were closed, sealed, and placed in an oven at 90 ± 2 °C for 7 days. Once cooled, the resulting solutions were sampled (filtered and acidified), then analyzed by ICP-AES. Normalized release rates were calculated based on the measured compositions using the average of the common logarithms of the leachate concentrations.

The viscosity of select glasses was measured following Procedure A of the ASTM C 965 standard.[13] Harrop and Orton high temperature rotating spindle viscometers were used with platinum crucibles and spindles. The viscometers were specially designed to operate with small quantities of glass to support measurements of radioactive glasses when necessary.[14,15] A well characterized standard glass was used to determine the appropriate spindle constants.[15,16] Measurements were taken over a range of temperatures from 1050 to 1250 °C in 50 °C intervals. Measurements at 1150 °C were taken at three different times during the procedure to provide an opportunity to identify the effects of any crystallization or volatilization that may have occurred during the test.

The liquidus temperatures (T_L) of select study glasses were determined using the isothermal heat treatment method.[17] The temperature profile of the furnace was carefully determined and periodically verified.[18] All thermocouples and temperature measurement devices were calibrated and periodically verified. A standard glass composition was incorporated into the test glass matrix as a method of verifying the measured data.[a] Polished samples of each quenched glass were observed via optical microscopy prior to T_L measurement to determine whether any preexisting crystals were present. Quenched glasses that were found to contain crystals were excluded from testing. The study glasses were ground, washed, dried, and heat treated in platinum boats with tight fitting lids. The glasses were air quenched after being removed from the furnace, then sectioned and polished for microscopy. Any bulk crystallization that occurred during the isothermal heat treatments was identified by optical microscopy. The procedure was repeated over various temperatures to determine the T_L to within a narrow range of tolerance.

Liquidus temperatures were determined for the KT01-series glasses. Liquidus temperatures for the KT02-series glasses were not measured due to their unrealistically high TiO_2 concentrations. Liquidus temperature measurements for the KT03 and KT04-series glasses are not yet complete.

---

[a] The glass standard is identified as 'Unknown Glass A' from the Pacific Northwest National Laboratory liquidus temperature round robin study.

RESULTS AND DISCUSSION

Homogeneity

The homogeneity of each glass sample was assessed via XRD. All of the quenched KT01-series glasses were found to be amorphous by XRD. XRD results identified crystallization in compositions KT01-HL and KT01-HF. Composition KT01-HL contained magnetite ($Fe^{2+}Fe_2^{2+}O_8$) and lithium silicate (Li_2SiO_3). Composition KT01-HF contained magnetite.

All of the KT02-series quenched glasses were amorphous by XRD. The XRD results for the CCC versions of the KT02-series glasses identified crystallization in all of the glasses except for compositions KT02-LA, KT02-LF, KT02-HN, and KT02-HK.

The KT02 compositions were selected to explore the impact of varying the concentrations of the major components of the glass on the retention of TiO_2. The XRD results for the KT02 glasses are summarized in Table 2, which offers some insight into the impact of compositional changes on the propensity for titanium to crystallize out of the glass. As mentioned earlier, all of the quenched glasses were XRD amorphous. Titanium-containing crystalline phases formed in all of the compositions except for the low Al_2O_3 concentration glass, the low Fe_2O_3 concentration glass, the high Na_2O concentration glass, and the high K_2O concentration glass.[a] In general, the impacts of these components will need additional investigation before drawing further conclusions (i.e., a larger number of compositions should be fabricated and characterized since the effects of an individual component are likely to be strongly influenced by overall composition).

XRD results for the KT03-series glasses indicated that all of the quenched glasses were amorphous. The XRD results for the CCC glasses showed that four of the compositions, KT03-LN, KT03-LB, KT03-LL, and KT03-HF, contained magnetite and trevorite ($NiFe_2O_4$). Two of the glasses were highly devitrified: KT03-HL contained trevorite ($NiFe_2O_4$) and KT03-HA contained magnetite with another unidentified phase.

XRD results for the KT04-series glasses indicated that all of the quenched glasses were amorphous, as well as all of the CCC versions.

---

[a] The CCC version of glass KT02-LS contained an unidentified phase that may or may not contain TiO_2. Further characterization of this glass is in progress to identify this phase.

Table 2. Summary of XRD Results for the KT02-Series Glasses.

Glass ID	Composition Note	Heat Treatment	XRD Results
KT02-HA	High Al_2O_3	quenched	amorphous
		CCC	$LiFeTiO_4$, unidentified phase
KT02-LA	Low Al_2O_3	quenched	amorphous
		CCC	amorphous
KT02-HB	High B_2O_3	quenched	amorphous
		CCC	$LiFeTiO_4$, unidentified phase
KT02-LB	Low B_2O_3	quenched	amorphous
		CCC	$LiFeTiO_4$, unidentified phase
KT02-HF	High Fe_2O_3	quenched	amorphous
		CCC	$LiFeTiO_4$, Fe_9TiO_{15}
KT02-LF	Low Fe_2O_3	quenched	amorphous
		CCC	amorphous
KT02-HL	High Li_2O	quenched	amorphous
		CCC	$LiFeTiO_4$, Li_2SiO_3
KT02-LL	Low Li_2O	quenched	amorphous
		CCC	$LiFeTiO_4$, unidentified phase
KT02-HN	High Na_2O	quenched	amorphous
		CCC	amorphous
KT02-LN	Low Na_2O	quenched	amorphous
		CCC	$LiFeTiO_4$, unidentified phase
KT02-HS	High SiO_2	quenched	amorphous
		CCC	Rutile (TiO_2), Hematite (Fe_2O_3)
KT02-LS	Low SiO_2	quenched	amorphous
		CCC	unidentified phase
KT02-HK	High K_2O	quenched	amorphous
		CCC	amorphous

Chemical Composition

A review of the ICP-AES data showed that there were no significant issues with meeting the targeted compositions for each of the study glasses.

Durability

All of the glasses had normalized leachate for boron (NL [B]) values that were well below the 16.695 g/L value of the benchmark EA glass.[11] No heat treatment effects were seen in any of the three sets of compositions. The predictability of the PCT responses was evaluated using the DWPF durability model.[19] The predicted PCT values, determined using the measured compositions of the glasses, were compared with the normalized PCT responses. Figure 1 provides plots of the DWPF model for B that relates the logarithm of the normalized PCT value to a linear function of a free energy of hydration term (ΔG_p, in kcal/100 g glass) derived from both of the heat treatments of the glasses. Prediction limits at a 95% confidence for an individual PCT result are also plotted along with the linear fits. The EA and ARM results are indicated on these plots as well. All but one of the KT01 glasses fall within the prediction limits of the DWPF model. Glass KT01-HK (the high K_2O concentration composition) has a PCT response that falls below the lower prediction limit; however, the PCT response of this glass remains considerably lower than that of the benchmark EA glass. All but two of the KT03 glasses fall within the prediction limits of the DWPF model. Glasses KT03-HK and KT03-MK have PCT responses that fall below the lower prediction limit; however, the PCT responses of these glasses remain considerably lower than that of the benchmark EA glass. As seen in the KT01 plot, there is an issue with the predictability of these high K_2O concentration glasses. However, this lack of model applicability may be of little practical importance since K_2O concentrations of this magnitude are unlikely for actual compositions to be processed at the DWPF. All of the KT04 glasses fall within the prediction limits of the DWPF model, regardless of compositional view or heat treatment, indicating good predictability for these compositions.

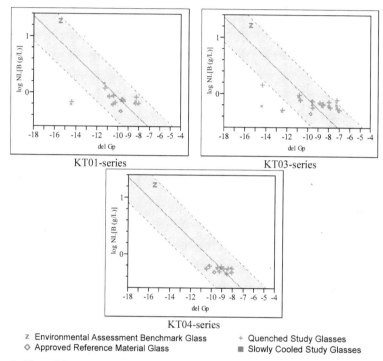

KT01-series   KT03-series

KT04-series

z   Environmental Assessment Benchmark Glass   +   Quenched Study Glasses
◇   Approved Reference Material Glass   ■   Slowly Cooled Study Glasses

Figure 1. Plots of the DWPF Durability Model Predictions versus Measured PCT Responses for the Three Series of Glasses.

Viscosity

Viscosity data were collected for all of the glasses in the KT01, KT03, and KT04-series. The measured viscosity at 1150 °C was determined by fitting the data for each glass to the Fulcher equation.[20,21] The results of the Fulcher fits were used to calculate a measured viscosity value for each glass at 1150 °C. The measured values are displayed graphically versus the model predicted values in Figure 2.

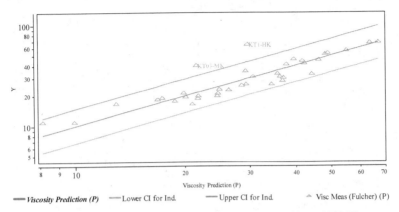

Figure 2. Measured Versus Predicted Viscosities at 1150 °C
for the Three Series of Glasses.

Figure 2 shows that all but two of the study glasses had measured viscosities that were predictable using the current DWPF viscosity model (based on the measured compositions). The two glasses with viscosities that fall above the upper confidence intervals for the model prediction are the high K_2O concentration glasses KT01-HK and KT03-MK. These glasses have measured K_2O concentrations of 16.4 and 9.9 wt %, respectively. These concentrations are above the DWPF viscosity model development range 0 to 5.73 wt % K_2O.[22] Interestingly, the KT03-HK composition, with a measured K_2O concentration of 17.6 wt %, had a measured viscosity that was well predicted by the current DWPF model. While these results point to a lack of applicability for the DWPF viscosity model for increased K_2O concentrations, it is important to note that these high K_2O concentration glasses were developed to determine potential impacts of K_2O on the retention of TiO_2 and that their K_2O concentrations are impractically high for actual DWPF processing. Overall, the measured viscosity values of the KT01, KT03, and KT04-series glasses are well predicted by the current DWPF viscosity model.

Liquidus Temperature
Liquidus temperatures were measured for the KT01-series glasses. Liquidus temperature estimates for the KT03 and KT04-series glasses are currently underway. Measured liquidus temperatures for the KT01-series glasses are presented in Figure 3 and compared with the predicted values from the DWPF model (based on the measured compositions of the glasses).

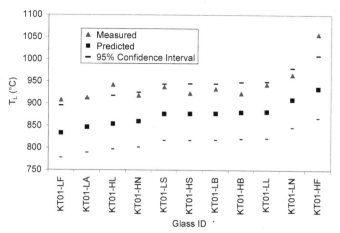

Figure 3. Predicted and Measured Liquidus Temperatures for the KT01-Series Glasses, Based on the Measured Compositions.

The results in Figure 3 are mixed with regard to the predictability of the T_L for each glass. All of the measured T_L values are higher than the model predicted values, although most fall within the 95% confidence intervals. Compositions KT01-LF, KT01-LA, KT01-HL, and KT01-HF have measured T_L values that are above the upper 95% confidence bounds on the predicted values. The concentrations of some of the components in these glasses, particularly TiO_2, fall outside the region of applicability of the current T_L model.[23] These results indicate that the model may need to be adjusted in order to more correctly predict the T_L of glasses when the SCIX streams are incorporated, although further data are necessary (and are currently being collected) for a more complete assessment.

SUMMARY

Four series of glass compositions were selected, fabricated, and characterized as part of a study to determine the impacts of the addition of CST and MST from the SCIX process on the DWPF glass waste form and the applicability of the DWPF process control models. The KT01 and KT02-series of glasses were chosen to allow for the identification of the influence of the concentrations of major components of the glass on the retention of TiO_2. The KT03 series of glasses was chosen to allow for the identification of these influences when higher Nb_2O_5 and ZrO_2 concentrations are included along with TiO_2. The KT04 series of glasses was chosen to investigate the properties and performance of glasses based on the best available projections of actual compositions to be processed at the DWPF (i.e., future sludge batches including the SCIX streams).

The glasses were fabricated in the laboratory without the radioactive components. They were characterized using XRD to identify crystallization, ICP-AES to verify chemical compositions, and the PCT to measure durability. The viscosity and liquidus temperature of several of the glasses were also measured.

The XRD results showed no titanium containing phases in the KT01-series glasses, regardless of whether they had been air quenched or slowly cooled. The target TiO_2 concentration had to be increased to 12 wt % in glass in the KT02-series before any compositional impacts on TiO_2 retention (e.g., the formation of titanium containing crystals) were apparent. Titanium containing crystalline phases formed in the slowly cooled versions of all of the KT02 compositions except for the low Al_2O_3 concentration glass, the low Fe_2O_3 concentration glass, the high Na_2O concentration glass, and the high K_2O concentration glass. The impacts of these components will need additional investigation before drawing further conclusions (i.e., a larger number of compositions should be fabricated and characterized). However, one should keep in mind that the 12 wt % TiO_2 concentrations in these glasses are unrealistic for DWPF processing with the SCIX streams and material from SWPF. Spinels were identified in some of the KT03 glasses after the CCC heat treatment. The spinels did not adversely impact the durability of the glasses but will be important for liquidus temperature determinations. The KT04-series glasses were amorphous regardless of heat treatment.

All of the glasses studied were considerably more durable than the benchmark EA glass. The measured PCT responses were compared with the predicted values from the current DWPF durability model. One of the KT01-series and two of the KT03-series glasses had measured PCT responses that were outside the lower bound of the durability model. All three of these glasses had intentionally high K_2O concentrations (to evaluate the impact of K_2O on TiO_2 retention), which may indicate a lack of applicability for the durability model in this composition region. However, this is likely of little practical importance since K_2O concentrations of this magnitude are unrealistic for actual compositions to be processed at the DWPF. All of the KT04 glasses had durabilities that were predictable regardless of heat treatment or compositional view.

All but two of the KT01, KT03, and KT04-series glasses had measured viscosities that were predictable using the current DWPF viscosity model (based on the measured compositions). The viscosities of two of the high K_2O concentration glasses, KT01-HK and KT03-MK, fell above the upper confidence intervals for the model prediction. These glasses have K_2O concentrations that are above the DWPF viscosity model development range for K_2O. Another high K_2O concentration glass, the KT03-HK composition, had a measured viscosity that was well predicted by the current DWPF model. While these results point to a lack of applicability for the DWPF viscosity model for increased K_2O concentrations, it is again important to note that these K_2O concentrations are unrealistic for actual DWPF processing. Overall, the measured viscosity values of the KT01, KT03, and KT04-series glasses are well predicted by the current DWPF viscosity model.

The results of T_L measurements for the KT01-series glasses were mixed with regard to the predictability of the T_L for each glass. All of the measured T_L values were higher than the model predicted values, although most fell within the 95% confidence intervals. Compositions KT01-LF, KT01-LA, KT01-HL, and KT01-HF had measured T_L values that were above the upper 95% confidence bounds on the predicted values. The concentrations of some of the components in these glasses, particularly TiO_2, fall outside the region of applicability of the current T_L model. These results indicate that the model may need to be adjusted in order to more correctly predict the T_L of glasses when the SCIX streams are incorporated, although further data are necessary for a more complete assessment. Liquidus temperature measurements for the KT03 and KT04-series glasses are underway.

FUTURE WORK

Overall, the results presented in this report show an ability to incorporate the anticipated SCIX streams into the DWPF-type glass compositions studied. Additional experiments are needed to determine whether to extend the validation range of the DWPF process control models or whether refitting of the models will be necessary. Liquidus temperature measurements are continuing for the KT03 and KT04 glasses, and should be performed for any additional compositions developed for this study.

Several additional sets of experimental glasses are being fabricated and characterized to provide further information on potential impacts of the SCIX streams on DWPF glass, including: a study of glass compositions previously shown to crystallize titanium containing phases at lower TiO_2 concentrations; a study of glass compositions covering a broader range of potential sludge compositions that remain acceptable for processing by the current DWPF process control models; a study of glass compositions incorporating noble metals that may serve as nucleation sites for titanium containing crystalline phases; and a study of glass compositions containing uranium and thorium that may impact the retention of TiO_2 or other glass properties.

At the completion of these studies, all of the data generated will be reviewed with regard to applicability of the DWPF PCCS models and recommendations will be made as to whether the validation ranges of the current models can be extended, or whether some or all of the models need to be refit to allow for the incorporation of the SCIX streams.

REFERENCES

1. Edwards, T. B., K. G. Brown and R. L. Postles, "SME Acceptability Determination for DWPF Process Control," *U.S. Department of Energy Report WSRC-TR-95-00364, Revision 5,* Washington Savannah River Company, Aiken, SC (2006).

2. Edwards, T. B. and J. R. Harbour, "Composition and Property Measurements for CST Phase 1 Glasses," *U.S. Department of Energy Report WSRC-TR-99-00245, Revision 0,* Westinghouse Savannah River Company, Aiken, SC (1999).

3. Edwards, T. B. and J. R. Harbour, "Composition and Property Measurements for CST Phase 2 Glasses," *U.S. Department of Energy Report WSRC-TR-99-00289, Revision 0,* Westinghouse Savannah River Company, Aiken, SC (1999).

4. Edwards, T. B. and J. R. Harbour, "Composition and Property Measurements for CST Phase 3 Glasses," *U.S. Department of Energy Report WSRC-TR-99-00291, Revision 0,* Westinghouse Savannah River Company, Aiken, SC (1999).

5. Edwards, T. B. and J. R. Harbour, "Composition and Property Measurements for CST Phase 4 Glasses," *U.S. Department of Energy Report WSRC-TR-99-00293, Revision 0,* Westinghouse Savannah River Company, Aiken, SC (1999).

6. Edwards, T. B., J. R. Harbour and R. J. Workman, "Summary of Results for CST Glass Study: Composition and Property Measurements," *U.S. Department of Energy Report WSRC-TR-99-00324, Revision 0,* Westinghouse Savannah River Company, Aiken, SC (1999).

7. Fox, K. M., T. B. Edwards, M. E. Stone and D. C. Koopman, "Paper Study Evaluations of the Introduction of Small Column Ion Exchange (SCIX) Waste Streams to the Defense Waste Processing Facility," *U.S. Department of Energy Report SRNL-STI-2010-00297, Revision 0,* Savannah River National Laboratory, Aiken, SC (2010).

8. Fox, K. M. and T. B. Edwards, "Impacts of Small Column Ion Exchange Streams on DWPF Glass Formulation: KT01, KT02, KT03, and KT04-Series Glass Compositions," *U.S. Department of Energy Report SRNL-STI-2010-00566, Revision 0,* Savannah River National Laboratory, Aiken, SC (2010).

9. Marra, S. L. and C. M. Jantzen, "Characterization of Projected DWPF Glass Heat Treated to Simulate Canister Centerline Cooling," *U.S. Department of Energy Report WSRC-TR-92-142, Revision 1,* Westinghouse Savannah River Company, Aiken, SC (1993).

10. ASTM, "Standard Test Methods for Determining Chemical Durability of Nuclear Waste Glasses: The Product Consistency Test (PCT)," *ASTM C-1285,* (2002).

11. Jantzen, C. M., N. E. Bibler, D. C. Beam, C. L. Crawford and M. A. Pickett, "Characterization of the Defense Waste Processing Facility (DWPF) Environmental Assessment (EA) Glass Standard Reference Material," *U.S. Department of Energy Report WSRC-TR-92-346, Revision 1,* Westinghouse Savannah River Company, Aiken, SC (1993).

12. Jantzen, C. M., J. B. Picket, K. G. Brown, T. B. Edwards and D. C. Beam, "Process/Product Models for the Defense Waste Processing Facility (DWPF): Part I. Predicting Glass Durability from Composition Using a Thermodynamic Hydration Energy Reaction Model (THERMO)," *U.S. Department of Energy Report WSRC-TR-93-672, Revision 1,* Westinghouse Savannah River Company, Aiken, SC (1995).

13. ASTM, "Standard Practice for Measuring Viscosity of Glass Above the Softening Point," *ASTM C-965*, (2007).

14. Schumacher, R. F. and D. K. Peeler, "Establishment of Harrop, High-Temperature Viscometer," *U.S. Department of Energy Report WSRC-RP-98-00737, Revision 0*, Westinghouse Savannah River Company, Aiken, SC (1998).

15. Schumacher, R. F., R. J. Workman and T. B. Edwards, "Calibration and Measurement of the Viscosity of DWPF Start-Up Glass," *U.S. Department of Energy Report WSRC-RP-2000-00874, Revision 0*, Westinghouse Savannah River Company, Aiken, SC (2001).

16. Crum, J. V., R. L. Russell, M. J. Schweiger, D. E. Smith, J. D. Vienna, T. B. Edwards, C. M. Jantzen, D. K. Peeler, R. F. Schumacher and R. J. Workman, "DWPF Startup Frit Viscosity Measurement Round Robin Results," *Pacific Northwest National Laboratory*, (Unpublished).

17. *"Isothermal Liquidus Temperature Measurement Procedure,"* Manual L29, ITS-0025, Savannah River National Laboratory, Aiken, SC (2009).

18. *"Furnace Functional Check,"* Manual L29, ITS-0096, Savannah River National Laboratory, Aiken, SC (2009).

19. Jantzen, C. M., K. G. Brown, T. B. Edwards and J. B. Pickett, "Method of Determining Glass Durability, THERMO™ (Thermodynamic Hydration Energy MOdel)," *United States Patent #5,846,278* (1998).

20. Fulcher, G. S., "Analysis of Recent Measurements of the Viscosity of Glasses," *Journal of the American Ceramic Society*, **8** [6] 339-355 (1925).

21. Fulcher, G. S., "Analysis of Recent Measurements of the Viscosity of Glasses, II," *Journal of the American Ceramic Society*, **8** [12] 789-794 (1925).

22. Jantzen, C. M., "The Impacts of Uranium and Thorium on the Defense Waste Processing Facility (DWPF) Viscosity Model," *U.S. Department of Energy Report WSRC-TR-2004-00311, Revision 0*, Westinghouse Savannah River Company, Aiken, SC (2005).

23. Brown, K. G., C. M. Jantzen and G. Ritzhaupt, "Relating Liquidus Temperature to Composition for Defense Waste Processing Facility (DWPF) Process Control," *U.S. Department of Energy Report WSRC-TR-2001-00520, Revision 0*, Westinghouse Savannah River Company, Aiken, SC (2001).

RADIATION RESISTANCE OF NANOCRYSTALLINE SILICON CARBIDE

Laura Jamison, Peng Xu, Kumar Sridharan, Todd Allen
University of Wisconsin-Madison
Madison, Wisconsin, USA

ABSTRACT

Silicon carbide is known for its superior thermomechanical performance, corrosion resistance, and radiation resistance. The radiation resistance of silicon carbide nanopowders with sizes of 10nm and 45nm-55 nm diameter, as well as bulk silicon carbide (grain size ~ 1μm), have been investigated *in situ* in the IVEM-Tandem facility at Argonne National Laboratory. The particles were irradiated with 1.0 MeV Kr$^+$ ions over a series of temperatures and doses. The radiation resistance was studied by observing changes in the critical amorphization dose. The 10nm particle size required a lower dose to amorphize at 100°C than either the 45-55nm particles or the bulk silicon carbide. This observation indicates a weakened radiation response with decreasing grain size at temperatures above 100°C. Below 100°C no difference in amorphization behavior was observed with changing grain size. *Ex situ* ion irradiations of thin film nanocrystalline silicon carbide using 2.0 MeV C^{2+} ions, up to 10 dpa at 600°C showed no evidence of amorphization of the original crystalline structure.

INTRODUCTION

The next generation nuclear reactors will demand exceptional material performance in extreme environments. One material that has shown promise for use in these environments is silicon carbide. Silicon carbide has been shown to have superior high temperature mechanical strength, corrosion resistance, and radiation resistance[1-4]. As such, it is of interest for use as a component in TRISO fuel, and as a structural component in nuclear reactors, such as cladding in a fission reactor and first wall material in a fusion system. Further improvement of these properties, for example by controlling the grain size, would further expand the use of silicon carbide in these aggressive environments. Modeling experiments have shown that decreasing the grain size of silicon carbide to the nanoscale can dramatically increase the toughness and strength of the material[5]. It has also been suggested that under lower energy irradiations, nanocrystalline silicon carbide has superior radiation resistance compared to silicon carbide with micron-scale grains[6]. In theory, smaller grain sizes will result in an improvement in radiation resistance due to the increased grain boundary density, which act as potential sinks for point defects produced in a radiation cascade. The annihilation of these point defects at the grain boundaries would be more pronounced in nanocrystalline silicon carbide due to the higher grain boundary density resulting in improved mechanical properties under radiation. This study reports results of preliminary investigation of the effect of grain size on radiation resistance in silicon carbide.

EXPERIMENTAL

Three types of silicon carbide were procured for this study. Silicon carbide nanopowder particles in the size range of 10 nm and 45 to 55nm were procured from the company NanoAmor. Bulk silicon carbide produced by the CVD process and with a grain size of 1μm was procured from the company, Rohm and Haas. Thin film nanocrystalline silicon carbide samples, 500nm thick deposited on silicon substrates were procured from North Carolina State

University for *ex situ* irradiation experiments. The silicon carbide film had an average grain size of 87nm. *In situ* irradiations of silicon carbide nanopowders and bulk silicon carbide were preformed at the IVEM-Tandem facility at Argonne National Laboratory using 1.0 MeV Kr^+ ions at temperatures ranging from 100K to 600K. TEM samples of the silicon carbide nanopowders were prepared by diluting the powders in methanol and then dispersing the particles on a carbon film TEM grid. TEM samples of the CVD silicon carbide procured from Rohm and Hass were prepared by mechanically polishing down to 30μm thickness and then ion milling to electron transparency. The thin film nanocrystalline silicon carbide samples were irradiated in the ion beam laboratory at the University of Wisconsin Madison, using 2.0 MeV C^{2+} ions.

RESULTS

To compare the radiation response of samples with various grain sizes, the dose to amorphization was selected as the criteria[7-9]. This also allowed for comparison to a model currently being developed[10] to study the response of silicon carbide to irradiation. Once the dose to amorphization was determined at various temperatures, a dose to amorphization versus temperature model (equations 1 and 2) was fit to the data, allowing the calculation of the critical amorphization temperature[11]. The lower the critical amorphization temperature, the more resistant the sample is to amorphization.

$$\ln(1 - \frac{D_o}{D}) = C - \frac{E}{kT} \tag{1}$$

$$T_c = \frac{E}{kC} \tag{2}$$

D_o is the dose at which the material amorphizes at 0K. D is the dose at which the sample amorphizes at a particular temperature, T. C is a constant dependent upon the ion flux and the damage cross section. E is the activation energy for recrystallization assumed to proceed under only one mechanism. T_C is the critical temperature above which no amorphization can take place for a particular sample. These equations assume that only one mechanism is responsible for amorphization.

To determine the dose at which the samples amorphized, the electron diffraction pattern was analyzed. Amorphization is indicated in an electron diffraction pattern when all diffraction spots and sharp lines have disappeared, leaving only a diffuse ring. The diffraction pattern was checked several times during the irradiation, to determine if the samples had amorphized in incremental dose steps. The dose to amorphization was determined to be the average of the dose at which the diffraction pattern was completely indicative of an amorphous state and the dose level just prior to this observation. The actual dose at which amorphization occurred could be at any level between these two doses, so the error in the dose to amorphization is taken as the difference between the highest dose achieved and the listed value.

In Situ Irradiation Results

During the *in situ* irradiation experiments performed at Argonne National Laboratory, observations of the effects of temperature and particle size on the amorphization dose were made. For the 10nm particle size samples, as the irradiation temperature increased, the dose to amorphization also increased (Figure 1). At 100K only 0.52dpa was needed to completely amorphize the sample, whereas at 600K even a dose of 4.35dpa was not enough to amorphize the sample.

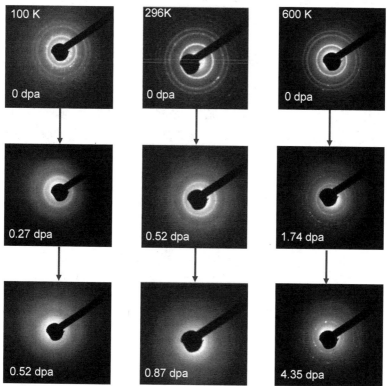

Figure 1: TEM electron diffraction patterns exhibiting the temperature dependence of the dose to amorphization for the 10nm particle size powder.

In addition to temperature dependence, the dose to amorphization also exhibited a particle size dependence at 100°C. For the 10nm particle size sample, only 1.04 dpa was needed to induce amorphization. Both 45-55nm particle size samples and bulk (micron-scale grains) samples required ~1.4dpa for amorphization (Figure 2). Further experiments are planned to determine if this trend continues at higher temperatures.

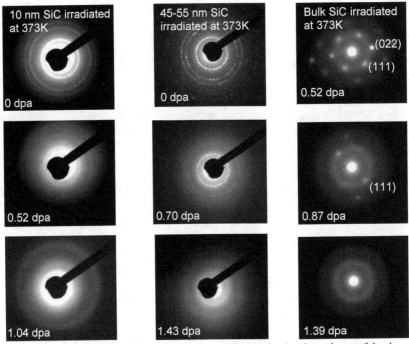

Figure 2: TEM electron diffraction patterns exhibiting the size dependence of the dose to amorphization at 373K.

Data collected during the *in situ* irradiation experiments was used to determine the critical amorphization temperature for both the 45-55nm and 10nm particle size powder TEM samples. This was found to be 610K for the 10nm particle size, and 485K for 45-55nm size, implying perhaps a lower radiation resistance for smaller particle sizes. Figure 3 summarizes the data collected and the fit of the dose to amorphization versus temperature plots. The values used to calculate these plots are shown in Table I. The vertical dashed lines in figure 3 indicate the critical amorphization temperatures. At lower temperatures, no difference in amorphization dose behavior is observed between either the nanocrystalline or micron scale samples. This is in agreement with a study performed using CVD nanocrystalline silicon carbide thin film[12], indicating that the same amorphization behavior is exhibited by nanocrystalline silicon carbide, regardless of the boundary type.

Table I: Fitting parameters and data from the *in situ* ion irradiation experiment.

Sample	Temperature (K)	Dose to Amorphization, D_C (dpa)	D_O	Activation Energy, E (eV)	C	Critical Amorphization Temperature, T_C (K)
10nm	100	0.48 ± 0.04		0.055	1.05	610
	173	0.57 ± 0.04				
	296	0.78 ± 0.09				
	373	0.96 ± 0.09				
	600	> 4.35				
45-55nm	296	0.8 ± 0.1	0.47	0.060	1.42	485
	373	1.3 ± 0.1				
	423	2.4 ± 0.2				
	473	> 4.35				
Micron CVD	296	< 0.87		N/A	N/A	N/A
	373	1.2 ± 0.2				

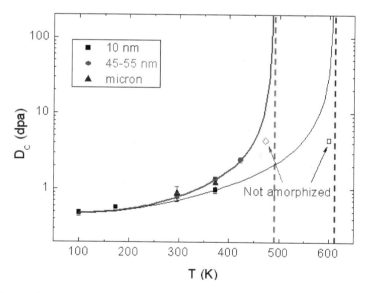

Figure 3: Dose to amorphization data plot as a function of irradiation temperature.

Ex Situ Irradiation Experiment

The University of Wisconsin-Madison ion beam lab was used to irradiate a thin film nanocrystalline silicon carbide sample to 10dpa at 600°C. TEM and electron diffraction showed no amorphization of the silicon carbide crystal structure. As shown in figure 4, the carbon ion implantation zone is well away from the silicon carbide thin film. The silicon carbide thin film does not deviate from stoichiometry due to implantation of the carbon ions in the film during irradiation. Future experiments are planned to determine the dose to amorphization at various temperatures such that the critical amorphization temperature can be calculated. The gathered data can then be compared to the *in situ* results to determine any ion species effect on amorphization dose.

Figure 4: Cross section of silicon carbide thin film sample irradiated at 600°C to 10dpa.

CONCLUSIONS

Experiments were conducted to investigate the response of silicon carbide to ion irradiation. *In situ* radiation experiments have shown that the dose to amorphization has both a temperature and particle size effect. Higher irradiation temperatures require a higher dose for amorphization to occur. At 100°C, smaller particles require a smaller dose for complete amorphization, suggesting that smaller particle sizes have lower radiation resistance. There was no substantial grain size dependence observed at lower temperatures. Similar behavior was observed in CVD nanocrystalline silicon carbide[12], suggesting that the boundary type, whether surface or grain boundary, does not change the result. The thin film sample irradiated at the ion beam lab exhibited no signs of amorphization after irradiation to 10dpa at 600°C.

Future work at the IVEM Tandem facility at Argonne National Laboratory will focus on gathering data on bulk (micron-scale) and nanocrystalline thin film silicon carbide samples. This will be compared to the results reported in this paper to determine if the size dependence of the critical amorphization temperature continues to follow the trend exhibited in these initial studies. Further experiments are also planned to expand the temperature window of the irradiation experiments to ensure all of the behavior is captured. Ion beam experiments will be continued at the University of Wisconsin-Madison to determine the critical amorphization temperature of nanocrystalline silicon carbide under carbon irradiation using the silicon carbide thin films.

ACKNOWLEDGEMENTS

This work is made possible by Department of Energy the Office of Basic Energy Sciences Grant Award Number DE-FG02-08ER46493. This research utilized NSF-supported shared facilities at the University of Wisconsin. This material is based upon work supported under a Department of Energy Nuclear Energy University Program Graduate Fellowship. Thank you to Professor Steve Shannon at North Carolina State University for supplying the CVD nanocrystalline silicon carbide thin films. Thank you to Edward Ryan and Paul Fuoss at Argonne National Laboratory for their expertise and assistance on the IVEM-Tandem.

REFERENCES

[1] J. Chen, P. Jung, and H. Klein, Production and Recovery of Defects in SiC After Irradiation and Deformation, *J. Nuc. Mat.,* **1803-1808**, 258-263 (1998).

[2] R. Jones, L. Giancarli, A. Hasegawa, Y. Katoh, A. Kohyama, B. Riccardi, L. Snead, and W. Weber, Promise and Challenges of SiCf/SiC Composites for Fusion Energy Applications, *J. Nuc. Mat.,* **307-311**, 1057-72 (2002).

[3] M. Osborne, J. Hay, L. Snead, D. Steiner, Mechanical- and Physical-Property Changes of Neutron-Irradiated Chemical-Vapor-Deposited Silicon Carbide, *J. Am. Ceram. Soc.* **82**, 2490-96 (1999).

[4] A. Ryazanov, A. Klaptsov, A. Kohyama, H. Kishmoto, Radiation Swelling of SiC Under Neutron Irradiation, *J. Nuc. Mat.,* **307-311**, 1107-11 (2002).

[5] Y. Mo, I. Szlufarska, Simultaneous Enhancement of Toughness, Ductility, and Strength of Nanocrystalline Ceramics at High Strain-Rates, *App. Phys. Let.,* **90**, 181926-3 (2007).

[6] Y. Leconte, I. Monnet, M. Levalois, M. Morales, X. Portier, L. Thomé, N. Herlin-Boime, C. Reynaud, Comparative Study of Structural Damage Under Irradiation in SiC Nanostructured and Conventional Ceramics, *Mat. Res. Soc. Symp. Proc.* **981**, JJ07-11 (2007).

[7] K. Kawatsura, N. Shimatani, T. Igarashi, T. Inoue, N. Terazawa, S. Arai, Y. Aoki, S. Yamamoto, K. Narumi, H. Naramoto, Y. Horino, Y. Mokuno, K. Fujii, Radiation-Induced Amorphization and Recrystallization of α-SiC Single Crystal, *J. Nucl. Mater.* **271-272**, 11-14 (1999).

[8] W. Bolse, J. Conrad, F. Harbsmeier, M. Borowski, T. Rödle, Long and Short Range Order in Ion Irradiated Ceramics Studied by IBA, EXAFS and Raman, *Mat. Sci. For.,* **248-249**, 319-326 (1997).

[9] W. Weber, N. Yu, In Situ and Ex Situ Investigation of Ion-Beam-Induced Amorphization in α-SiC, *Nuc. Inst. Meth. Phys. Res. B,* **127-128**, (1997).

[10]N. Swaminathan, P. Kamenski, D. Morgan, I. Szlufarska, Effects of Grain Size and Grain Boundaries on Defect Production in Nanocrystalline 3C-SiC, *Acta Mat.*, **58**, 2843-53 (2010).
[11]W. Weber, Models and Mechanisms of Irradiation-Induced Amorphization in Ceramics, *Nuc. Inst. Meth. Phys. Res. B*, **166-167**, 98-106 (2000).
[12]W. Jiang, H. Wang, I. Kim, I. Bae, G. Li, P. Nachimuthu, Z. Zhu, Y. Zhang, W. Weber, Response of Nanocrystalline 3C Silicon Carbide to Heavy-Ion Irradiation, *Phys. Rev. B*, **80**, 161301 (2009).

PERFORMANCE OF A CARBON STEEL CONTAINER IN A CANADIAN USED NUCLEAR FUEL DEEP GEOLOGICAL REPOSITORY

Gloria M. Kwong
Nuclear Waste Management Organization, Toronto, Ontario, Canada.

Steve Wang and Roger C. Newman
University of Toronto, Toronto, Ontario, Canada

ABSTRACT

Experimental work was undertaken to evaluate the long-term corrosion behaviour of a carbon steel used fuel container in postulated Canadian deep geological repository environments. Specifically, atmospheric corrosion testing was conducted in anoxic atmospheres at 30, 50 and 70°C, over a wide range of relative humidity (30-100% RH). Hydrogen evolved from steel corrosion was initially monitored with a high sensitivity pressure gauge system (the sensitivity of such measurements is ca. 0.005 $\mu m\ y^{-1}$) and later using a solid-state electrochemical hydrogen sensor (the hydrogen sensor can detect the pressure increase down to a limit of ca. 0.1 Pa corresponding to a corrosion rate as low as ca. 0.0001 $\mu m\ y^{-1}$). Results observed in this study indicate that it is highly likely that the corrosion rate will fall below 1 nm y^{-1} within a year once an anoxic condition is attained. In parallel with the corrosion experiments, corrosion product surface analyses were performed using (i) scanning electron microscopy coupled with energy dispersive X-ray analysis (SEM/EDX) to determine the morphology of the corrosion product films, (ii) X-ray photoelectron spectroscopy (XPS) to identify the chemical composition of the films, and (iii) Raman and Fourier Transform Infrared spectroscopy (FTIR) to study bonding. Oxides formed on steel surfaces were found to consist mostly of Fe_3O_4, with some Fe(III) species from traces of air exposure; carbonate was detected on the NaCl contaminated surfaces which had been subject to a degree of prior aerobic corrosion. A high humidity (100%) environment produces more loose surface oxide than a lower humidity environment (75%). The experimental results support the use of carbon steel as a potential used fuel container material in a Canadian deep geological repository located in sedimentary rock.

INTRODUCTION

Steel corrosion in postulated deep geological repository environments has been studied, but has mostly focused in two main areas: (i) aerobic or oxygen containing environments (both in vapour and liquid phases); and (ii) anaerobic, solution environments [1]. The long-term behaviour of carbon steels in humid, non-immersed, anaerobic conditions has not been studied. The lack of such data may be due to the relatively slow corrosion rates and difficulties in precise measurements. This condition (i.e., humid and anoxic) has nevertheless been reported in various Canadian sedimentary host rock studies to potentially span from thousands to millions of years after repository closure [1,2]. The need for addressing the anaerobic steel corrosion issue in humid, vapour phase is evident.

CORROSION EXPERIMENTS

The initial purposes of this study were to evaluate the optimal corrosion cell setup, hydrogen detection techniques and testing protocols. Hydrogen generation from carbon steel has been the primary means to monitor the corrosion behaviour of the material under anaerobic unsaturated conditions. Two approaches were taken to monitor the hydrogen evolved from anaerobic corrosion of steel – (i) a high sensitivity pressure gauge corrosion system (sensitivity of ca. $0.005 \ \mu m \ y^{-1}$) to directly measure the amount of hydrogen evolved; and (ii) a solid-state electrochemical hydrogen sensor with a detection limit of ca. 0.1 Pa (equivalent to a corrosion rate of $0.0001 \ \mu m \ y^{-1}$ based on the measurement procedure). In parallel with the corrosion experiments, the structures and chemical compositions of the oxides formed during the anaerobic corrosion studies were analyzed using various surface analysis techniques.

Test conditions studied in the series of corrosion experiments are summarized in Table 1.

Table 1 Test conditions evaluated in the high sensitivity pressure gauge system

Test condition		Duration (h)
Wire precorroded with 0.5M NaCl		
Test #1	68% RH, 32 °C	1347
Test #2	71% RH, 32 °C	935
Test #3	75% RH, 32°C	1730
Test #4	100% RH,32 °C	1270
Test #5	30% RH, 50°C	1962
Test #6	51% RH, 50°C	1053
Test #7	75% RH, 50°C	450
Test #8	100% RH, 50°C (I)	237
Test #9	100% RH, 50°C (II)	1725
Test #10	100% RH, 70°C	213
Wire precorroded with 0.05M NaCl		
Test #11	51% RH, 50°C	1867
Test #12	75% RH, 50°C	2326
Test #13	75% RH, 32°C	3355
As-cleaned surfaces		
Test #14	100% RH, 32°C	379
Test #15	100% RH, 50°C (I)	591
Test #16	100% RH, 50°C (II)	615
Pickled surfaces		
Test #17	75% RH, 32°C	1171
Test #18	75% RH, 50°C	331
Test #19	75% RH, 70°C	1121
Test #20	100% RH, 32°C	1337
Test #21	100% RH, 50°C	1002
Test #22	100% RH, 50°C	688
Test #23	100% RH, 70°C	1006

PROCEDURE

"Iron" wire of 0.25 mm diameter and 99.5% purity, supplied by Goodfellow Cambridge, was used in this study. Its nominal composition is listed in Table 2 indicating that it can be considered to be a steel with very low carbon content. A total length of 130 m of steel wire, corresponding to $0.1 \ m^2$ surface area was used for each exposure. The wire was cut into 50 mm lengths and 100 pieces of the cut wires were tied with two PTFE rings to form a bundle, 26 bundles were placed into a glass cell. Preparations of the wire specimens and experimental procedures for the conducted tests are provided in more detail in a separate report [3].

Table 2 Nominal composition of iron wire [ppm]

Fe	Al	Ca	Co	Cr	Cu	Mn	Ni	Sn	C	P	S
balance	30	30	100	100	100	4000	200	50	<800	<600	<600

Following degreasing and cleaning of the wire bundles, one of three surface finishes was applied, namely (i) salt deposit or pre-corroded with NaCl; (ii) as-cleaned; and (iii) pickled. Weight gains on the wires with salt deposit ranged from 0.18 g (i.e. for wires pre-corroded with 0.05M NaCl) to 1.14 g (for wires pre-corroded with 0.5M NaCl).

Leak Test

To ensure air tightness and to evaluate potential H_2 leakages of the pressure gauge system, several leak tests were carried out. Detailed leak test procedure and test results are documented in Reference [3].

Figure 1 shows the results of one of the leak tests confirming the glass cell is air-leak tight and able to maintain a stable pressure for more than one month.

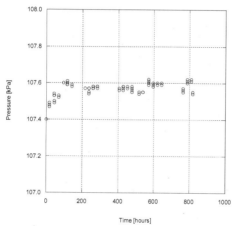

Figure 1. Typical leak test results

H_2 measurement using a solid-state electrochemical hydrogen sensor

In cases where low rates of hydrogen generation are expected (i.e. pickled wire surfaces at various relative humidities and temperatures), hydrogen monitoring was conducted using a solid state electrochemical hydrogen sensor. This technology of monitoring hydrogen in the gas phase using a solid-state sensor containing a ceramic proton-conducting electrolyte (protic salt, prototype was hydrogen uranyl phosphate, HUP) began in the 1980's [4]. The extraordinary sensitivity of the original invention, which can operate as a Nernstian potentiometric sensor and detect hydrogen partial pressures down to the 10^{-6} bar range, was found to suffer from problems with the oxidizing nature of the HUP. The hydrogen probe used in this study was manufactured by Ion Science Ltd, model Hydrosteel 6000 and operates by drawing a stream of ambient air at quite a high rate (30 ± 1 mL/min) past the electrochemical sensor. The test-gas stream is confined within a specially designed collector plate contained in the probe assembly. The collected hydrogen is conveyed through narrow bore tubing into the measurement analyzer, and across the face of a very sensitive amperometric detector. This suction approach removes the necessity for tight sealing around any barnacle type of attachment on a pipe or vessel, which is a significant advantage in practice despite the dilution of the effusing hydrogen.

Initial measurements using this electrochemical hydrogen probe were planned to be carried out in a semi-continuous manner, i.e. by using a copper corrosion chamber and effusing the produced hydrogen through a palladium membrane. Extensive test trials later concluded that this effusion method did not work well particularly for the near-ambient temperature cases. Rather than the semi-continuous monitoring of hydrogen effusion, the study switched to a batch mode of monitoring hydrogen in the copper cell. The test conditions evaluated in this series of experiments are tabulated in Table 3. The calibration of the hydrogen probe and the procedure for measuring hydrogen are described in detail in Reference [3].

Table 3 Test conditions evaluated in the copper cell – H_2 sensor system

Test Conditions	Duration (h)
Pickled wire, 75% RH, 50°C	2733
Pickled wire, 85% RH, 50°C	1826
Pickled wire, 100% RH, 50°C	779

Surface Analyses and Microscopy

The following surface analytical measurements were carried out to determine the surface oxide morphology, composition and where possible, chemical state, as a result of exposure to various environmental conditions:

1. Preliminary optical microscopy examinations were performed using a low power (10X – 80X) Zeiss Discovery V8 stereozoom microscope and a high power Zeiss Axioplan optical microscope.
2. Scanning Electron Microscopy coupled with Energy Dispersive X-ray Analysis (SEM/EDX) was performed using a Leo 440 conventional SEM equipped with a Quartz Xone EDX system. Secondary electron micrographs and EDX spectra were obtained from two different areas of interest per sample.
3. X-ray Photoelectron Spectroscopy (XPS) analyses were performed using a Kratos AXIS Ultra X-ray photoelectron spectrometer. Monochromatic X-rays produced by an Al (Kα) source were bombarded on the specimen and the resulting photoelectrons were collected using a spherical mirror analyser. Survey scan spectra were collected from an analysis area of 300 x 700 microns with a pass energy of 160 eV. High resolution XPS analyses were carried out on an analysis area of 300 x 700 microns with a pass energy of 20 eV.
4. Laser Raman Spectroscopy measurements were performed using a Renishaw Model 2000 Raman Spectrometer. The measurements were made using a 633 nm laser. The sampling area for this analysis is approximately 2 μm in diameter.
5. Finally Fourier Transform Infrared Spectroscopy (FTIR) was performed to give further information on bonding. FTIR measurements were performed using a Bruker IFS55 instrument under reflection mode using a microscope attachment.

RESULTS AND DISCUSSION
Glass cell - Pressure Gauge System

Results from the glass cell – pressure gauge system are graphed as "corrosion rates" derived from differentiation of the raw, temperature-corrected pressure data. Table 4 summarizes the results obtained and Figures 2 to 6 show the corrosion behaviours for the NaCl pre-corroded wires. The estimated corrosion rates at the end of the tests were found to be in the range of 0.01 to 0.8 $\mu m\ y^{-1}$ for test duration of 935 to 1725 hours. Higher transient corrosion rates in the range of 1 to 5 $\mu m\ y^{-1}$ were found for tests with shorter test durations of 213 to 450 hours. At 32°C, the final corrosion rates were estimated to be about 0.15 to 0.45 $\mu m\ y^{-1}$ for RH in the range of 68% to 100% (Table 4). The corrosion rates appear to be slightly higher at lower (68% to 75%) RH than at 100% RH (Figure 7), which may be explained by a more concentrated thin-film electrolyte at lower RH. This was, however, not a general result, as testing at 50°C showed the reverse trend.

Table 4. Glass cell – pressure gauge experimental results

Test Conditions	Duration	Estimated Corrosion Rate
Wire pre-corroded with 0.5M NaCl		
68% RH, 32 °C	1347 h	$< 0.15\ \mu m\ y^{-1}$
71% RH, 32 °C	935 h	$\sim 0.2\ \mu m\ y^{-1}$
75% RH, 32°C	1730 h	$< 0.4\ \mu m\ y^{-1}$
100% RH,32 °C	1269 h	$< 0.4\ \mu m\ y^{-1}$
30% RH, 50°C	1962 h	$< 0.01\ \mu m\ y^{-1}$
51% RH, 50°C	1053 h	$< 0.01\ \mu m\ y^{-1}$
75% RH, 50°C	450 h	$< 1\ \mu m\ y^{-1}$
100% RH, 50°C (I)	237 h	$< 3\ \mu m\ y^{-1}$
100% RH, 50°C (II)	1725 h	$0.8\ \mu m\ y^{-1}$
100% RH, 70°C	213 h	$< 5\ \mu m\ y^{-1}$
Wire pre-corroded with 0.05M NaCl		
51% RH, 50°C	1867 h	$< 0.01\ \mu m\ y^{-1}$
75% RH, 50°C	2326 h	$\sim 0.08\ \mu m\ y^{-1}$
75% RH, 32°C	3355 h	$< 0.01\ \mu m\ y^{-1}$
As-cleaned surfaces		
100% RH, 32°C	379 h	$< 0.001\ \mu m\ y^{-1}$
100% RH, 50°C	591 h	$< 0.001\ \mu m\ y^{-1}$
100% RH, 50°C	615 h	$< 0.001\ \mu m\ y^{-1}$
Pickled surfaces		
75% RH, 32°C	1171 h	---
75% RH, 50°C	331 h	$< 0.001\ \mu m\ y^{-1}$
75% RH, 70°C	1121 h	$< 0.001\ \mu m\ y^{-1}$
100% RH, 32°C	1337 h	$< 0.01\ \mu m\ y^{-1}$
100% RH, 50°C	1002 h	$< 0.01\ \mu m\ y^{-1}$
100% RH, 50°C	688 h	$< 0.01\ \mu m\ y^{-1}$
100% RH, 70°C	1006 h	$< 0.01\ \mu m\ y^{-1}$

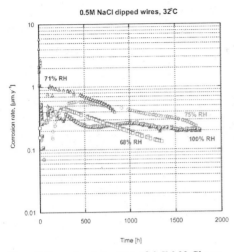

Figure 2. Corrosion behaviour of 0.5M NaCl pre-corroded wires tested at 32°C and various RH

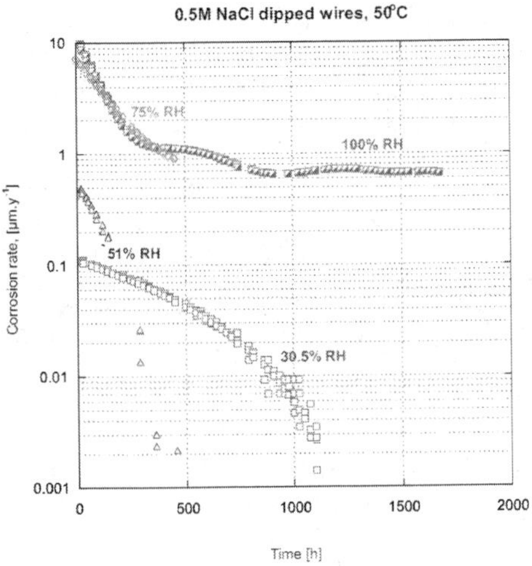

Figure 3. Corrosion behaviour of 0.5M NaCl pre-corroded wires tested at 50°C and various RH

At a higher temperature of 50 °C, the estimated corrosion rates were 1 and 0.8 μm y^{-1} at 75% and 100% RH, respectively (Table 5). The test at 75% is still in its rapid transient stage since the test duration is short (450 hours) (Figure 3). At 30.5 and 50% RH, the estimated final corrosion rates were much lower, ca. 0.01 μm y^{-1}, which was still within the resolution range of the pressure gauge method.

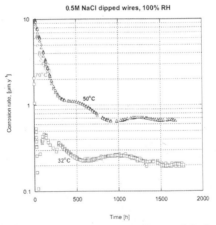

Figure 4. Corrosion behaviour of 0.5M NaCl pre-corroded wires tested at 100% RH and various temperatures

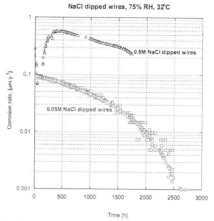

Figure 5. Corrosion behaviour of NaCl pre-corroded wires tested at 75% RH and 32°C

The tests at 32 °C and 75% RH indicate that soaking the wires in 0.05M NaCl solution was much less aggressive than soaking in 0.5M NaCl solution (Figure 5). With 0.05M NaCl soaking, the metal loss tended to level out with time, indicating that chloride is more than just a catalyst for corrosion – the thickness of liquid layer and absolute quantity of chloride in that layer are important. These test results appear to indicate that corrosion rates may increase quite sharply with increasing concentration of the NaCl solution and increasing amount of NaCl deposited on the wire surface.

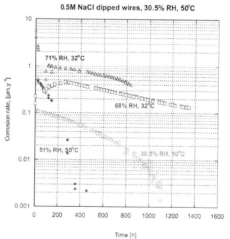

Figure 6. Corrosion behaviour of 0.5M NaCl pre-corroded wires tested at
RH ≤ saturated NaCl equilibrium RH

The test results also show that steel corrosion would not stop as the relative humidity was reduced to, or slightly below, the value that is in equilibrium with saturated NaCl (i.e. 75.5% RH at 30 °C and 74.5% RH at 50 °C) (Figure 6). This is well-known finding for atmospheric corrosion in ordinary aerated environment – that 60-70% RH provides sufficient moisture to support electrochemical reactions – but this is probably the first such study in an anoxic environment. Various explanations for this are possible, including simple gas-phase reactions with water vapour, reactions with multilayer adsorption of water, hydrous corrosion products, capillary condensation, and involvement of hygroscopic iron salts ($FeCl_2 \cdot xH_2O$). Nevertheless, there was a steep decline in corrosion rate as the RH was reduced to 51% and 30.5% (Figure 6). At these low RH values, corrosion is not well sustained.

The time dependence of corrosion rates for the NaCl pre-corroded tests that showed sustained corrosion were then fitted to enable extrapolation to longer times. The initial attempt to fit the curve by straight-line decay approximates to the following:

$$R = R_0 \exp\frac{-t}{\tau}$$

with $R_0 \simeq 6 \ \mu m \ y^{-1}$ and $\tau \simeq 870$ h. R, R_0 and τ are defined as corrosion rate at time t, "initial" corrosion rate (after curve fitting) and decay time constant, respectively. Such a decay law would result in the corrosion rate falling to 0.5 nm y^{-1} after only 1 year. Alternatively, curve fittings by power laws led to more pessimistic predictions for long-term corrosion rates. For illustrative purposes, Figure 7 shows a rather conservative fit for the 0.5M NaCl pre-corroded wires exposed to 70 °C and 100% RH; this approach resulted in a corrosion rate at 0.24 $\mu m \ y^{-1}$ after 1 year, and 3.7 nm y^{-1} after 1,000 years.

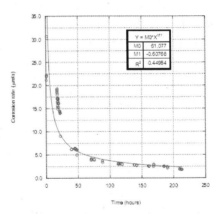

Figure 7. Power law fit of corrosion rates – 0.5M NaCl pre-corroded wires, exposed to 100% RH and 75°C

Tests using wires without salt deposits form the basis for the use of the hydrogen sensor as the pressure gauge method began to approach its limit of sensitivity for the detection of hydrogen. In cases when hydrogen was detectable by the pressure gauge method, the hydrogen generation was mostly transient in nature. The pickled wires tested at 32 °C, 50 °C and 70 °C, 100% RH showed similar behaviour and these tests indicated no long-term sustained corrosion. The question of whether hydrogen could be absorbed from the pickling, then effuse during the test was studied. It was concluded as if that were the case, it should happen in every test, which it did not. Testing with degreased wires ("as-cleaned") showed even less corrosion since as expected, these wires have a well-formed air oxide prior to exposure to the tested conditions.

Copper cell – H$_2$ sensor experiments

Table 5 summarizes the results obtained with the hydrogen sensor. The reason for the increase in corrosion rate with time is not apparent. It is unclear at present whether this is a real phenomenon, or an artifact of the extremely low amount of incremental oxidation that occurs in each batch stage of the reaction, combined with the perturbation of removing the hydrogen formed during the previous stage. Nevertheless, these results are important in that they demonstrate the lowest corrosion rate that has been measured in such an environment. Regarding the repository, the observed corrosion rates would not pose any significant threat to the design life of the used fuel container.

Table 5. H$_2$ sensor experimental results

Test Condition	Interval between Measurement time increment	Estimated Corrosion Rate (average over increment)
Pickled wires, 85% RH, 50°C		
1st measurement	164 hrs	1.1×10^{-4} µm y^{-1}
2nd measurement	259 hrs	1.0×10^{-4} µm y^{-1}
3rd measurement	313 hrs	1.6×10^{-4} µm y^{-1}
4th measurement	235 hrs	2.5×10^{-4} µm y^{-1}
5th measurement	262 hrs	2.0×10^{-4} µm y^{-1}
6th measurement	332 hrs	2.4×10^{-4} µm y^{-1}
7th measurement	221 hrs	3.1×10^{-4} µm y^{-1}
Pickled wires, 75% RH, 50°C		
1st measurement	353 hrs	7.0×10^{-5} µm y^{-1}
2nd measurement	359 hrs	1.8×10^{-4} µm y^{-1}
3rd measurement	310 hrs	3.1×10^{-4} µm y^{-1}
4th measurement	216 hrs	4.8×10^{-4} µm y^{-1}
5th measurement	283 hrs	---
6th measurement	260 hrs	---
7th measurement	408 hrs	3.3×10^{-4} µm y^{-1}
8th measurement	212 hrs	6.3×10^{-4} µm y^{-1}
9th measurement	218 hrs	4.9×10^{-4} µm y^{-1}
10th measurement	91 hrs	1.01×10^{-3} µm y^{-1}
Pickled wires, 100% RH, 50°C		
1st measurement	473 hrs	---
2nd measurement	93 hrs	6.7×10^{-3} µm y^{-1}
3rd measurement	93 hrs	2.5×10^{-3} µm y^{-1}

Surface Analyses

Table 6 lists the samples selected for surface analysis and microscopy. Due to the large volume of analysis results, only summaries of the analyses are presented below. More detailed surface analysis results and discussion are provided in Reference [3].

Table 6. 15 samples examined by surface analyses

Sample #	Condition	RH	Temp	Duration
1	Control, no previous exposure, wires degreased in acetone	--	--	--
2	Control, no previous exposure, wires coated with 0.5M NaCl (no rinse before analyses)	--	--	--
3	Degreased wires (w/ acetone), exposed to deaerated environment (using N_2)	100%	50°C	615 hrs
4	Pickled wires (w/ HCl)	100%	50°C	1001 hrs
5	Pickled wires (II)	100%	50°C	688 hrs
6	0.5M NaCl dipped wires, no rinse before analyses	75%	50°C	370 hrs
7	0.5M NaCl dipped wires, rinsed before analyses	75%	50°C	370 hrs
8	Degreased wires, exposed to anoxic environment (deaerated w/ N_2 / H_2)	100%	50°C	591 hrs
9	0.5M NaCl dipped wires, no rinse before analyses	100%	50°C	237 hrs
10	0.5M NaCl dipped wires, rinsed before analyses	100%	50°C	237 hrs
11	Pickled wires	100%	70°C	1006 hrs
12	Pickled wires	74.1%	70°C	1121 hrs
13	Raw steel wires for XPS sputtering experiment	--	400°C	Heated in air, 3 hrs
14	0.5M NaCl dipped wires, no rinse before analyses	100%	50°C	1725 hrs
15	0.5M NaCl dipped wires, rinsed before analyses	100%	50°C	1725 hrs

XPS

Table 7 summarizes the apparent near-surface atomic compositions obtained from XPS survey spectra of the examined samples, without any sputter etching except for sample 13. This sample was used to determine the elemental ratios that would exist on a well-characterized, oxidized iron surface after sputter etching - a 2:3 Fe:O ratio is found after 1 minute sputtering, but after 5 minutes there is evidence of film reduction. Table 8 shows the results of fitting of high-resolution XPS spectra.

Table 7. Composition data from XPS survey scan spectra

Sample		Composition (at. %)									
		Fe	C	O	Na	Cl	F	Mn	Si	N	Ca
1	Area 1	13	42.4	40.6	0.3	0.5		0.5	1.2	0.7	0.8
2	Area 1	22.4	19.8	49.1	3.0	4.5		0.9		0.3	0.1
3	Area 1	4.3	48.6	39.9		0.5		0.9	3.9	0.9	1.2
4	Area 1	10.8	39.3	44.7		0.2			3.9	1.1	
5	Area 1	12.2	42.1	43.2		0.7			1.7		
6	Area 1	20.5	27.0	49.5	0.8	0.9			1.2		
7	Area 1	18.1	27.6	51.8	1.1	0.8			0.7		
8	Area 1	3.4	49.4	40.3					6.1		0.8
9	Area 1	13.7	34.0	47.5	3.0	1.9					
10	Area 1	18.1	28.8	51.1		0.5			1.5		
11	Area 1	12.5	40.7	44.6			2.2				
12	Area 1	13.2	40.0	41.3		0.6			3.7	1.2	
13	As-heated	14.0	42.2	43.3	0.6						
	After 1 min sputtering	36.3	11.5	51.8	0.4						
	After 5 min sputtering	31.9	7.9	59.9	0.4						
14	Area 1	20.4	29.3	42.8	0.3	4.2					
15	Area 1	15.0	40.8	43.9		0.4					

Table 8. Summary of XPS high-resolution spectra data

Sample	XPS Summary
1	Fe metal 4.3%, Fe_2O_3: 70.4%; FeO: 25.3%; film thickness: ~ 6 nm
2	Fe_2O_3: 60.6%; Fe_3O_4: 2.3%; FeOOH: 37.1%
3	Fe metal: 6.1%; Fe_2O_3: 47.9%; Fe_3O_4: 19.5%; FeOOH: 26.4%; film thickness: ~ 5.7 nm
4	Fe metal: 2.1%; Fe_2O_3: 30.6%; Fe_3O_4: 48.6%; FeOOH: 18.7%; film thickness: ~ 7.5 nm
5	Fe metal: 1.8%; Fe_2O_3: 29.6%; Fe_3O_4: 41.8%; FeOOH: 26.9%; film thickness: ~ 7.8 nm
6	$FeCO_3$: 63.9%; Fe_2O_3: 11.2%; FeO: 24.7%
7	$FeCO_3$: 64.6%; Fe_2O_3: 15.7%; FeO: 19.7%
8	Fe metal: 4.8%; Fe_3O_4: 26.2%; FeOOH: 10.3%; Fe_2O_3: 58.9%
9	$FeCO_3$: 60.1%; Fe_2O_3: 3.9%; FeO: 35.9%
10	Fe metal: 1.9%; Fe_3O_4: 50.4%; FeOOH: 5.2%; Fe_2O_3: 42.6%
11	Fe_3O_4: 60.9%; Fe_2O_3: 39.1%
12	Fe_3O_4: 49.3%; FeOOH: 29.4%; Fe_2O_3: 21.5%
13	Steel wire oxidized in air at 400°C and found to contain a mixture of Fe_2O_3/Fe_3O_4 in XPS analysis. Upon sputtering, the well-defined oxide (mixture) is reduced and is sufficiently damaged to make curve-fitting of the resulting spectra difficult. Peak widths are broadened significantly and metallic iron is formed.
14	Appears to be a mixture of Fe_2O_3, Fe_3O_4 and $FeCO_3$. FeOOH may also be present.
15	Appears to be a mixture of Fe_2O_3, Fe_3O_4 and $FeCO_3$. $FeCO_3$ levels are much lower than sample 14. FeOOH may also be present.

SEM/EDX

EDX results of all examined specimens exhibit some indication of the spatial variability of average composition within the analysis depth, which is much greater than that of XPS. Oxygen levels give a broad indication of which films have substantial thickness, of the order of at least 0.1 micron. Not much information was obtained for specimens that were exposed to anoxic conditions without a salt deposit, although oxygen was detected in some cases.

Raman Spectroscopy and FTIR

Table 9 lists the findings of Raman spectroscopy analysis. While it is realized that this technique may not be sensitive to $Fe(OH)_2$ [5], to the extent that Fe_3O_4 is positively identified, it clarifies the nature of the films formed in anoxic conditions. The formation of some Fe(III) oxide or oxyhydroxide, or carbonates, by exposure to traces of oxygen does not invalidate this basic finding. The thin corrosion product films and the poorly crystalline $Fe(OH)_2$ did not allow the relative amounts of Fe_3O_4 and $Fe(OH)_2$ to be determined. The lack of positive evidence for OH in the FTIR data suggests preponderance of Fe_3O_4.

Table 9. Summary of Raman and FTIR spectra data

Sample	Spectra data
1	Fe_3O_4, could be FeO and amorphous carbon
2	Fe_3O_4, α-FeOOH, likely carbonate species, amorphous carbon and other iron species such as maghemite (γ-Fe_2O_3) and ferrihydrite (generally presented as $Fe_2HO_8 \cdot 4H_2O$, also written as 5 $Fe_2O_3 \cdot 9H_2O$ or as $Fe_2O_3 \cdot 2FeOOH \cdot 2.6H_2O$
3	Raman analysis did not find any evidence of oxide species on this sample.
4	Typical of Fe_3O_4.
5	Typical of Fe_3O_4. No other oxide species were detected.
6	Amorphous carbon, Fe_3O_4, Fe_2O_3 and gamma and alpha Fe_2O_3.
7	Fe_3O_4, carbonate, beta-FeOOH(Cl), and gamma-Fe_2O_3.
8	Raman: Amorphous carbon and Fe_3O_4. FTIR: Weak with no assignable peaks
9	Raman: Fe_3O_4, carbonate and amorphous carbon. FTIR: Likely mixed carbonate/hydroxide species.
10	Raman: Fe_3O_4, amorphous carbon, alpha FeOOH, Fe_2O_3, gamma-Fe_2O_3 and carbonate. FTIR: Likely mixed carbonate/hydroxide species.
11	Raman: spectra are consistent with Fe_3O_4 with some contribution from amorphous carbon. FTIR: spectra are very weak and undifferentiated. No evidence of hydroxide.
12	Raman: spectra are consistent with Fe_3O_4 with some contribution from amorphous carbon. FTIR: No evidence of hydroxide in the FTIR spectra.
13	Sample not examined using Raman
14	Raman: presence of mainly Fe_3O_4 with some areas showing the presence of gamma-Fe_2O_3 (maghemite). FTIR: mainly showed the presence of broad undifferentiated peaks which were not assigned.
15	Raman: presence of Fe_3O_4, with some areas also showing the presence of a peak likely arising from carbonates. FTIR: spectra were weak in some areas and strong in others. The strong spectra were consistent with the spectra observed previously and proposed as arising from a mixed carbonate/hydroxide species.

CONCLUSIONS

Corrosion tests of very low carbon content steel wire show that corrosion is persistent with salt deposition, but likely to stabilize in the range of $0.01 - 0.1$ $\mu m\ y^{-1}$. In cases where salt deposition is reduced or absent, extremely slow corrosion rates were observed even at 75-100% RH and 50-70°C. To apply the obtained final corrosion rates as benchmark for a deep geological repository, current results indicate that it is highly likely that the corrosion rate will fall below 1 $nm \cdot y^{-1}$ within a year once an anoxic condition is attained.

Results of this study suggest that chloride was not purely catalytic, as it is in phenomena such as filiform corrosion of coated steel [6], but became much less influential for an order of magnitude reduction in the surface loading of chloride. Further research will provide information on the mechanism of this chloride effect.

Contrary to the literature, the Raman spectroscopy results have shown that magnetite is a predominant low-temperature product. This crystallization of magnetite on the surface will be favourable for protecting the carbon steel container in the repository, particularly when compared to a looser $Fe(OH)_2$ film. Despite the wire surfaces being hydroxylated and contaminated with carbon in some cases, XPS results nevertheless support the presence of significant hydroxide in the corrosion product. In future, sample transfer protocols will be improved to further eliminate the potential exposure to oxygen and to enhance the quality of these surface analyses. Further combined use of Raman and XPS (with sputter etch profiling) is recommended, including spatially resolved studies such as examination of corrosion at sulfide inclusions.

This research points towards the following priorities for future study of the anoxic, atmosphere corrosion of carbon steel:

(i)   Longer-term testing investigating the effect of the areal loading of various salts on anoxic corrosion, and its mechanism in terms of sequestration of anions by corrosion products;

(ii)   Evaluation of the electrochemical hydrogen sensor for estimation of very low corrosion rates; and

(iii)   Exploitation of the sensitive performance of the Raman microscope and XPS, specifically in studies of film development including the potential use of the "vacuum suitcase" technology.

REFERENCES

[1] F. King. 2007. Overview of a Carbon Steel Container Corrosion Model for a Deep Geological Repository in Sedimentary Rock, NWMO TR-2007-01.

[2] M. Mazurek. 2004. Long-Term Used Nuclear Fuel Waste Management – Geoscientific Review of the Sedimentary Sequence in Southern Ontario. Background Paper prepared by Rock-Water Interactions, Institute of Geological Sciences, University of Bern for the Nuclear Waste Management Organization. (Available at www.nwmo.ca).

[3] R. Newman, S. Wang and G. Kwong. 2010. Anaerobic Corrosion Studies of Carbon Steel Used Fuel Containers. NWMO Technical Report, TR-2010-07 (to be published).

[4] S.B. Lyon and D.J. Fray. 1983. Hydrogen Measurements using Hydrogen Uranyl Phosphate Tetrahydrate, Solid State Ionics, 9 and 10, pg. 1295.

[5] M.S. Odziemkowski, T.T. Schuhmacher, R.W. Gillham, and E.J. Reardon. 1998. Mechanism of Oxide Film Formation on Iron in Simulating Groundwater Solutions: Raman Spectroscopic Studies. Corrosion Science, Vol. 40, No. 2/3, pg. 371-389.

[6] G. Williams, H.N. McMurray. 2003. The Mechanism of Group(I) Chloride Initiated Filiform Corrosion on Iron. Electrochemistry Communications. Vol. 5, Issue 10, pg. 871-877.

DEVELOPMENT OF CERAMIC WASTE FORMS FOR AN ADVANCED NUCLEAR FUEL
CYCLE

A. L. Billings, K. S. Brinkman, K. M. Fox and J. C. Marra
Savannah River National Laboratory
Aiken, SC, USA

M. Tang and K. E. Sickafus
Los Alamos National Laboratory
Los Alamos, NM, USA

ABSTRACT
    A series of ceramic waste forms were developed and characterized for the immobilization of a
Cesium/Lanthanide (CS/LN) waste stream anticipated to result from nuclear fuel reprocessing. Simple
raw materials, including Al_2O_3 and TiO_2 were combined with simulated waste components to produce
multiphase ceramics containing hollandite-type phases, perovskites (particularly $BaTiO_3$), pyrochlores
and other minor metal titanate phases. Three fabrication methodologies were used, including melting
and crystallizing, pressing and sintering, and Spark Plasma Sintering (SPS), with the intent of studying
phase evolution under various sintering conditions. X-Ray Diffraction (XRD) and Scanning Electron
Microscopy coupled with Energy Dispersive Spectroscopy (SEM/EDS) results showed that the
partitioning of the waste elements in the sintered materials was very similar, despite varying
stoichiometry of the phases formed. Identification of excess Al_2O_3 via XRD and SEM/EDS in the first
series of compositions led to a Phase II study, with significantly reduced Al_2O_3 concentrations and
increased waste loadings. The Phase II compositions generally contained a reduced amount of
unreacted Al_2O_3 as identified by XRD. Chemical composition measurements showed no significant
issues with meeting the target compositions. However, volatilization of Cs and Mo was identified,
particularly during melting, since sintering of the pressed pellets and SPS were performed at lower
temperatures. Partitioning of some of the waste components was difficult to determine via XRD.
SEM/EDS mapping showed that those elements, which were generally present in small concentrations,
were well distributed throughout the waste forms.

INTRODUCTION
    Efforts being conducted by the United States Department of Energy (DOE) under the Fuel
Cycle Research and Development (FCR&D) program are aimed at demonstrating a proliferation-
resistant, integrated nuclear fuel cycle. The envisioned fuel reprocessing technology would separate
the fuel into several fractions, thus, partitioning the waste into groups with common chemistry. With
these partitioned waste streams, it is possible to treat waste streams separately or combine waste
streams for treatment when it is deemed appropriate. A trade study conducted in 2008 concluded that
it was beneficial from a cost perspective to combine waste streams and treat them using existing waste
form technologies.[1] A borosilicate glass was identified as the preferred waste form for the Cs/Sr (CS),
lanthanide (LN) and transition metal fission product (TM) combined waste stream. Unfortunately,
several fission products (e.g. noble metals and molybdenum) have limited solubility in borosilicate
glasses. Therefore, the use of borosilicate glass may simplify waste form processing but result in
significant increases in waste form volumes. This would defeat a major advanced fuel cycle
reprocessing objective of minimizing high level waste form volumes. In this work, experimental work
to develop crystalline ceramics to immobilize a combined CS and LN waste stream is discussed with a
goal of obtaining high waste loadings and establishing baseline compositions for future CS/LN/TM
combined waste stream work.

Titanate ceramics have been thoroughly studied for use in immobilizing nuclear wastes (e.g., the SYNROC family) due to their natural resistance to leaching in water.[2,3] Assemblages of several titanate phases have been successfully demonstrated to incorporate radioactive waste elements, and the multiphase nature of these materials allows them to accommodate variation in the waste composition.[4] While these materials are typically densified via hot isostatic pressing (HIP), recent work has shown that they can also be produced from a melt. For example, demonstrations have been completed using the Cold Crucible Induction Melter (CCIM) technology to produce several crystalline ceramic waste forms, including murataite-rich ceramics,[5] zirconolite/pyrochlore ceramics,[6] Synroc-C (zirconolite, hollandite, perovskite),[7] aluminotitanate ceramics, and zirconia.[8] This production route is advantageous since melters are already in use for commercial and defense high level waste (HLW) vitrification in several countries, and melter technology greatly reduces the potential for airborne contamination as compared to powder handling operations.

EXPERIMENTAL PROCEDURES

Waste Stream Composition and Chemical Additives
The CS/LN waste stream is a combination of the Cs/Sr separated stream and the Trivalent Actinide - Lanthanide Separation by Phosphorous reagent Extraction from Aqueous Komplexes (TALSPEAK) waste stream consisting of lanthanide fission products. In this testing, a projected composition for this combined stream was used (Table I).
Ceramic host systems for this study were selected based on the objectives of forming durable titanate and aluminate phases, using a minimum of additives to form the desired phases (i.e., achieving high waste loadings), and fabrication from a melt. Target compositions for the CS/LN waste form are given in Table II. The only additives used were Al_2O_3 and TiO_2, with targeted waste loadings of 50-60 wt %. The alkali and alkaline earth elements in the waste were anticipated to partition to aluminotitanate phases approximating hollandite or $(Ba,Cs,Rb)(Al,Ti)_2Ti_6O_{16}$. The lanthanides were anticipated to partition to aluminate perovskites ($LnAlO_3$) and the strontium to a titanate perovskite ($SrTiO_3$). Following characterization of the phase I samples, it was realized that titanate phases preferentially formed over aluminate phases (titanate phases thermodynamically more stable). Therefore, Phase II compositions were formulated with lower alumina concentrations (i.e. minimum alumina levels required to form the hollandite phase). The Phase II compositions are also given in Table II.

Table I. Composition of the CS/LN Waste Stream

Oxide	Wt %
Rb_2O	2.22
SrO	5.14
Y_2O_3	0.49
Cs_2O	15.08
BaO	11.55
La_2O_3	8.21
Ce_2O_3	15.28
Pr_2O_3	7.51
Nd_2O_3	27.11
Pm_2O_3	0.08
Sm_2O_3	5.58
Eu_2O_3	0.89
Gd_2O_3	0.84
Tb_2O_3	0.02

Table II. Target Compositions for the Phase I and Phase II CS/LN Ceramic Waste Forms.

Phase I			
Component	CS/LN-03	CS/LN-04	CS/LN-05
Waste	52	50	50
Al$_2$O$_3$	18	15	20
Phase II			
Component	CS/LN-06	CS/LN-08	CS/LN-09
Waste	50.0	55.0	60.0
Al$_2$O$_3$	4.0	7.3	8.0
TiO$_2$	46.0	37.7	32.0

Fabrication Methods

Simulated waste material and the ceramic forming additives were blended in the appropriate ratios via ball milling. Three different fabrication methods were used to densify the ceramic waste forms, including melting and crystallizing, cold pressing and sintering for extended periods, and rapid heating under pressure via spark plasma sintering. The intent was to provide insight into the phase assemblage that is formed at varying proximity to equilibrium conditions.

Samples of each of the ceramic materials were melted in an electric resistance heated furnace to simulate melter production. The blended and dried powders were placed into Pt/Rh alloy crucibles and melted at 1500°C for 1 hour. Power to the furnace was then turned off with the crucibles remaining inside to cool slowly (furnace cooling) to roughly approximate the slow cooling conditions experienced by a waste form poured from a melter into a canister. The temperature of the furnace fell below 200°C after 6.5 hours of cooling. The crucibles were removed from the furnace once cooled and photographed to document the degree of melting that occurred for each composition.

Sintering of pellets for longer periods was used to approximate equilibrium conditions and allow the more stable phases to form within each composition. The blended and dried powders were cold pressed into pellets using a steel die and uniaxial hydraulic press. Thermal analysis, conducted as a screening tool for crystallization and melting point determination, indicated that a majority of the compositions displayed initial melting point endotherms in the range of 1280-1300°C. The pellets were then sintered in an electric resistance heated furnace at 1200°C in air for 25 hours and furnace cooled.

Spark plasma sintering (SPS) is a powder densification process similar to hot pressing, except that heating occurs directly by passing a pulsed DC current through the die and powder rather than using resistive heating elements. This allows for very rapid heating of the powder and enhanced sintering due to high localized temperatures at particle interfaces, electric field assisted diffusion, and plastic deformation. Materials can typically be sintered via SPS at temperatures much lower than those of other sintering techniques and in much shorter times. SPS was explored for the ceramic waste forms to gauge the impacts of sintering at conditions far from equilibrium on the resulting phase assemblages. One sample (CS/LN-03) was sintered via SPS. Approximately 2 g of the sample powder was placed in a graphite die of approximately 20 mm diameter lined with graphite foil. The loaded die set was placed into the SPS chamber and a pressure of 3 MPa was applied. The chamber was evacuated and then backfilled with argon to 9.4x10^{-4} Torr. The sample was then heated from room temperature to 600°C at a rate of 150°C per minute using a DC pulse duration of 25 ms on and 5 ms off. The pressure was increased from 3 to 10 MPa during this time. The sample was held at 600°C for 15 seconds. Heating then continued at 100°C per minute to 1300°C, with the pressure being increased from 10 to 100 MPa during this time. The sample was held at 1300°C and 100 MPa for 15 minutes. The pressure was then relieved and the current was turned off to allow the sample to cool.

Sample Characterization
Representative samples of select compositions were characterized to confirm that the as-fabricated ceramics met the target compositions. A lithium-metaborate fusion was used to dissolve the samples. The resulting solutions were analyzed by Inductively Coupled Plasma – Atomic Emission Spectroscopy (ICP-AES). Two measurements were taken for each element of interest, and the average of these two measurements was reported as the measured value. Inductively Coupled Plasma – Mass Spectroscopy (ICP-MS) was used to measure the concentrations of Cs, Pd, Rh, and Ru. Only one measurement was performed for each of these elements.

Scanning Electron Microscopy (SEM) and Energy Dispersive Spectroscopy (EDS) analyses were performed on select samples. Specimens were cut, ground, and polished with alumina lapping films to produce a flat surface for imaging and analysis. All of the samples were final polished using 40 nm colloidal silica slurry to remove mechanical polishing damage. Secondary electron and backscattered electron imaging were used to identify grain size and morphology, as well as general homogeneity of the specimen. EDS elemental mapping was used to identify partitioning of the waste elements among the various phases.

Representative samples of each ceramic waste form were analyzed using X-ray diffraction (XRD) to assess phase development. Samples were ground prior to analysis. In some cases, the surfaces of unground samples were first analyzed to determine any differences between surface and bulk phase assemblage.

RESULTS AND DISCUSSION

Fabrication of the Ceramic Waste Forms
Photographs of the Phase I CS/LN waste forms fabricated from 1500°C melts are shown in Figure 1. Each of the three compositions appeared to have melted completely, with some void spaces after cooling and a mottled appearance that may be indicative of their multiphase composition.

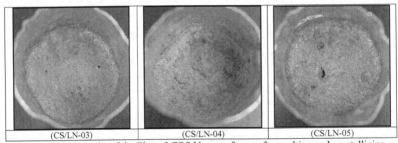

| (CS/LN-03) | (CS/LN-04) | (CS/LN-05) |

Figure1. Photographs of the Phase I CS/LN waste forms after melting and crystallizing.

Photographs of the Phase II CS/LN waste forms fabricated from 1500°C melts are shown in Figure 2. The CS/LN-06 and -08 compositions appeared to have melted completely. The CS/LN-09 composition, which has a higher concentration of waste oxides and Al_2O_3 in place of TiO_2, appears to have become too refractory to melt completely at 1500°C.

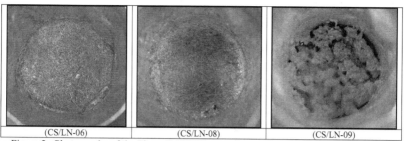

| (CS/LN-06) | (CS/LN-08) | (CS/LN-09) |

Figure 2. Photographs of the Phase II CS/LN waste forms after melting and crystallizing.

Chemical Composition Measurements

The chemical compositions of CS/LN waste forms fabricated by melting and crystallizing were measured and compared to the target compositions. In general, most of the components were present at concentrations close to their targeted values. The Cs_2O concentrations were about 20-30% below their targeted values, which was likely due to volatilization during melting. There may have also been some volatilization of La_2O_3.

Microstructure and Elemental Analysis

Samples of compositions CS/LN-05 fabricated by melting and crystallizing and composition CS/LN-09 fabricated by pressing and sintering were analyzed using SEM/EDS. A backscattered electron micrograph of composition CS/LN-05 is shown in Figure 3a while a secondary electron micrograph of CS/LN-09 is shown in Figure 3b. The differences in contrast in Figure 3a indicated at least four crystalline phases were present in the CS/LN-05 sample prepared by melting and crystallizing, with varying grain sizes and morphology. Porosity is visible as the black area at the left of the micrograph. The more angular black areas may indicate grain pullout during polishing.

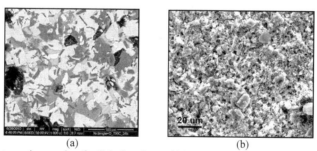

(a)   (b)

Figure 3. Electron micrographs of polished surfaces of (a) composition CS/LN-05 fabricated by melting and crystallizing (backscattered image) and (b) composition CS/LN-09 fabricated by pressing and sintering (secondary electron image).

The results of EDS elemental mapping for this sample are shown in Figure 4. Observations of these maps showed that:

- Unreacted Al_2O_3 was readily apparent as high aspect ration, or needle-like grains.
- Ba appeared to partition mainly to a titanate phase, but is present in all of the phases except for the unreacted Al_2O_3.
- Ce appeared to partition most strongly to a different titanate phase, with additional Ba and Nd.
- Cs and La appeared to partition to the same phase, although La is also distributed in other phases.
- Nd and Pr appeared to partition to the same phases.
- O is dispersed throughout the material as expected, although higher concentrations appeared with Al.
- Rb and Sr were distributed fairly uniformly, although some of the Sr appeared to partition to the phase containing Nd and Pr.

The secondary electron micrograph of composition CS/LN-09 (Figure 3b) was obtained from a rough cut surface of this sample since its low density made polishing difficult. This Phase II composition contained a lower concentration of Al_2O_3, and did not appear to contain the elongated grains of excess Al_2O_3 seen in the Phase I sample. The average grain size appeared to be smaller than that of the previous sample, with most grains measuring less than 10 μm in diameter. Porosity continued to be apparent. The secondary electron image did not allow for an estimate of the number of phases present based on z-contrast. However, the image suggested at least two phases based on the morphology of the grains: a phase with high aspect ratio platelet or needle-like grains, and a phase with larger diameter and smaller aspect ratio grains.

The results of EDS elemental mapping for the CS/LN-09 sample are shown in Figure 5. Observations of these maps showed that:

- There were still a small number of unreacted Al_2O_3 grains present, although they were not the high aspect ratio grains seen in the previous samples. The Al_2O_3 grains did not incorporate any of the other elements analyzed.
- Ba and Ti were distributed throughout the sample except for the Al_2O_3 grains, but were in the highest concentrations together.
- Ce was distributed throughout the sample, with a few grains of higher concentration. The results for Cs were similar, although the grains with higher Cs concentrations were different from those with higher Ce concentrations.
- La, Nd, Pr, and Sm, were fairly evenly distributed throughout the sample.
- Sr appeared evenly distributed among some of the phases and depleted in others.

The results of the EDS mapping were used to aid in the identification of the crystalline phase assemblages, along with the XRD data discussed in the following section.

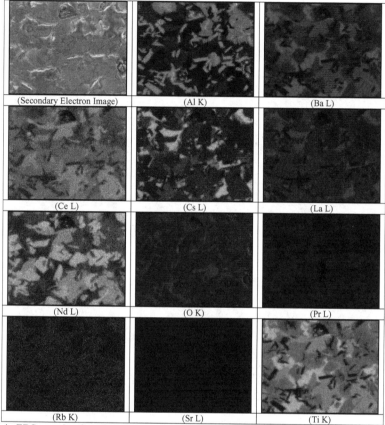

Figure 4. EDS mapping for select elements in composition CS/LN-05 fabricated by melting and crystallizing. The element and emission line are given for each image.

Phase Analysis

XRD data for the Phase I CS/LN compositions prepared via the three fabrication methods are summarized in Table III. XRD scans of the pellet surfaces and ground samples of the same material showed no texturing effects. The XRD data showed that some of the expected phases in the CS/LN system formed, while others did not. There was some dependence on the type of fabrication method used. The hollandite-type phases formed in each of the compositions as predicted. For the press and sinter method, $(Ba,Cs,Rb)Al_2Ti_5O_{14}$ was identified. The higher temperature processes produced $CsTiAlO_4$ and $(Ba,Cs)(Al_2Ti_6)O_{16}$ phases approximating hollandite. All of the methods produced the predicted $SrTiO_3$ perovskite phase, as well as $BaTiO_3$.

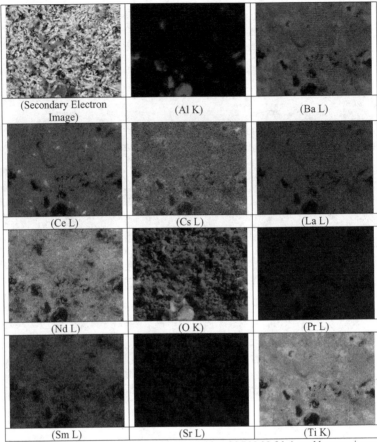

Figure 5. EDS mapping for select elements in composition CS/LN-09 fabricated by pressing and sintering. The element and emission line are given for each image.

The LnAlO$_3$ phases predicted to host the lanthanides for the CS/LN waste composition did not form. Instead, a Ba$_4$(Sr$_2$Sm$_8$)(TiO$_3$)$_{18}$ phase was detected in all of the compositions, as well as a Nd$_2$Ti$_2$O$_7$ pyrochlore phase potentially with all the fabrication methods except for the SPS sample. The Nd$_2$Ti$_2$O$_7$ pyrochlore phase in the samples produced from melts was difficult to positively identify by XRD but the SEM and EDS results served to confirm the presence of this phase (see Figure 4). A brief review of thermodynamic data for these phases shows that the lanthanide titanate phases should indeed be more stable than the lanthanide aluminate phases originally predicted to form. The most prevalent lanthanide in the CS/LN waste stream is Nd, and the free energy of formation of NdAlO$_3$ is approximately -41 kJ/mol,[9] as compared to approximately -120 kJ/mol for Nd$_2$Ti$_2$O$_7$.[10] Although the conditions used to fabricate these compositions were likely not at equilibrium, the lower free energy of formation of $Nd_2Ti_2O_7$ is in agreement with the XRD results.

Partitioning of the waste components Ce, La, and Pr was not clear from the XRD data. Based on the EDS results (and XRD results for the Phase II compositions discussed below), it was likely that these components were present as substitutional cations in the perovskite phase. Corundum (Al_2O_3) was identified in all of the samples, indicating that an excess of Al_2O_3 was added to the compositions and remained unreacted after melting or sintering. This finding led to the reduced Al_2O_3 concentrations targeted in the Phase II compositions (see Table II).

XRD data for the Phase II CS/LN compositions prepared via two of the fabrication methods (the Phase II compositions were not sintered by SPS) are summarized in Table IV. For most of the compositions fabricated by pressing and sintering, the predicted hollandite-type phases were formed, as well as Ba perovskite and Nd pyrochlore phases. CS/LN-06, which had the lowest target Al_2O_3 concentration, had no unreacted Al_2O_3 detectable by XRD. Titanate phases differing from a typical hollandite stoichiometry formed in the CS/LN-08 and -09 compositions, but may have similar structures.

For most of the Phase II CS/LN compositions fabricated by melting and crystallizing, hollandite-type and pyrochlore phases again formed. No perovskites or unreacted Al_2O_3 were detected in these samples via XRD. However, the EDS data (Figure 5) showed unreacted Al_2O_3 in composition CS/LN-09. Cs partitioned to an aluminotitanate phase in most of these samples. The alkaline earths, transition metals and some of the lanthanides also partitioned to titanate phases. The predicted $LnAlO_3$-type phases did not form using either fabrication method, which again may be due to a larger free energy of formation of these phases as compared to the titanates.

Partitioning of the Ce, Cs, La, and Pr in the Phase II CS/LN compositions was unclear from the XRD data. The Ce, La, and Pr may have partitioned to the perovskite phase as identified for some of the Phase I compositions. The EDS data showed these elements to be fairly well distributed throughout the material (see Figure 5).

Table III. Summary of XRD Data for Phase I CS/LN Waste Forms Prepared by Three Methods

Phases	Press and Sinter			Melt and Crystallize			SPS
	CS/LN-03	CS/LN-04	CS/LN-05	CS/LN-03	CS/LN-04	CS/LN-05	CS/LN-03
$BaTiO_3$/$SrTiO_3$ perovskite	X	X	X	X	X	X	X
$CsTiAlO_4$ hollandite type				X	X	X	X
$(Ba,Cs)(Al_2Ti_6)O_{16}$ hollandite type							X
$(Ba,Cs,Rb)Al_2Ti_5O_{14}$ hollandite	X	X	X				
$Ba_4(Sr_2Sm_8)(TiO_3)_{18}$	X	X	X	X	X	X	X
$Nd_2Ti_2O_7$ pyrochlore	X	X	X	?	?	X (SEM/EDS)	
Unreacted Al_2O_3	X	X	X	X	X	X	X

Table IV. Summary of XRD Data for Phase II CS/LN Waste Forms Prepared by Two Methods

Phases	Press and Sinter			Melt and Crystallize		
	CS/LN-06	CS/LN-08	CS/LN-09	CS/LN-06	CS/LN-08	CS/LN-09
$Sr_{0.34}Nd_{2.44}Ti_4O_{12}$				X	X	
$BaNd_2Ti_4O_{12}$			X	X		X
$BaAlTi_5O_{14}$ Hollandite	X	X		X	X	
$Nd_2Ti_2O_7$ Pyrochlore	X	X	X		X	X
$Cs_2Ti_2Al_2O_8$					X	X
$BaTiO_3$ perovskite	X	X	X			
Unreacted Al_2O_3		X	X			X (SEM/EDS)
$Ba_4(Sr_2Sm_8)(TiO_3)_{18}$		X	X			

CONCLUSIONS

A series of ceramic waste form compositions for the immobilization of CS/LN waste streams anticipated to result from nuclear fuel reprocessing were developed. Three fabrication methodologies were used, including melting and crystallizing, pressing and sintering, and SPS, with the intent of studying phase evolution under various sintering conditions. XRD and SEM/EDS results showed that the partitioning of the waste elements in the materials was very similar, despite varying stoichiometry of the phases formed. The Phase II compositions generally contained a reduced amount of unreacted Al_2O_3 as identified by XRD. They also had phase assemblages that were closer to the initial targets. Chemical composition measurements showed no significant issues with meeting the target compositions. However, volatilization of Cs and La was identified, particularly during melting, since sintering of the pressed pellets and SPS were preformed at lower temperatures. Partitioning of some of the waste components was difficult to determine via XRD. EDS mapping showed that these elements, which were generally those present in smaller concentrations, were well distributed throughout the waste forms.

ACKNOWLEDGEMENTS

The authors would like to thank Professor Serge Stefanovsky of SIA Radon Institute for his insight and suggestions into potential compositions for the host ceramic phases, David Best, David Missimer, Irene Reamer, Phyllis Workman, Pat Simmons, Whitney Riley, and Curtis Johnson for their assistance with sample preparation and characterization, Dr. Daniela Fredrick and Robert Aalund at Thermal Technology LLC for providing the spark plasma sintered samples.

This manuscript has been authored by Savannah River Nuclear Solutions, LLC under Contract No. DEAC09-08SR22470 with the U.S. Department of Energy. This work was funded by the Department of Energy Office of Nuclear Energy Fuel Cycle Research and Development Program. The authors gratefully acknowledge this financial support.

REFERENCES
[1] D. Gombert, S. Piet, T. Trickel, J. Carter, J. D. Vienna and W. Ebert, "Combined Waste Form Cost Trade Study," U.S. Department of Energy Report GNEP-SYSA-PMO-MI-DV-2009-000003, Idaho National Laboratory, (2008).

[2] A. E.Ringwood, E. S. Kesson, N. G. Ware, W. Hibberson and A. Major, "Geological Immobilisation of Nuclear Reactor Wastes," *Nature,* **278** 219 (1979).

[3] A. E. Ringwood, E. S. Kesson, K. D. Reeve, D. M. Levins and E. J. Ramm, "Synroc," pp. 233-334 in *Radioactive Waste Forms for the Future,* W. Lutze and R. C. Ewing, eds. Elsevier, North-Holland, Amsterdam, Netherlands (1988).

[4] D. S. Perera, B. D. Begg, E. R. Vance and M. W. A. Stewart, "Application of Crystal Chemistry in the Development of Radioactive Wasteforms," *Advances in Technology of Materials and Materials Processing,* **6** [2] 214-217 (2004).

[5] S. V. Stefanovsky, A. G. Ptashkin, O. A. Knyazev, S. A. Dmitriev, S. V. Yudintsev and B. S. Nikonov, "Inductive Cold Crucible Melting of Actinide-bearing Murataite-based Ceramics," *Journal of Alloys and Compounds,* **444-445** 438-442 (2007).

[6] A. V. Demine, N. V. Krylova, P. P. Polyektov, I. N. Shestoperov, T. V. Smelova, V. F. Gorn and G. M. Medvedev, "High Level Waste Solidification Using a Cold Crucible Induction Melter"; pp. 27-34 in Mater. Res. Soc. Symp. Proc., Vol. 663, *Scientific Basis for Nuclear Waste Management XXIV.* Edited by K. P. Hart and G. R. Lumpkin. Warrendale, PA, 2001.

[7] T. Advocat, G. Leturcq, J. Lacombe, G. Berger, R. A. Day, K. Hart, E. Vernaz and A. Bonnetier, "Alteration of Cold Crucible Melter Titanate-based Ceramics: Comparison with Hot-Pressed Titanate-based Ceramic"; pp. 355-362 in Mater. Res. Soc. Symp. Proc., Vol. 465, *Scientific Basis for Nuclear Waste Management XX.* Edited by W. J. Gray and I. R. Triay. Pittsburgh, PA, 1997.

[8] G. Leturcq, T. Advocat, K. Hart, G. Berger, J. Lacombe and A. Bonnetier, "Solubility Study of Ti Zr-based Ceramics Designed to Immobilize Long-lived Radionuclides," *American Mineralogist,* **86** [7-8] 871-880 (2001).

[9] J. E. Saal, D. Shin, A. J. Stevenson, G. L. Messing and Z.-K. Liu, "First-Principles Calculations and Thermodynamic Modeling of the Al_2O_3-Nd_2O_3 System," *Journal of the American Ceramic Society,* **91** [10] 3355-3361 (2008).

[10] K. B. Helean, S. V. Ushakov, C. E. Brown, A. Navrotsky, J. Lian, R. C. Ewing, J. M. Farmer and L. A. Boatner, "Formation Enthalpies of Rare Earth Titanate Pyrochlores," *Journal of Solid State Chemistry,* **177** [6] 1858-1866 (2004).

DETERMINATION OF STOKES SHAPE FACTOR FOR SINGLE PARTICLES AND AGGLOMERATES

J. Matyáš, M. Schaible, and J.D. Vienna
Pacific Northwest National Laboratory
Richland, WA, USA

ABSTRACT

   Large octahedral crystals of spinel can precipitate from glass during the high-level waste vitrification process and potentially block the glass discharge riser of electrically heated ceramic melters. To help predict the settling behavior of spinel in the riser, the settling of single particles and agglomerates was studied in stagnant and transparent viscosity oils at room temperature with a developed optical particle-dynamics-analyzer. The determined dimensions and terminal settling velocities of particles were used to calculate their Stokes shape factors. The calculated shape factor for the glass beads was almost identical with the theoretical shape factor of 2/9 for a perfect sphere. The shape factor for a single spinel crystal was about 7.6% higher compared to the theoretically predicted value for an octahedron. The Stokes shape factor of irregularly shaped multi-particle agglomerates was lower than that of the glass beads and individual spinel crystals because of the higher surface drag caused by the larger surface area-to-volume ratio.

INTRODUCTION

   The high-level radioactive waste (HLW) from the Hanford and Savannah River Sites is being vitrified in stable borosilicate glass for long-term storage and disposal. A major concern of the vitrification process is the formation and settling of large spinel crystals in the glass discharge riser of the HLW melter.[1,2] During numerous and extended melter idling periods from 20 to 100 days[3,4], while new feed is being incorporated into the melt, the temperature of molten glass in the riser can drop to ~ 850°C. At this temperature, a significant volume of large octahedral crystals of spinel [Fe,Ni,Mn,Zn][Fe,Cr]$_2$O$_4$ can precipitate in the glass.[5] Matyas et al.[6] demonstrated that these crystals rapidly settle, forming a thick sludge layer with rates up to 0.6 mm/day. These rates are fast enough to form a few cm thick plug that can partially or completely block the riser during idling, therefore, preventing discharge of molten glass during normal operation. This is supported by the fact that the spinel sludge cannot be dissolved because the temperature in the riser is relatively low, and the sludge cannot be easily disturbed.[7] To assess the crystal accumulation during melter operation, detailed knowledge of spinel settling behavior in high-viscosity Newtonian liquids is needed.

   The settling velocity of spherical particles suspended in a stagnant Newtonian fluid can be estimated from the well-known Stokes equation:

$$v = \frac{2(\rho_p - \rho_f)gr^2}{9\eta} \qquad (1)$$

where ρ_p is the particle density, ρ_f is the fluid density, η is the fluid dynamic viscosity, g is the gravitational acceleration, and r is the particle effective radius. This equation can be modified by considering non-spherical particles:

$$v = k_s \frac{(\rho_p - \rho_f)gr^2}{\eta} \qquad (2)$$

195

where k_s is the Stokes shape factor. The shape factor of the non-spherical particle in diluted suspensions can be calculated from experimentally measured terminal settling velocities using the equation:

$$k_s = \frac{v'}{v} = \frac{v'\eta}{\left(\rho_p - \rho_f\right)gr_v^2} \tag{3}$$

where v' is the experimentally measured terminal settling velocity, and r_v is the particle effective radius. A review of the various methods to calculate the drag on non-spherical particles was performed by Chhabra et al.,[8] and a more recent study concentrating on isometric particles, including octahedrons, was undertaken by Hazzab et al.[9] Williams et al. reviewed a number of different techniques commonly employed to study sedimentation, such as X-ray attenuation and nuclear magnetic resonance (NMR) imaging.[10] However, optical monitoring was used in this study because it allowed settling particles to be directly observed and collected data to be quickly evaluated.

To improve predictions of the accumulation rate of spinel crystals in the glass discharge riser during melter idling, the settling of single particles of different sizes and agglomerates in stagnant transparent liquids at room temperature was studied with a developed optical particle-dynamics-analyzer.

EXPERIMENTAL
Materials
Settling experiments were performed with two different sizes of spherical barium calcium silicate and soda-lime-silica glass beads from Mo-Sci Corporation, and octahedral spinel crystals that were suspended in Brookfield 0.48, 0.98, 1.025, 4.84, and 5.08 Pa.s standard viscosity oils of specific gravity 0.97 g/cm^3. Two sets of barium calcium silicate glass beads had a density 4.1741 g/cm^3, and their average diameters were 72 and 161 µm. Two sets of soda-lime-silica glass beads had a density 2.4842 g/cm^3, and their average diameters were 67 and 137 µm. The spinel crystals were produced by precipitation in the waste simulant glass during 10 days of heat-treatment at 850°C. The glass was dissolved overnight in continuously mixed 25% HNO$_3$ solution at 60°C. The collected mixture of crystals and silica gel was then treated with 5% HF to obtain the gel-free crystals that were separated into distinct size groups by sieving. The crystals have a density 5.3954 g/cm^3, as-measured with the gas pycnometer, and contained about 31.4 mass % of Ni, 48.3 mass % of Fe, 1.3 mass % of Cr, and 19 mass% of O, as-determined with scanning electron microscopy and energy dispersive spectroscopy (SEM-EDS). Figure 1 shows the SEM images of the 161-µm glass beads and spinel crystals.

A   B

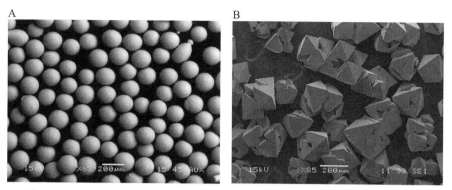

Figure 1. SEM images of barium calcium silicate glass beads (A) and spinel crystals (B).

Methods

To obtain terminal settling velocities of individual beads and crystals, a few particles were sprinkled with a sieve onto the surface of viscosity oil in a 2.5×2.5×15-cm clear quartz cuvette. The cuvette was tall enough for particles to reach their terminal settling velocities and wide enough that the wall effect on free-settling rates could be neglected. Figure 2 shows the experimental setup of the optical particle-dynamics-analyzer for measuring the free-settling velocity of individual glass beads and spinel crystals. An infinitely corrected lens 20×/0.42 with a field of view 415×515 μm was mounted on a digital camera and focused at the center of the cuvette approximately 4 cm above the bottom. PAX-it imaging software was used to record images of settling particles at 5-s intervals and measure their size and vertical distances. Figure 3 shows an example of time-sequence images collected during the settling of spinel crystals, skipping every second image.

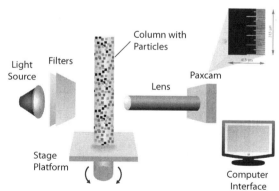

Figure 2. Optical particle-dynamics-analyzer.

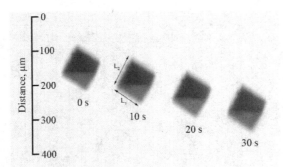

Figure 3. Time sequence images for spinel with edge lengths 69 μm (L_1) and 79 μm (L_2).

The settling velocity was calculated by dividing the difference in distance between successive images by the known time interval. While the effective radius of the glass beads was measured directly from collected images, the non-spherical shape of the spinel crystals required the radius to be calculated for an equivalent volume sphere. Therefore, the average length of the crystal edge was determined as $a = (L_1+L_2)/2$, and the volume of an isometric octahedron was calculated from the equation:

$$V_s = \tfrac{1}{3} \sqrt{2} a^3 \tag{5}$$

Then, the equivalent sphere radius was acquired from the equation:

$$R_{eq} = \left(\frac{3V_s}{4\pi} \right)^{1/3} \tag{6}$$

In the case of agglomerates, the radius of an equivalent area was obtained from the measured boundary area of the agglomerates A_{ag} by calculating the radius of a sphere with an identical cross-sectional area:

$$R_{eq} = \sqrt{\frac{A_{ag}}{\pi}} \tag{7}$$

To determine the Stokes shape factors, the measured terminal settling velocities were plotted against $[(\rho_p-\rho_f)gR^2]/\eta$. The slopes of the linear regression curves corresponded to the unknown Stokes shape factors.

RESULTS AND DISCUSSION

The baseline settling experiments were performed with individual glass beads to evaluate our experimental method for determining particle shape factors. Table I shows the calculated settling velocities and shape factors for glass beads of different size and density in oils of three different viscosities. The shape factor was about constant, with an average value of 0.2219 and relative standard deviation (RSD) of ~ 1.5 %. Figure 3 shows a linear fit of settling velocity vs. $[(\rho_p-\rho_f)gR^2]/\eta$ for glass beads. The shape factor of 0.2227 which was determined from fitting was almost identical with the tabulated Stokes shape factor for a perfect sphere, 2/9. The excellent agreement in shape factors indicated that a developed optical particle-dynamics-analyzer can be efficiently used to study the settling of particles in the Stokes Regime (Re <0.5) and to determine their shape factors.

Table I. Settling velocities and shape factors for glass beads of different size and density in different viscosity oils.

#	Bead Size (μm)	Bead Density (g/cm³)	Oil Viscosity (Pa.s)	Settling Velocity (μm/s)	Shape Factor
1	190	4.1741	4.840	13.3500	0.2278
2	186	4.1741	4.840	12.6000	0.2243
3	188	4.1741	4.840	12.8000	0.2231
4	168	4.1741	4.840	10.0000	0.2182
5	142	2.4842	1.025	16.3333	0.2236
6	144	2.4842	1.025	16.5000	0.2196
7	148	2.4842	1.025	17.6875	0.2229
8	124	2.4842	1.025	12.4000	0.2226
9	82	4.1741	1.025	11.7500	0.2279
10	74	4.1741	1.025	9.4857	0.2260
11	78	4.1741	1.025	9.6000	0.2193
12	76	4.1741	1.025	10.2286	0.2168
13	126	2.4842	1.025	12.7000	0.2185
14	120	2.4842	1.025	11.7000	0.2219
15	64	2.4842	0.480	7.0000	0.2209
16	70	2.4842	0.480	8.5000	0.2242
17	70	2.4842	0.480	8.3250	0.2196
18	84	2.4842	0.480	12.1000	0.2217
19	68	2.4842	0.480	8.1500	0.2278
20	72	2.4842	0.480	8.9000	0.2219

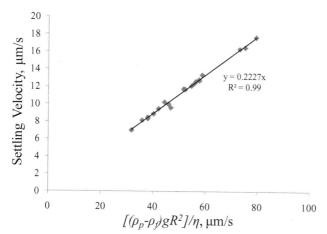

Figure 3. Linear fit of settling velocity vs. $[(\rho_p-\rho_f)gR^2]/\eta$ for glass beads.

Figure 4 shows the linear fit of settling velocity vs. $[(\rho_p-\rho_f)gR^2]/\eta$ for individual spinel crystals and spinel agglomerates. The fitted shape factor for individual crystals was slightly higher compared to glass beads. Happel and Brenner[11] used a Stokes shape factor correlated to the particle sphericity through the equation:

$$k' = 0.843 \times \log \frac{\psi}{0.065} \tag{8}$$

where the sphericity ψ is defined as the ratio of the surface area of a sphere to the surface area of a particle with the same volume. If multiplied by 2/9, this shape factor becomes the Stokes shape factor k_s from Equation (3). The obtained spinel shape factor of 0.2246 was 7.6% higher than the theoretical Stokes shape factor of octahedrons 0.2087 (ψ =0.846).[12] This difference can be attributed to the higher scatter of data (smaller R^2 value) resulting from irregularities in crystal shape (e.g., crystal elongation and surface defects) and differences in crystal orientation during settling. Variations in crystal geometry cause a divergence from the sphericity of an isometric octahedron.

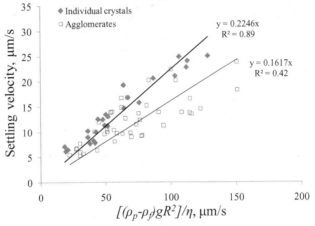

Figure 4. Linear fit of settling velocity vs. $[(\rho_p-\rho_f)gR^2]/\eta$ for single and agglomerated spinel crystals.

The data in Figure 4 show that the calculated shape factors for spinel agglomerates were widely scattered. Case-to-case variations in the number of agglomerated crystals, the degree of overlap, and the agglomerate orientation can explain this variance. In addition, multi-crystal agglomerates likely contained holes and inter-pockets that slowed down their settling velocity. These features could not be detected from the shallow three-dimensional images but were observed during SEM analysis of the precipitated spinel crystals (Figure 1). Also, when only two- or three-particle agglomerates were observed, the determined shape factors were within 10% of that of the single spinel crystals. Figure 5 shows examples of a two-particle agglomerate and a multi-particle agglomerate with Stokes shape factors ~ 0.2442 and ~ 0.1879, respectively. The fitted shape factor for multiparticle agglomerates was more than 25% lower compared to glass beads. This can be attributed to larger drag forces caused by a higher surface area-to-volume ratio. Spheres have the smallest surface area for all shapes of a given volume and at low

Reynolds numbers experience a minimum amount of resistance as they travel through a fluid. A decrease in the sphericity, corresponding to an increase in the surface area-to-volume ratio, results in an increase in the surface drag and a decrease in velocity for an object travelling through a given fluid.

A   B

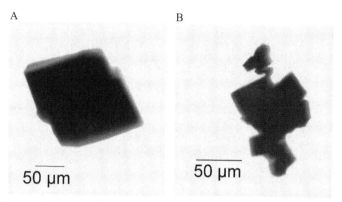

Figure 5. Two (A) and multiple-particle (B) agglomerates of spinel crystals.

CONCLUSION

The free-settling experiments with spherical glass beads produced a Stokes' shape factor very close to the predicted theoretical value. This provided confidence in the experimental setup and confirmed the appropriate use of the developed methodology to study the free settling of single particles and agglomerates. The experimental shape factor for the spinel crystals was about 7.6% higher compared to the non-spherical shape factor calculated from the equation used by Happel and Brenner.[11] In contrast, the Stokes shape factor for agglomerates was smaller than that of the beads because of the higher surface drag caused by the larger surface area-to-volume ratio. The obtained shape factors will be used to study hindered settling of particles in concentrated suspensions.

ACKNOWLEDGEMENT

The authors would like to thank Mike Schweiger for laboratory support, Carol Burnett for supplying glass beads, and Brian Riley for technical support. Micah Schaible would like to thank the Office of Science for his internship opportunity at Pacific Northwest National Laboratory. This work was funded by the U.S. Department of Energy's Environmental Management Program EM31. Pacific Northwest National Laboratory is operated by Battelle for the U.S. Department of Energy under Contract DE-AC05-76RL01830.

REFERENCES
[1]J. Matyáš, J. Kloužek, L. Němec, and M. Trochta: Spinel settling in HLW melters, *The 8-th International Conference Proceedings (ICEM'01)*, Bruges, Belgium, 2001.
[2]P. R. Hrma: Impact of particle size and agglomeration on settling of solids in continuous melters processing radioactive waste glass, DOE-ORP-39901, Richland, WA, December 2008.

[3]M.E. Smith, D.F. Bickford, The behavior and effects of the noble metals in the DWPF melter system, WSRC-TR-97-00370, Aiken, SC, March 1998.

[4]N.D. Hutson, D.C. Witt, D.F. Bickford, S.K. Sundaram, On the issue of noble metals in the DWPF melter, WSRC-TR-2001-00337, Aiken, AC, August 2001.

[5]P. Izak, P. Hrma, B.W. Arey, and T.J. Plaisted, Effect of feed melting, temperature history, and minor component addition on spinel crystallization in high-level waste glass, *Journal of Non-Crystalline Solids*, vol. 289, 17-29, 2001.

[6]J.Matyáš, J.D. Vienna, A. Kimura, M. Schaible, and R.M. Tate: Development of crystal-tolerant waste glasses, Advances in Materials Science for Environmental and Nuclear Technology edited by K Fox, E Hofmann, N Manjooran, G Pickrell, *Ceramic Transactions*, vol. 222, 41-51, 2010.

7M. Mika, P. Hrma, M.J. Schweiger, Rheology of spinel sludge in molten glass, *Ceramics-Silikaty*, vol. 44 (3), 86-90, 2000.

[8]R. P. Chhabra, L. Agarwal, N. K. Sinha: Drag on non-spherical particle: an evaluation of the available methods, *Powder Technology*, vol. 101, 288-295, 1999.

[9]A. Hazzab, A. Terfous, A. Ghenaim: Measurement and modeling of the settling velocity of isometric particles, *Powder Technology*, vol. 184, 105-113, 2008.

[10]R.A. Williams, C.G. Xie, R. Bragg, W.P.K. Amarasinghe, Experimental techniques for monitoring sedimentation in optically opaque suspensions, *Physics of Fluids A*, vol 4 (12), 1992.

[11]J. Happel, H. Brenner, Low Reynolds number hydrodynamics: with special applications to particulate media, 1-st pbk. ed edn. M. Nijhoff (Distributed by Kluwer, Boston), The Hague Boston, Hingham, MA, USA, pp. 411, 1983.

[12]M. Hartman, O. Trnka, and K. Svoboda: Free settling of nonspherical particles, *Ind. Eng. Chem. Res.*, vol. 33, 1979-1983, 1994.

GLASSY AND GLASS COMPOSITE NUCLEAR WASTEFORMS

Michael I. Ojovan
Immobilisation Science Laboratory, University of Sheffield,
Sir Robert Hadfield Building, Mappin Street, Sheffield, S1 3JD, United Kingdom

William E. Lee
Department of Materials, Imperial College London,
South Kensington campus, London, SW7 2AZ, United Kingdom

ABSTRACT

The current status of vitrification techniques for immobilisation of nuclear wastes in glasses and glass composite materials (GCM) are reviewed. One-stage and two-stage processes are described along with the compositions and properties of the resulting glassy wasteforms. Techniques for assessing wasteform durability are discussed along with the mechanisms of corrosion on contact with groundwater.

INTRODUCTION

Raw radioactive waste usually contains mobile contaminants therefore immobilisation is used to convert the waste into a solid and stable wasteform which can be handled, stored and disposed of safely and conveniently, significantly reducing any potential release of radionuclides into the environment. One of the most developed waste immobilisation technologies is vitrification which has been used on an increasing scale for over a half of century. Vitrification results in glassy wasteforms which comprise both homogeneous glasses and glass composite materials (GCM)[1-7]. Waste vitrification provides:

(i)   Reliable immobilisation of a wide range of elements;

(ii)   Small volume of the resulting glassy wasteform;

(iii)   High chemical and radiation durability of wasteform;

(iv)   Versatile immobilisation technology.

Almost all elements can be incorporated into the glass structure either directly as a constituent of the glass network or in physically encapsulated particles. The volume of vitrified waste is typically several times smaller than the initial waste volume resulting in a reduced costs of waste transportation, storage and disposal. The high chemical resistance of glass allows it to remain stable in corrosive environments at very low alteration rates (of the order of fractions of a millimetre per million years) which ensures a high degree of environmental protection. The relatively high initial investment and then operational costs of vitrification are fully justified when account is taken of transportation and disposal expenses. Vitrification provides the highest currently achievable degree of safety on waste transportation, storage and disposal.

NUCLEAR WASTE VITRIFICATION

Vitrification technology comprises several stages, starting with evaporation of excess water from liquid radioactive waste, followed by batch preparation, calcination, glass melting, and ending with pouring and cooling of vitrified waste blocks with potentially small amounts of secondary waste. Thin film evaporators are typically used and the remaining salt concentrate is mixed with the necessary additives and, depending on the type of vitrification process, is directed to one or another process apparatus. There are two types of vitrification processes: One-stage process in which both waste calcination and melting occur in the melter and two-stage process in which the waste is calcined prior to melting. One-stage vitfication is currently used in USA, Russia, India, Germany, Japan and Slovakia whereas two-stage vitrification is currently used in France and the UK. In the one-stage vitrification process (Fig. 1) glass forming additives are mixed with concentrated liquid wastes and so a glass-forming batch is formed (often as a paste). This batch is then fed into the melter where further water evaporation occurs, followed by calcination and glass melting which occur directly in the melter.

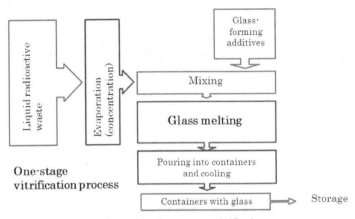

Figure 1. Schematic of a one-stage vitrification process.

In the two-stage vitrification process with separate calcination (Fig. 2) the waste concentrate is fed into the calciner. After calcination the required glass-forming additives (usually as a glass frit) together with the calcine are fed into the melter.

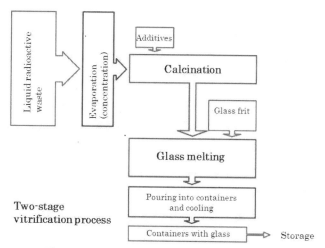

Figure 2. Schematic of a two-stage vitrification process.

In both vitrification processes two streams come from the melter: the glass melt containing most of radioactivity and the off gas flow, which contains off gases and aerosols. Two types of melters are currently used at waste vitrification plants: Joule heated ceramic melters (JHCM) and induction-heated melters which can either be hot (IHC) or cold e.g. cold crucible melters (CCM, Fig. 3)[4, 8].

Figure 3. An induction-heated CCM.

The melt waste glass is poured into containers (canisters) made of stainless steel when immobilising high level radioactive waste (HLW, Fig. 4)[4] or carbon steel for vitrified low and intermediate level radioactive waste (LILW)[8]. These may or may not be slowly cooled in an annealing furnace to avoid accumulation of mechanical stresses in the glass. When annealing is not used, cracking occurs resulting in a large surface area being potentially available for attack by water in a repository environment. Despite the higher final surface areas of non-annealed glasses these are sufficiently durable to ensure a suitable degree of radionuclide retention. Hence in many cases annealing is not used in vitrification facilities.

Figure 4. Stainless steel canisters for HLW glass.

The second stream from the melter goes to the gas purification system, which is usually a complex system that removes from the off gas not only radionuclide but also chemical contaminants. Operation of this purification system leads to generation of a small amount of secondary waste. For example, the distribution of beta gross activity at the PAMELA waste vitrification plant in Belgium was (%): >99.88 in waste glass, and the rest in secondary waste, e.g. <0.1 in intermediate level waste, <0.01 in cold waste and <0.01 in off gas[9]. Table I summarises data on radioactive waste vitrification facilities worldwide.

Table I. Operational Data on Vitrification.

Facility	Waste type	Melting process	Operational period	Performance
R7/T7, La Hague, France	HLW	IHC[1]	Since 1989/92	6,811 tonnes (237.9 10^6 Ci in 17206 canisters) to 2009
AVM, Marcoule, France	HLW	IHC	Since 1978	857.5 tonnes in 2412 canisters
R7, La Hague, France	HLW	CCM[2]	Since 2003	GCM: U-Mo glass
WVP, Sellafield, UK	HLW	IHC	Since 1991	>5,000 canisters to 2009
DWPF, Savannah River, USA	HLW	JHCM[3]	Since 1996	5000 tonnes in 2845 canisters to 2009
WVDP, West Valley, USA	HLW	JHCM	Since 1996	~500 tonnes in 275 canisters to 2002
EP-500, Mayak, Russia	HLW	JHCM	Since 1987	~8000 tonnes to 2009 (900 10^6 Ci)
CCM, Mayak, Russia	HLW	CCM	Pilot plant	18 kg/h by phosphate glass
PAMELA, Mol, Belgium	HLW	JHCM	1985-1991	~500 tonnes in 2200 canisters
VEK, Karlsruhe, Germany	HLW	JHCM	2010	~60 m^3 of HLW (24 10^6 Ci), to be completed in 2010
Tokai, Japan	HLW	JHCM	Since 1995	> 100 tonnes in 241 canisters (110 l) to 2007
Radon, Russia	LILW	JHCM	1987-1998	10 tonnes
Radon, Russia	LILW	CCM	Since 1999	> 30 tonnes
Radon, Russia	ILW	SSV[4]	2001-2002	10 kg/h, incinerator ash
VICHR, Bohunice, Slovakia	HLW	IHC	1997-2001, upgrading to restart work	1.53 m^3 in 211 canisters
WIP, Trombay, India	HLW	IHPT[5]	Since 2002	18 tonnes to 2010 (110 10^3 Ci)
AVS, Tarapur, India	HLW	IHPT	Since 1985	
WIP, Kalpakkam, India	HLW	JHCM	Under testing & commissioning	
WTP, Hanford, USA	LLW	JHCM	Pilot plant since 1998	~ 1000 tonnes to 2000

[1]IHC - Induction, hot crucible, [2]CCM – Cold crucible induction melter, [3]JHCM – Joule heated ceramic melter, [4]SSV - Self-sustaining vitrification, [5]IHPT – Induction heated pot type melter.

GLASSY WASTEFORMS

Two main glass types are currently used for nuclear waste immobilisation: borosilicates and phosphates. The exact compositions of nuclear waste glasses are tailored for easy preparation and melting, avoidance of phase separation and uncontrolled crystallisation, and acceptable chemical durability, e.g. leaching resistance. Vitrification can be performed efficiently at temperatures below 1500 K (1200 °C) because of the volatility of the fission products, notably Cs and Ru, so avoiding excess radionuclide volatilisation and maintaining viscosities below 10 Pa·s to ensure high throughput and controlled pouring into canisters. A more fluid glass is preferred to minimise blending problems. Phase separation on melting is most important for waste streams containing glass-immiscible constituents however these can be immobilised in form of isolated and phase separated disperse phase (in GCM's). The leaching resistance of nuclear waste glasses is a paramount criterion as it ensures low release rates for radionuclides on any potential contact with water.

Vitrification involves melting of waste materials with glass-forming additives so that the final vitreous product incorporates the waste contaminants in its macro- and micro-structure. Hazardous waste constituents are immobilised either by direct chemical incorporation into the glass structure or by physical encapsulation. In the former, waste constituents are dissolved in the glass structure with Si, B, P being included in the glass network on cooling while others such as Cs, K, Na, Li, Ca, Pb, Mg act as modifiers. A number of glass compositions have been designed for nuclear waste immobilisation however few are used in practice[1-8]. Table II gives compositions of several nuclear waste glasses.

Table II. Compositions of Some Nuclear Waste Glasses, Mass%.

Plant, Waste, Country	SiO_2	P_2O_5	B_2O_3	Al_2O_3	CaO	MgO	Na_2O	Misc	Waste loading
R7/T7, HLW, France	47.2	-	14.9	4.4	4.1	-	10.6	18.8	≤28
DWPF, HLW, USA	49.8	-	8.0	4.0	1.0	1.4	8.7	27.1	≤38
WVP, HLW, UK	47.2	-	16.9	4.8	-	5.3	8.4	17.4	≤25[1]
PAMELA, HLW, Germany-Belgium	52.7	-	13.2	2.7	4.6	2.2	5.9	18.7	<30
Mayak, HLW, Russia	-	52.0	-	19.0	-	-	21.2	7.8	≤33[2]
Radon, LILW, Russia	43	-	6.6	3.0	13.7	-	23.9	9.8	<35

[1]It was recently demonstrated that waste loading can be increased up to 35-38%[10];
[2]≤10 for Fission Products and Minor Actinide oxides.

High waste loadings and high chemical durability can be achieved in both borosilicate and aluminophosphate glasses. Moreover such glasses can immobilise large quantities of actinides, for example, borosilicate glasses can accommodate up to 7.2, mass% of PuO_2[11]. In contrast to borosilicate melts molten phosphate glasses are highly corrosive to refractory furnace linings, behaviour which has limited their application. Currently, this glass is used only in Russia, which has immobilised HLW from nuclear fuel reprocessing in alumina-phosphate glass since 1987[12]. It should be emphasized that nuclear waste glasses are never completely homogeneous vitreous materials but contain significant amounts of bubbles, foreign inclusions such as refractory oxides and other immiscible components.

Encapsulation is applied to elements and compounds which have low solubility in the glass melt and do not fit into the glass microstructure either as network formers or modifiers. Immiscible constituents which do not mix easily into the molten glass are typically sulphates, chlorides and molybdates as well as noble metals such as Rh and Pd, refractory oxides with high liquidus temperatures such as PuO_2, noble metal oxides and spinels. Encapsulation is carried out either by deliberate dispersion of insoluble compounds into the glass melt, immiscible phase separation on cooling or by sintering of glass and waste powders so that the waste form produced is a GCM. However, this requires a more complex melter supplied with a stirrer[8].

Although developed initially for HLW, vitrification is being used currently for immobilisation of LILW such as from operation and decommissioning of nuclear power plants[13, 14]. Vitrification is one of the technologies that have been chosen to solidify 18,000 tonnes of ore mining tailings at the Fernald site, Ohio, USA[15]. Plans are in place to vitrify vast volumes of waste; for example the vitrification of the low level radioactive waste at Hanford, USA is expected to produce over 160,000 m^3 of glass[16]. The USA Department of Energy (DOE) plans to vitrify 54 million gallons of mixed radioactive waste stored at its Hanford site in eastern Washington State, which represents 60% of the United States' volume of radioactive waste[17]. The world's largest waste vitrification plant (Waste Treatment Project (WTP)) is now under construction at Hanford. Borosilicate glass will be used for immobilisation of Hanford's low-activity waste (LAW). The vitrified LAW will be disposed of in a shallow land-burial facility. The proposed disposal system has been shown to adequately retain the radionuclides and prevent contamination of the surrounding environment. Release of radionuclides from the waste form via interaction with water is the prime threat to the environment surrounding the disposal site; the two major dose contributors in Hanford LAW glass that must be retained are ^{99}Tc and ^{129}I[18]. A number of glasses were developed to immobilise Hanford low activity wastes with composition ranges that will meet the desired performance of the Hanford site burial facility[18]. It is planned that the WTP will vitrify 99% of Hanford's waste by 2028. The WTP melter chosen to vitrify HLW is a JHCM which has nickel–chromium alloy electrodes which heat the waste and glass-forming additives to ~1450 K (1150 °C). The glass melt is stirred by convection and by bubbler elements and then poured into carbon steel canisters to cool. Canisters with vitrified HLW are sealed and decontaminated. It was planned that the vitrified HLW would be disposed of in the Yucca Mountain geological repository although due to the change in the US Government policy this HLW will need to be stored. Current plans also provide for the vitrified LAW to be stored on site. Moreover at Hanford it is planned to use a Bulk Vitrification process in which liquid waste is mixed with controlled-composition soil in a disposable melter[17]. The process of Bulk Vitrification involves mixing LAW with Hanford's silica-rich soil and surrounding it with sand and insulation in a large steel box. Electrodes are inserted to vitrify the mixture and when cooled the melter, its contents and the embedded electrodes will be buried as LLW in an on-site burial ground.

Vitrification of L ILW was studied intensively in Russia in the mid 1970's[8]. A number of glass compositions were developed for immobilisation of liquid waste containing mainly sodium nitrate. Various boron-containing minerals as well as sandstone were tested as glass-forming additives. Datolite $CaBSiO_4(OH)$ was found to be most suitable fluxing agent. Other systems such as Na_2O (LILW oxides) - $2CaO$ B_2O_3 - SiO_2 were studied and glass forming regions, melt viscosity and resistivity, leach rate of sodium (and ^{137}Cs for actual waste), density, radiation stability, and compressive strength were measured. Suitable glass composition areas were established[8]. The most important properties of these glasses are given in Table III.

Table III. Properties of Vitrified LILW.

Properties	Borosilicate glasses		GCM
	High sodium waste	Operational WWER[1] waste	Glass immiscible (high sulphate) waste
Waste oxide content, mass %	30-35	35-45	30-35 and up to 15vol. % of immiscible waste[2]
Viscosity, Pa s, at 1500 K (1200 °C)	3.5-5.0	2.5-4.5	3.0-6.0 (for vitreous phase)
Resistivity, Ω m, at 1500 K (1200 °C)	0.03-0.05	0.02-0.04	0.03-0.05
Density, g/cm^3	2.5-2.7	2.4-2.6	2.4-2.7
Compressive strength, MPa	80-100	70-85	50-70
Normalised leach rate, g/(cm^2 day) ^{137}Cs	10^{-5}-10^{-6}	~10^{-5}	10^{-4}-10^{-5}
^{90}Sr	10^{-6}-10^{-7}	~10^{-6}	10^{-6}-10^{-7}
Cr, Mn, Fe, Co, Ni	~10^{-7}-10^{-8}	~10^{-7}	10^{-7}-10^{-8}
REE, An	~10^{-8}	~10^{-8}	~10^{-8}
Na	10^{-5}-10^{-6}	~10^{-5}	10^{-4}-10^{-5}
B	<10^{-8}	<10^{-8}	≤10^{-8}
SO_4^{2-}	~10^{-6} (when present)	-	10^{-4}-10^{-5} at content <15vol. %

[1]WWER or VVER, water-water energetic reactor, Russian analogue of western PWR, pressurised water reactor. [2]e.g. yellow phase[19].

Loam and bentonite clays were also used as glass forming additives. Up to 50% of either loam clay or bentonite in the batch was substituted for sandstone. This substitution increases the chemical durability of glass and, moreover, such batches containing 20-25 wt% of water form homogeneous pastes which are stable for long times without segregation and are transportable in pipes over long distances[8]. Sodium nitrate is the major component of both institutional liquid LILW and nuclear power plant (NPP) operational wastes from RBMK (channel type uranium-graphite) reactors. NPP wastes from WWER reactors contain boron although the major components of this waste are sodium nitrate and sodium tetrahydroxyl borate NaB(OH)$_4$. As a result there is no need for boron-containing additives when vitrifying WWER waste. Silica, loam or bentonite clay or their mixtures are suitable as glass forming additives. WWER waste glasses are in the Na$_2$O-(Al$_2$O$_3$)- B$_2$O$_3$-SiO$_2$ system for which glass forming regions are well-known. Long-term tests of vitrified LILW have been carried out in a shallow ground experimental repository since 1987[20]. These show a low and diminishing leaching rate of radionuclides. Boron-free aluminosilicate glasses in the Na_2O-CaO-Al_2O_3-SiO_2 system for immobilisation of institutional and RBMK wastes were also produced from waste, sandstone and loam clay (or bentonite).

Some liquid waste streams contain sulphate and chloride ions which because of the low sulphate and chloride solubility (~1%) in silicate and borosilicate melts, limits the waste oxide content to 5-10 wt% and LILW vitrification becomes inefficient. Excess sulphates/chlorides segregate as separate phases floating on the melt surface due to the immiscibility of silicate and sulphate (chloride) melts. The same phenomenon occurs for molybdate- and chromate-containing wastes, where the separate phase is coloured and named yellow phase[19]. Vitrification of this waste can be done by using vigorous melt agitation followed by rapid cooling to the upper annealing temperature to fix the dispersed sulphate-chloride phase into the host borosilicate glass. Sulphate-chloride-containing GCM have only a slightly diminished chemical durability compared to sulphate-chloride free aluminosilicate and borosilicate glasses (see Table III); sufficiently high for them to be used for waste immobilisation. GCM produced using a thermochemical technique based on exothermic self-sustaining reactions are also composed of vitreous and crystalline phases, mainly silicates and aluminosilicates[21].

DURABILITY OF GLASSY WASTEFORMS

A set of standard tests to determine the water durability of vitrified waste and other wasteforms was developed at the Materials Characterization Centre (MCC) of the Pacific Northwest National Laboratory, USA. These MCC tests are now the internationally-approved standards used worldwide. The most important tests are given in Table IV[22].

Table IV. Standard Tests of Immobilisation Reliability.

Test	Conditions	Use
ISO 6961, MCC-1	Deionised water. Static. Monolithic specimen. Sample surface to water volume (S/V) usually 10 m^{-1}. Open to atmosphere. Temperature 298 K (25 °C) for ISO test, 313 K (40 °C), 343 K (70 °C) and 363 K (90 °C) for MCC-1 test	To compare different waste forms.
MCC-2	Deionised water. Temperature 363 K (90 °C). Closed.	Same as MCC-1 but at high temperatures.
PCT (MCC-3)	Product consistency test. Deionised water stirred with glass powder. Various temperatures. Closed.	For durable waste forms to accelerate leaching.
SPFT (MCC-4)	Single pass flow through test. Deionised water. Open to atmosphere.	The most informative test.
VHT	Vapour phase hydration. Monolithic specimen. Closed. High temperatures.	Accelerates alteration product formation.

As the cost saving incentive is to increase the waste loading in a wasteform the optimal glassy wasteform compositions are tailored as a compromise between waste loading and final glass durability accounting also for processing parameters on vitrification[7]. Table V gives typical data on parameters of HLW borosilicate and phosphate glasses[5,23].

Table V. Typical Properties of HLW Glasses.

Glass	Density g/cm^3	Compressive strength, MPa	NR, 28-th day, in 10^{-6} g/cm^{2}d	Thermal stability[1], K (°C)	Damaging dose[2], Gy
Borosilicate	2.7	22 – 54	0.3 (Cs); 0.2 (Sr).	≥ 823 (550)	>10^9
Phosphate	2.6	9 – 14	1.1 (Cs); 0.4 (Sr).	≥ 723 (450)	>10^9

[1]Thermal stability is the temperature above which the radionuclide normalised leaching rates increase >10^2 times. [2]The irradiation has a small impact on glasses and the damaging dose is the absorbed dose above which the radionuclide normalised leaching rates increase several (2 – 5) times.

Vitrified radioactive waste is chemically durable and the glass reliably retains radioactive species. Typical normalised leaching rates of vitrified waste forms are below 10^{-5} – 10^{-6} g/cm^2 day. Corrosion durability of vitrified waste is the most important acceptance parameter for disposal[24]. The release of radioactive species, which in nuclear waste glasses are invariably cations, can be caused by corrosion of the glass in contact with groundwater. However, the potential contact of water with glass is deferred in actual disposal systems to times after the waste container has been breached. For vitrified HLW containers, which are made of stainless steel, these times are expected to be of the order of many hundreds or even thousands of years. High temperatures and radiation dose rates are likely only for the first few hundred years after HLW vitrification so that container temperatures will be close to those of the ambient rock by the expected time of contact with groundwater. Moreover the role of βγ-radiolysis will also become negligible because of low radiation dose rates. Vitrified LILW is almost invariably at the ambient temperature of a repository environment. In addition this type of waste is expected to be disposed of in near-surface repositories which are often characterised by near-neutral groundwaters and relatively low host rock temperatures. Hence the temperatures of nuclear waste glasses at the times of expected contact with groundwater are likely to be close to those of the surrounding repository environment.

Aqueous corrosion of nuclear waste glasses is a complex process which depends on many parameters such as glass composition and radionuclide content, time, temperature, groundwater chemical composition and pH. Corrosion of silicate glasses, including nuclear waste borosilicate glasses, involves two major processes – diffusion-controlled ion exchange and glass network hydrolysis. Diffusion-controlled ion exchange reactions lead to selective leaching of alkalis and protons entering the silicate structure to produce a hydrated alkali-deficient layer on the glasses. Hydrolysis being a near-surface reaction of hydroxyl ions with the silicate network leads to its destruction resulting in congruent dissolution of glass constituents and subsequent precipitation of hydrous silica-gel layers as secondary alteration products. Table VI summarises the three most important mechanisms of glass corrosion in non-saturated aqueous solutions[20].

Table VI. Main Characteristics of Corrosion Mechanisms.

Mechanism, rate behaviour	Instantaneous surface dissolution	Ion exchange	Hydrolysis
Time[1]	Short-term effect $\propto \exp(-kt)$	Diminishes $\propto t^{-1/2}$	Independent[1]
Temperature	Arrhenian	Arrhenian, Universal activation energy	Arrhenian, One high activation energy
pH	Dependent	Decreasing $\propto 10^{-0.5pH}$	Increasing $\propto 10^{0.5pH}$
Saturation effects[2]	Unlikely	Unlikely	Impeded $\propto (1-C_{Si}/C_{Si\ saturation})$
Selectivity	Selective	Selective	Congruent

[1]Time behaviour may be affected by saturation effects; [2]Changes in solution chemistry may affect solution pH.

Ion-exchange occurs preferentially in acidic and neutral solutions but diminishes quickly with increase of pH, whereas hydrolysis occurs preferentially in basic solutions but diminishes quickly with decrease of pH. It is normally considered that for pH < 9 – 10 ion exchange dominates glass corrosion whereas hydrolysis reactions are significant when pH exceeds 9[25,26]. In acidic media below pH=6 the water concentration of protons (or hydronium ions) is high resulting in a high rate of ion exchange. The role of glass network dissolution in this process is insignificant and cation leaching is ion-selective with different leaching rates for different cations. Above pH=9 the role of ion exchange becomes insignificant due to the high water concentration of hydroxyl ions and thus the glass network begins to dissolve rapidly. In such basic media the release of cations becomes congruent as destruction of the glass network results in practically complete dissolution of all glass constituents. The pH dependence of corrosion rate has a U-form curve with typical minimal changes in the near-neutral water solutions. Fig. 5 illustrates the pH dependence of corrosion rate for borosilicate glasses.

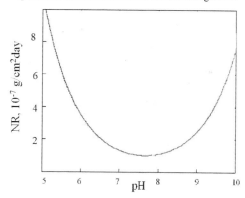

Figure 5. pH dependence of Cs normalised leaching rate calculated for glass K-26[20].

The mass release from glass follows simple power laws only below pH=6 and above pH=9 whereas in the interval 6<pH<9 the dependence is a more complex function with a changing slope when pH changes and with minimal corrosion rates achieved close to but not at pH=7. Because of the time dependence of ion-exchange rates in corroding glasses the minimum rates drift with time to lower values of pH. Therefore attempts to model the pH dependences by simple power laws separated at pH=7 will inevitably result in smaller values of exponent terms m and η. For example, the exponent terms for UK Magnox waste glass based on data from the pH ranges 2<pH<7 and 7<pH<10 were m=0.39 for boron, m=0.43 for silicon, and η=0.43[27] which are somewhat smaller than the theoretical value of m=0.5.

The ion exchange reaction of glass with water leads to gradual diminution of cation content in the near-surface glass layers. Because of this depletion in the glass near-surface layers over time the rate of ion-exchange reduces. In contrast the rate of glass hydrolysis, although small in near-neutral conditions, remains constant. Hence hydrolysis will eventually dominate, once the near-surface glass layers have become depleted in cations. Depending on glass composition and the conditions of aqueous corrosion, as well as on time, the contribution of the basic mechanisms to the overall corrosion rate can be different. E.g. in dilute solutions ion exchange controls the initial corrosion stage. Moreover at expected disposal temperatures (below several tens of $^\circ$C) the corrosion of glasses will occur via ion exchange for very long times even in contact with non-silica saturated groundwater, although ion exchange controls corrosion of glasses over geological timescales when the contacting groundwater is silica-saturated and the hydrolytic dissolution of the glass network is impeded.

The time required for silicate glasses to reach the hydrolysis stage in near-neutral solutions depends mainly on glass composition and temperature assuming unchanged water parameters so there is no coupling between water chemistry and the corroding glass. More highly polymerised glasses are hydrolytically decomposed more slowly. Thus glasses with higher silica contents require longer times before hydrolysis becomes dominant compared to high sodium content glasses. The corrosion regimes of silicate glasses should be characterised in terms of time-temperature parameters as the higher the temperature the sooner hydrolysis becomes dominant. As corrosion progresses the impact of hydrolysis becomes significant with comparable contributions from both ion exchange and hydrolytic reactions. Finally, glass corrosion in deionised water is fully controlled by hydrolysis.

CONCLUSIONS

The physical and chemical durability of glasses combined with their high tolerance to compositional changes makes glasses irreplaceable when highly toxic wastes such as long-lived and highly-radioactive wastes need reliable immobilisation for safe long-term storage, transportation and consequent disposal. Vitrification is attractive because of its flexibility, the large number of elements which can be incorporated in the glass, its high corrosion durability and the reduced volume of the resulting wasteform. Hazardous waste constituents are immobilised either by direct incorporation into the glass structure or by encapsulation when the final glassy material can be in the form of a glass composite material (GCM). Glasses and GCM are highly corrosion resistant, their high nuclide retention is expected to last for many millennia.

REFERENCES

1. C.M. Jantzen, K.G. Brown and J.B. Pickett. Durable glass for thousands of years. *International Journal of Applied Glass Science* **1**, 38–62 (2010).

2. I.W. Donald. *Waste Immobilisation in Glass and Ceramic Based Hosts.* Wiley, Chippenham (2010).

3. D. Caurant, P. Loiseau, O. Majérus, V. Aubin Chevaldonnet, I. Bardez and A. Quintas. *Glasses, Glass-Ceramics and Ceramics for Immobilization of Highly Radioactive Nuclear Wastes*. Nova, New York (2009).

4. M.I. Ojovan and W.E. Lee. *New Developments in Glassy Nuclear Wasteforms*. Nova, New York (2007).

5. W.E. Lee, M. I. Ojovan, M.C. Stennett and N.C. Hyatt. Immobilisation of radioactive waste in glasses, glass composite materials and ceramics. *Advances in Applied Ceramics*, **105**, 3-12 (2006).

6. M.I. Ojovan and W.E. Lee. *An Introduction to Nuclear Waste Immobilisation*. Elsevier, Amsterdam (2005).

7. I.L. Pegg and I. Joseph. Vitrification. In *Hazardous and Radioactive Waste Treatment Technologies Handbook*, by C. Ho Oh, 4.2.1-27. CRC Press, Boca Raton (2001).

8. I.A. Sobolev, S.A. Dmitriev, F.A. Lifanov, A.P. Kobelev, S.V. Stefanovsky and M.I. Ojovan. Vitrification processes for low, intermediate radioactive and mixed wastes. *Glass Technology*, **46**, 28-35 (2005).

9. V. Friedrich, J.B. Morris, G. Roth, D.W. Cleland, W. Baehr, W. Hebel, M. Klein, E.J. Moskal, T. Ogushi, C.T. Randall, M. Yoneya and J. Zhang. *Design and Operation of Off-gas Cleaning Systems at High Level Liquid Waste Conditioning Facilities*. IAEA TRS-291, Vienna (1988).

10. N.R. Gribble, R. Short, E. Turner and A.D. Riley. The impact of increased waste loading on vitrified HLW quality and durability. *Mat. Res. Soc. Symp. Proc.*, **1193**, 283-290 (2009).

11. J.K. Bates, A.J.G. Ellison, J.W. Emery and J.C. Hoh. Glass as a waste form for the immobilisation of plutonium. *Mat. Res. Soc. Symp. Proc.*, **412**, 57-64 (1996).

12. A.A. Vashman, A.V. Demine, N.V. Krylova, V.V. Kushnikov, Yu.I. Matyunin, P.P. Poluektov, A.S. Polyakovand E.G. Teterin. *Phosphate Glasses with Radioactive Waste*. CNIIAtominform, Moscow (1997).

13. F.A. Lifanov, M.I. Ojovan, S.V. Stefanovsky and R. Burcl. Cold crucible vitrification of NPP operational waste. *Mat. Res. Soc. Symp. Proc.* **757**, II5.13.1-6 (2003).

14. M.-J. Song. The vitrified solution. *Nuclear Engineering International*, **2**, 22-26 (2003).

15. C.M. Jantzen, J.B. Pickett and R.S. Richards. *Vitrification of simulated Fernald K-65 silo waste at low temperatures*. WSRC-MS-97-00854, Rev.1 (1997).

16. B.P. McGrail, D.H. Bacon, J.P. Icenhower, F.M. Mann, R.J. Schaef, H.T. Puigh and S.V. Mattigod. Near-field performance assessment for a low-activity waste glass disposal system: laboratory testing to modelling results. *J. Non-Cryst. Solids*, **298**, 95-111 (2001).

17. R. Alvarez. Reducing the risks of high-level radioactive wastes at Hanford. *Science and Global Security*, **13**, 43-86 (2005).

18. J. D. Vienna, P. Hrma, A. Jiricka, D. E. Smith, T. H. Lorier, R. L. Schulz and I. A. Reamer. Hanford immobilized LAW product acceptance testing: Tanks focus area results. PNNL-13744 (2001).

19. J.A.C. Marples. The preparation, properties, and disposal of vitrified high level waste from nuclear fuel reprocessing. *Glass Technol.*, **29**, 230-247 (1988).

20. M.I. Ojovan, R.J. Hand, N.V. Ojovan and W.E. Lee. Corrosion of alkali-borosilicate waste glass K-26 in non-saturated conditions. *J. Nucl. Mater*. **340**, 12-24 (2005).

21. M.I. Ojovan and W.E. Lee. Self sustaining vitrification for immobilisation of radioactive and toxic waste. *Glass Technology*, **44** (6) 218-224 (2003).

22. D.M. Strachan. Glass dissolution: testing and modelling for long-term behaviour. *J. Nucl. Mat.*, **298**, 69-77 (2001).

23. N.V. Krylova and P.P. Poluektov. Properties of solidified forms of high level wastes as one of barriers in the disposal system. *At. Energy*, **78**, 93-98 (1995).

24. C.M. Jantzen, D.I. Kaplan, N.E. Bibler, D.K. Peeler and M.J. Plodinec. Performance of a buried radioactive high level waste (HLW) glass after 24 years. *J. Nucl. Mater.*, **378**, 244-256 (2008).

25. B.M.J. Smets and M.G.W. Tholen. The pH dependence of the aqueous corrosion of glass. *Phys. Chem. Glasses*, **26**, 60-63 (1985).

26. W.L. Ebert. The effect of the leachate pH and the ratio of glass surface area to leachant volume on glass reactions. *Phys. Chem. Glasses*, **34** 58-65 (1993).

27. P.K. Abraitis, F.R. Livens, J.E. Monteith, J.S. Small, D.P. Triverdi, D.J. Vaughanand R.A. Wogelius. The kinetics and mechanisms of simulated British Magnox waste glass dissolution as a function of pH, sicilic acid activity and time in low temperature aqueous systems. *Applied Geochemistry*, **15**, 1399-1416 (2000).

ADVANCES IN MATERIALS CORROSION RESEARCH IN THE YUCCA MOUNTAIN PROJECT

Raul B. Rebak
GE Global Research
1 Research Circle, Schenectady, NY 12309, USA

ABSTRACT

Several countries are currently considering disposing nuclear waste in stable geologic repositories. Engineered barriers are planned between the waste and the geologic formation. One of these barriers is the container for the waste. In most of the planned repositories the proposed materials for the container are carbon steel and copper. For the repository that was planned in the US, some of the most corrosion resistant commercially available alloys, such as nickel and titanium, were characterized. More than 20 years of research in the US program helped to understand the environmental behavior of the materials. For example, immersion corrosion testing determined that alloy C-22 was free of crevice corrosion after more of 9 years immersion in highly concentrated acidified aerated ground water while alloy 825 suffered crevice corrosion in similar conditions. Electrochemical testing showed the effect of chloride concentration, temperature, applied potential and presence of inhibitors on the crevice corrosion resistance of C–22. Several oxyanions such as nitrate, sulfate, phosphate and carbonate were identified as inhibitors of the crevice corrosion induced by chlorides.

INTRODUCTION

Recent concerns about global warming and the release of greenhouse gases by the fossil fuel power industry have re-ignited the consideration of alternative sources of energy such as wind, solar, fuel cells and nuclear power. In the last decade, many reports have appeared discussing a nuclear renaissance. But is there a nuclear renaissance? Figure 1 shows the number of new commercial nuclear power reactors connected to the grid for the last 50 years according to the International Atomic Energy Agency. [1] The highest number of commercial reactors connected was 33 a year in 1984 and 1985. Since then this number has decreased markedly and for the last decade the average annual number of new reactors was only 3.3. Figure 1 also shows that, for the first time in 50 years, there were no new reactors connected to the grid in 2008. Figure 2 shows that in 2010 there were more than 50 new nuclear power reactors under construction in the world, the majority of them in Asia. [1] During 2007-2009 there were 26 applications to the Nuclear Regulatory Commission to build new nuclear power reactors in the United States. [2] However; there may be a period of 10 years between the application to the operation of a new reactor. [3]

NUCLEAR WASTE DISPOSITION

Even though nuclear power has been used for more than 60 years, the issue of the toxic radioactive waste generated during energy production still needs to be resolved. Radioactive materials are not only pertinent to nuclear power since they are also used worldwide in other fields including medical applications and production of weapons. Once the radioactive materials lose their commercial value, they are considered radioactive waste, and they need to be isolated from the environment until the radioactive decay has reduced its toxicity to innocuous levels for plants, animals, and humans. Different types of radioactive waste are produced during commercial and defense nuclear fuel cycles. One type of waste, denoted high-level waste (HLW), contains the highest concentration of radiotoxic and heat-generating species. Because of this factor, the most stringent standards for disposing of radioactive wastes are being placed worldwide on HLW, and the majority of the radioactive waste management effort is being directed toward the HLW problem. One of the most common and most voluminous types of HLW is the spent fuel (SF) from commercial nuclear reactors for power generation.

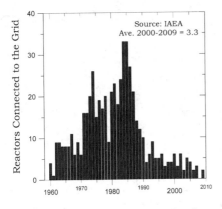

Figure 1. New Reactors Connected to the Grid since 1960

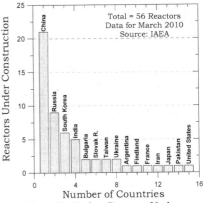

Figure 2. Nuclear Reactors Under Construction in 2010

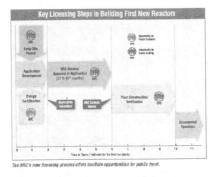

Figure 3. Timeline from Application to Operation of a New Power Reactor[3]

NATIONAL PROGRAMS FOR NUCLEAR WASTE DISPOSAL

More than 30 countries are currently studying the options for disposing of HLW in deep stable geologic formations, which will be the primary barrier for accomplishing this isolation. It is postulated that by the very nature of these geological sites, they will contain the waste for long times, limiting their spread, for example, through water flow. All the repository designs also plan to delay the release of radionuclides to the environment by the construction of engineered barrier systems (EBS) between the waste and the geologic formation. These barriers will be installed to limit water reaching the repository and to restrict radionuclide migration from the waste. The principal engineered component in this multibarrier approach is the waste package, which includes the waste itself, possibly a stabilizing matrix for the waste such as glass, and a metallic container that encloses the wasteform. Beyond the metallic containers, other secondary barriers could be added to attenuate the impact of the emplacement environment on the containers. The secondary barriers may include a drip shield such as in the USA design or a backfilling with bentonite such as in the Canadian and other designs (Table I).[4,5]

Table I. Plans for Nuclear Waste Repositories for Selected Countries

Country	Possible environment, host rock	Scheduled start operations	Container materials being studied
Belgium	Reducing or anoxic, clay	2035	Carbon steel – cement – stainless steel
Canada	Reducing or anoxic, granite	Unspecified	Carbon steel insert - Copper
Finland	Reducing or anoxic, granite	2020	Cast iron insert - Copper
France	Reducing or anoxic, clay	Unspecified	Carbon steel
Germany	Salt dome layer	2030	Carbon steel
Japan	Reducing or anoxic, granite + bentonite buffer	Late 2030s	Carbon steel, titanium
Sweden	Reducing or anoxic, granite + clay	2023	Cast iron insert - Copper
Switzerland	Reducing or anoxic, clay	Unspecified	Carbon steel
US of America	Oxidizing, non-saturated, volcanic tuff	Initially 1998, later 2020, now on hold	Ni-Cr-Mo Alloy C-22, Titanium Gr 7, 28 & 29

Twenty years ago most of the repository designs specified lifetimes from 300 to 1000 years. Currently, the minimum length of time specified for some repositories has increased to 10,000, 100,000 and even to 1,000,000 years.[6,7] The viability of extrapolating degradation data from short term testing to long time performance has been debated for some time and addressed by some investigators and the American Society for Testing and Materials (ASTM). [8]

One of the most advanced studies for a repository corresponded to the USA, which was planning to locate its nuclear waste at a remote desert site in Nevada. [9] However, since March 2010 the Yucca Mountain Project (YMP) may be on indefinite hold. See next section on the Yucca Mountain Project. The United Kingdom has also recognized that the geological disposal of the waste in a mined repository is the best available approach. In June 2008 the UK government issued the white paper "Managing Radioactive Waste Safely," where a framework for implementing geological disposal is outlined. The location for the repository will be defined through geological screening and community engagement. To complement the permanent repository studies, the UK Committee for Radioactive Waste Management has also recommended a robust program of interim storage.

The Finnish repository will be located in crystalline bedrock at Olkiluoto island on the western coast of Finland. The waste containers will be made using nodular cast iron with a 50 mm thick over pack of copper.[10] The repository in Finland should start operations in 2020 and will continue for approximately 100 years.

Sweden has elected a site for their underground repository in the municipality of Östhammar, 500 m below ground in crystalline rock.[11] The waste will packed in cast iron baskets inside thick copper canisters, surrounded by bentonite clay. Each container is a double walled cylinder of approximately 1 m diameter and 5 m long. The Swedish repository is scheduled to open in 2023 and it is designed to contain the waste for 100,000 years.

The Japanese Final Disposal Plan calls for a repository that will start operating in the late 2030s.[12] The final site for the repository has not been selected yet but two underground research laboratories have been selected, one 1000 m-deep in crystalline rock in the presence of fresh water and the second in sedimentary rock, 500 m-deep in the presence of saline water. In the final Japanese repository the metal containers will be surrounded by bentonite buffer material. The final material for the container has not been selected yet, but it is reported that a thick steel container surrounded by a bentonite buffer overpack would be a robust design.

The waste disposal for the French nuclear industry has been outlined in the document Dossier 2005 and calls for the commissioning of a disposal facility by the year 2025.[13,14] An important concept in the design of the French repository is its reversibility or design evolution at all steps for at least 100 years. The cylindrical containers for the high level waste (type C in a glass matrix) will be made of standard steel 5 cm thick, 60 cm diameter and approximately 1.5 m long. It is estimated that this container will remain leak-proof for 4,000 years. The spent fuel container may have a wall thickness of over 10 cm and last 10,000 years.

Germany is exploring the possibility of a repository in salt dome at Gorleben probably using steel containers for the nuclear waste. The waste repository should be stable for a million years and the containers should be retrievable for the entire operation time.

Most nations around the world have specified common materials for their nuclear waste containers, notably carbon steel or copper. The Swedish, Finnish and Canadian repositories have copper as their outer barrier. In the deep anoxic dilute aqueous environments, copper should stable or immune to corrosion according to the Pourbaix diagram.

The repositories in Finland and Sweden seem the most advanced currently, with a strong support of the local communities. It is apparent that nuclear waste repositories may not be constructed and operated anywhere if the public has the perception that they are unsafe.

THE YUCCA MOUNTAIN PROJECT

The Nuclear Waste Policy Act of 1982 created the Office of the Civilian Radioactive Waste Management (OCRWM) within the Department of Energy (DOE) and established that OCRWM-DOE would be responsible for finding a site, building, and operating an underground nuclear waste disposal facility. The Act of 1982 also established the Nuclear Waste Fund, by which consumers of nuclear generated electricity contribute one-tenth of a cent/kilowatt-hr into the waste fund or about $750 million per year. Between 1982 and 1987 OCRWM-DOE investigated several sites for the repository including Hanford, Washington, Deaf Smith County, Texas and Yucca Mountain, Nye County, Nevada. In 1987 the US Congress amended the Act of 1982 and designated Yucca Mountain to be the only place that will be characterized for the future US nuclear waste repository. The characterization studies were divided in several steps called (1) Viability Assessment (1998), (2) Site Recommendation (2002) and License Application (2008).

After more than two decades of scientific investigations, the US Department of Energy submitted a license application to build the repository on 03-Jun-2008 and the Nuclear Regulatory Commission (NRC) accepted this application on 08-Sep-2008.[15] It was anticipated that after a formal review process of three to four years construction would start in late 2011, and the first waste emplacement would not occur until 2020.[16] On 01-Feb-2010 the Las Vegas Review Journal reported that the funding for the Yucca Mountain Project will be zeroed out. On 03-March-2010, DOE officially filed a motion with the NRC Atomic Safety and Licensing Board to withdraw with prejudice the License Application to build the repository. In January 2010 the Blue Ribbon Commission on America's Nuclear Future was established within DOE to conduct a comprehensive review of policies for managing the back end of the nuclear fuel cycle, including all alternatives for the storage, processing, and disposal of civilian and defense used nuclear fuel, high-level waste, and materials derived from nuclear activities. The Commission will produce a final report with its recommendations in January 2012.

MATERIALS RESEARCH IN THE YUCCA MOUNTAIN PROJECT

The design of the waste package for the Yucca Mountain repository has evolved in the last fifteen years. In previous versions of the container, a thick layer of carbon steel was specified for the outer shell of the container and a corrosion resistant material as the inner shell. However, since 1998, the design of the engineered barriers has not changed significantly and it specified a double walled cylindrical container covered by a titanium alloy drip shield. The outer shell of the container would be a Ni-Cr-Mo alloy (N06022) (Table II), with an inner shell of nuclear grade austenitic Type 316 stainless steel (S31600). The function of the outer barrier was to resist corrosion and the function of the inner barrier was to provide mechanical strength and a shield to radiation. The drip shield would be made of Ti Gr 7 and a higher strength Ti alloy (Ti Gr 29) would be used for the internal ribs of the shields. The function of the drip shield was to deflect rock fall and early water seepage on the container.

At this moment the Yucca Mountain Project for the permanent US Nuclear Waste Repository is on hold until the courts and the Blue Ribbon Commission decide what is going to be the next step. However, even though the Yucca Mountain Project may not be materialized as initially planned, much knowledge has been gained through twenty years of meticulous research and documentation. For example, in the area of corrosion behavior or materials for the waste package (container and drip shield) there was a large increase in the knowledge mainly for the alloys Hastelloy C-22 and Ti Gr 7. The YMP was supported by DOE funded scientific research in several national laboratories including Lawrence Berkeley, Lawrence Livermore National Laboratory (LLNL), Los Alamos National Laboratory, Sandia National Laboratories and the U.S. Geological Survey. Each laboratory was specialized in different areas of the project, and LLNL has historically been the primary site for corrosion testing, phase stability and metallurgical studies. DOE also funded materials research at the University of Nevada Reno. At the same time, other institutions were also performing research in the area of materials and corrosion, notably including the Southwest Research Institute – Center for Nuclear Waste

Regulatory Analyses (SwRI-CNWRA) funded by NRC, the Electric Power Research Institute and the Argentina Commission of Atomic Energy. One group that deserves a special mention was the cooperative started by Prof. Joe Payer from Case Western Reserve University. Prof. Payer was able to obtain funds from the OCRWM Office of Science and Technology to organize a large consortium of universities that would perform basic research to understand the fundamental mechanisms of materials degradation and to give confidence to the information already obtained by YMP. The Payer group was called Corrosion and Materials Performance Cooperative and included Arizona State University, Case Western Reserve, the University of Virginia, Pennsylvania State University, The Ohio State University, The University of Western Ontario, the University Of Toronto, the University of Minnesota, and the University of California Berkeley.

Most if not all research was regarding the material of the outer barrier of the container (Hastelloy C-22) and almost no research on Ti alloys or other metals. And most of the research on Hastelloy C-22 was on crevice corrosion since very little attention was devoted to general passive corrosion and to environmentally assisted cracking. Localized corrosion was always considered a mechanism that could limit the lifetime performance of the containers. Also, the area of crevice corrosion initiation and repassivation was incredibly rich since there were a myriad of interlinked affecting variables to explore.

MATERIALS RESEARCH AT LAWRENCE LIVERMORE NATIONAL LABORATORY

In the time period 2002-2006 extensive materials research has been conducted at Lawrence Livermore National Laboratory (LLNL) to support the DOE Yucca Mountain repository. A large variety of alloys were tested; however, most of the work was on Hastelloy C-22 and Ti Grade 7, the materials for the outer barrier of the container and the drip shield, respectively. Very little research has been conducted regarding the resistance of alloy C-22 to stress corrosion cracking (SCC). However, LLNL was the first laboratory to report stress corrosion cracking of C-22 under anodic applied potential in simulated concentrated water using the slow strain rate test technique. [17, 18] LLNL materials research was mainly concentrated in two areas; thermal stability and corrosion performance. The corrosion characterization included both standard immersion test for weight loss as well as U-bend specimens for SCC pass / does not pass criteria, and electrochemical testing. The corrosion testing was carried out to evaluate the response to three mechanisms of degradation: (1) general or uniform corrosion, (2) localized corrosion such as crevice corrosion, and (3) environmentally assisted cracking. A short summary is given for three major areas where LLNL contributed significantly to the advance of corrosion testing: (a) The long-term corrosion test facility (LTCTF), (b) Long-term monitoring of the corrosion potential (E_{corr}), and (c) Measurement of the repassivation or critical potential (E_{crit}) for crevice corrosion.

The Long Term Corrosion Test Facility (LTCTF)

The LTCTF started its operations in 1996 to provide corrosion-engineering data on a variety of candidate materials for the fabrication of the nuclear waste package. It was designed to allow an evaluation of all the forms of corrosion and to permit successive withdrawals of test specimens to obtain corrosion rates and behavior over different time intervals. It initially included 28 polymeric vessels containing six different waters at different temperatures that were prepared to simulate the bounding chemical and thermal conditions in the repository. Each vessel contained up to six racks, and each rack could hold up to 246 specimens. More than 20,000 specimens were exposed in the vessels during the decade long experiment. The tested materials included corrosion allowance materials (e.g. carbon steels, alloy steels), intermediate corrosion resistant alloys (copper-based), and corrosion resistant alloys (nickel and titanium alloys) (Table II). Type of specimens included weight (mass) loss, localized (crevice) corrosion, stress corrosion cracking and galvanic corrosion. The specimens were welded and non-welded. Many specimens were designed to determine the amount of hydrogen absorbed by titanium alloys when coupled to carbon steel. The volume of each vessel was approximately 2,000 L and they were filled with 1,000 L of electrolyte solution, that is about half of the specimens were exposed to the vapor phase and half to the liquid phase. There were three types of electrolytes in the vessels, all variations of the basic composition of well water (J-13) at the Yucca Mountain site. The aqueous solutions were: (1) simulated dilute water (SDW), which is 10 times more concentrated than J-13, pH ~ 10, (2) Simulated concentrated water (SCW), which is 1000 times more concentrated than J-13 water, pH ~ 10, (3) Simulated acidified water (SAW), which is 1000 times more concentrated than J-13 water and it is acidified to pH 2.8 to simulate metal hydrolysis and microbial activity, and (4) Simulated cement modified water (SCMW), which would represent water in contact with concrete. The ingress of air to the vessels was not restricted, that is, the electrolytes were naturally aerated. The testing temperatures were 60°C and 90°C. The temperature, pH and volume of the electrolyte was monitored as a function of time; however the E_{corr} or redox potential of the system was not monitored. The operation of the LTCTF started in the time interval September 1996 to September 1997 and all the vessels were shut down and the specimens removed, catalogued and stored in August to September 2006. The longest exposure time was approximately 9-10 years. Most of the 20,000 coupons removed from the vessels were never examined, data not acquired nor analyzed. [19]

Table II. Alloys that were tested in the LTCTF at LLNL 1996-2006

Prefix Letter	Alloy - ASTM	UNS	Approx. Composition
A	Incoloy 825 B 424	N08825	42Ni + 22Fe + 21Cr + 3Mo + 2Cu + 1Ti
B	Hastelloy G-3 B 582	N06985	Ni + 22Cr + 19Fe + 7Mo + 2Cu + 5Co*
C	C-4 B 575	N06455	65Ni + 16Cr + 16Mo + 3Fe* + 2Co*
D	Hastelloy C-22	N06022	57Ni + 22Cr + 13Mo + 3W + 5Fe*
E	Ti Gr 12 B 265	R53400	Ti + 0.8Ni + 0.3Mo + 0.3Fe*
F	Ti Gr 16 B 265	R52402	Ti + 0.05Pd
G	Monel 400 B 127	N04400	67Ni + 31Cu + 2.5Fe + 2Mn
H	CDA715 B 171	C71500	70Cu + 30Ni
I	Alloy Steel A387 Gr 22	K21590	Fe + 2.25Cr + 1Mo + 0.5Si*
J	Carbon Steel A516 Gr 55	K01800	Fe + 0.8Mn + 0.25C + 0.2Si
K	Cast Steel A27 Gr 70-40	J02501	Fe + 1Mn* + 0.35C*
L	Inconel 625 B 443	N00625	58Ni + 21Cr + 9Mo + 3.5(Nb+Ta) + 5Fe*
M	Inconel 686 B 575	N06686	Ni + 21Cr + 16Mo + 4W
N	Ti Gr 7 B 265	R52400	Ti + 0.15Pd

* Maximum

Results from the LTCTF showed that none of the five nickel alloys listed in Table II suffered stress corrosion cracking under the tested conditions. [19]

Flat coupons from the LTCTF were used to obtain general corrosion of alloy C-22. General corrosion (or passive corrosion) is the uniform thinning of the container alloy at its rest, open circuit potential or corrosion potential (E_{corr}). In the presence of aerated multi-ionic brines, such as those that may be present at the repository site, alloy C-22 is expected to remain passive at its E_{corr}. The passive corrosion rates of alloy C-22 after 5 years immersion in multi-ionic solutions simulating concentrated ground waters from pH 2.8 to 10 are extremely low and in the order of 10 nm/year. [4] This low corrosion rate was measured at 60°C and 90°C for welded and non-welded alloy C-22 at an E_{corr} range from −100 mV to +400 mV SSC (saturated silver chloride electrode). The low corrosion rates or passive behavior of Alloy C-22 is because of the formation of a protective inner chromium rich oxide film between the alloy (metal) and the surrounding electrolyte. This passive film is even stable in the presence of strong mineral acids at temperatures below 60°C. It has been shown that the thickness of this passive film formed in concentrated hot electrolyte solutions could be only in the range of 5 to 6 nm. The long-term extrapolation of the corrosion rate of Alloy C-22 has been modeled considering that the dissolution rate is controlled by the injection of oxygen vacancies at the oxide film/solution interface. From this modeling it has been concluded that it is unlikely that catastrophic failure of the container may occur due to long-term passive film dissolution. That is, the passive dissolution of Alloy C-22 was not considered to be the limiting factor for the life performance of the waste container.

The Long Term Monitoring of the Corrosion Potential

In order to determine the susceptibility of an alloy to localized corrosion such as crevice corrosion, two parameters are important. One is the potential at which the alloy may suffer localized corrosion (critical potential or E_{crit}) and the other is the rest or corrosion potential (E_{corr}). If the rest potential E_{corr} is equal or higher than E_{crit} then the alloy may suffer localized corrosion under natural conditions. That is, if

$$E_{crit} - E_{corr} \rangle 0 \qquad \text{no localized corrosion}$$

Therefore it was necessary to build a bank of data on the corrosion potential behavior of C-22 in a variety of aqueous electrolytes both as a function of the solution composition (mainly chloride and nitrate composition) and the temperature. In the development of localized corrosion models, the data on E_{corr} was later compared with the values of E_{crit} obtained not only at LLNL but also in numerous universities and other research institutions. The monitoring of E_{corr} for times as long as 3 years was a unique feature of LLNL. From 2001 to 2006, 37 bench top cells were installed at LLNL to monitor the corrosion potential of C-22 and platinum as a function of time. [19] The specimens were immersed in a variety of electrolytes while the open circuit potential or free corrosion potential (E_{corr}) was acquired every hour. All the electrolytes were aerated by flowing air above the level of the solution in the cells. The air exited the cells through a condenser that avoided the evaporation of the solutions. The electrolyte solutions included highly concentrated 18 molal $CaCl_2$ + 9 molal $Ca(NO_3)_2$ at 155°C and 0.1 molar oxalic acid at ambient temperature. C-22 was tested under different metallurgical and surface conditions including non-welded rods, as-welded rods, as-welded plus high-temperature aged (173 h at 700°C) rods, and high-temperature air-oxidized (20 min at 1,121°C) specimens.

Measuring the Repassivation Potential for Crevice Corrosion

C-22 is highly resistant to pitting corrosion but it may suffer crevice corrosion, generally at anodic applied potentials and at temperatures higher than 70°C. There are a large number of interconnected factors affecting the susceptibility of C-22 to crevice corrosion. [20] They can be divided into (1) Internal or metallurgical factors and (2) External or environmental factors. The internal factors include: wrought vs. welded microstructure, presence of second phase precipitates, thermal exposure treatment including temperature and time, surface stress mitigation such as laser shock peening or burnishing. The external factors include: chloride concentration, temperature, applied potential, presence of inhibitors such as nitrate, sulfate, carbonate, etc., volume of electrolyte (e.g. fully immersed vs. thin film), crevice former geometry or tightness of the crevice gap, type of crevicing material, etc.

There are various methods for measuring the repassivation potential of alloy C-22, including the traditional cyclic potentiodynamic polarization (CPP in ASTM G 61) and the THE method developed as part of the YMP, which is now the ASTM standard G 192.

Figure 4 shows that repassivation potential for alloy C-22 decreased 300 mV when the chloride concentration increased from 0.0005 M to 1-4 M, both at 60°C and 90°C. [21] The same relationship between the repassivation potential and the chloride concentration occurred at 60°C and at 90°C; however, the repassivation potential at each chloride concentration was approximately 100 mV lower at 90°C than at 60°C. Figure 5 shows that the repassivation potential for alloy C-22 in 5 M CaCl₂ solution decreased as the temperature increased from 30 to 120°C. Above 75°C there is a linear relationship between the repassivation potential and the temperature using the CPP method, showing that at T<75°C it was difficult to obtain a reproducible value of the repassivation potential using CPP. However, using the THE method the relationship between the repassivation potential and the temperature was linear in the entire temperature tested range. The THE method applies the current to the specimen in a more gentle manner than the CPP method and therefore gives rise to more reproducible values of repassivation potential. [21]

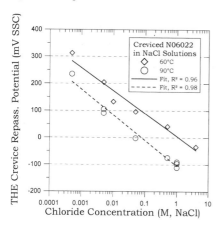

Figure 4. Effect of the chloride concentration on the repassivation potential of C-22

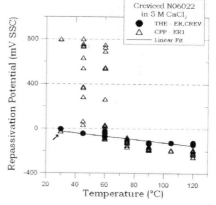

Figure 5. Effect of the temperature on the repassivation potential of C-22

There are several anions (mainly oxyanions) that were shown to inhibit the crevice corrosion process induced by chloride anions. The most effective inhibitor is nitrate since it may counter act the aggressive effect of chloride via several mechanisms. [22] Other inhibitors of crevice corrosion include sulfate, [23] carbonate/bicarbonate, [24] fluoride [25] and phosphate. [26] The presence of an inhibitor is generally stated using the ratio R = [inhibitor]/[Cl⁻]. Figure 6 shows the cyclic potentiodynamic polarization of Alloy C-22 in three different electrolyte solutions at 110°C. The electrolytes are: (1) pure 8 molal chloride (4 molal NaCl + 4 molal KCl), (R = 0) and (2) 8 molal chloride with added sodium and potassium nitrate to obtain R = 0.1 and (3) to obtain R = 0.5. Figure 6 shows that as R increased the repassivation potential increased, that is, for R = 0, E_{crit} = −210 mV SSC, for R = 0.1, E_{crit} = −50 mV SSC and for R = 0.5, E_{crit} = 337 mV SSC. For R = 0.5 there was a total inhibition of crevice corrosion (no hysteresis in the reverse scan in Figure 6). [27]

Constant potential laboratory tests have shown that crevice corrosion in Alloy C-22 often repassivates after initiation due to a stifling mechanism. [28] Figure 7 shows the crevice corroded area under one crevice former tooth (ASTM G 192) after a week long test in 3.5 m NaCl + 0.175 m KNO₃ solution (R = 0.05) at 100°C at a constant applied potential of +100 mV SSC. The test was conducted at approximately 200 mV higher than the crevice repassivation potential of Alloy C-22 (-110 mV SSC) in the same conditions. [29] Current measurements showed that the crevice corrosion nucleated 10 minutes after the potential was applied and it progressed with increasing anodic currents for the next 14 hours, after which the anodic current started to decrease becoming cathodic a the hour 79 even though the potential was maintained at +100 mV SSC during the entire test. Figure 7 also shows that crevice corrosion occurred under the entire footprint of the crevice former and that the depth of attack was shallow and even both for the base metal (lower part of the image) and the weld metal (upper part of the image). In the weld metal the attack was interdendritic and in the base metal the attack was intergranular.

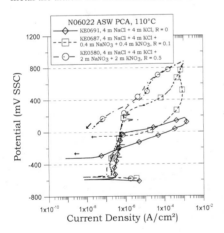

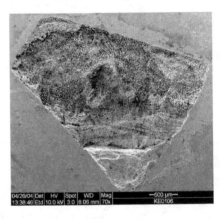

Figure 6. Effect of the addition of nitrate to a chloride solution on the repassivation potential of C-22

Figure 7. Crevice corrosion formed in alloy C-22 under constant applied potential in a Cl⁻ + NO₃⁻ solution, R = 0.05, at 100°C

Environmentally Assisted Cracking of Alloy C-22

Wrought mill annealed (MA) Alloy C-22 is highly resistant to environmentally assisted cracking (EAC) in most environments, including acidic concentrated and hot chloride solutions. Welded and non-welded U-bend specimens of Alloy C-22 and five other nickel based alloys (Table II) exposed for more than 5 years to multi-ionic solutions that represent concentrated ground water of pH 2.8 to 10 at 60°C and 90°C were free from EAC. [30] Even though Alloy C-22 is resistant to EAC in concentrated hot chloride solutions, it may be susceptible under other severe environmental conditions. Slow strain rate tests were performed using MA Alloy C-22 specimens in Simulated Concentrated Water (SCW) and other solutions as a function of the temperature and applied potential. [31] SCW has a pH 8–10, and it is approximately 1000 times more concentrated than ground water. Alloy C-22 was found susceptible to EAC in hot SCW solutions and bicarbonate plus chloride solutions at anodic applied potentials approximately 300–400 mV more positive than E_{corr}. The occurrence of EAC was related to the presence of an anodic peak in the polarization curve of the alloy in SCW environments. For example, at ambient temperatures, the peak is not present and EAC does not take place. [31] It was demonstrated that the most aggressive species for EAC in SCW was bicarbonate, but that the presence of chloride in the bicarbonate solution enhances the aggressiveness of the environment. [18]

Corrosion Behavior of Titanium Alloys

Titanium grade 7 (Ti Gr 7 or R52400) was selected to fabricate the detached drip shield for the repository in Yucca Mountain. Other Ti alloys of higher strength such as Ti Gr 29 were also going be used for the structural parts of the drip shield. The presence of the drip shield would deflect early water seepage from the containers. This drip shield would also deflect rock fall from the containers. Ti Gr 7 belongs to a family of Ti alloys especially designed to withstand aggressive chemical environments (Table II). [32] The superior corrosion resistance of Ti and Ti alloys is due to a thin, stable and tenacious oxide film that forms rapidly in air and water, especially under oxidizing conditions. The presence of fluoride in the ground water may render Ti Gr 7 more susceptible to general and crevice corrosion under anodic polarization. [33]

Weight-loss, creviced, and U-bend specimens of Ti Gr 7, 12 and 16 were exposed to three different aerated electrolyte solutions simulating concentrated ground water for over five years both at 60°C and at 90°C in the vapor and liquid phases of these solutions. [34] Ti Gr 7 generally exhibited the lowest corrosion rates irrespective of temperature or solution type while Ti Gr 12 generally exhibited the highest corrosion rates. Titanium and Ti alloys may be susceptible to environmentally assisted cracking (EAC), such as hydrogen embrittlement (HE). Embrittlement by hydrogen is a consequence of absorption of atomic hydrogen by the metal to form hydrides. This may happen in service when the Ti alloy is coupled to a more active metal in an acidic solution. A critical concentration of hydrogen in the metal may be needed for HE to occur. Results from up to 5-year immersion testing at 60°C and 90°C of U-bend specimens made of wrought and welded Ti Gr 7 and Ti Gr 16 alloys showed that these alloys were free from environmentally assisted cracking (EAC) in multi-ionic solutions that could be representative of concentrated ground water. [35] Welded Ti Gr 12 U-bend specimens suffered EAC in SCW liquid at 90°C. Under the same conditions, non-welded Ti Gr 12 was free from cracking. [19, 35]

SUMMARY AND CONCLUSIONS

The consensus around the world is that high-level nuclear waste should be deposited in stable geological repositories and several countries are currently developing these repositories. Most of the repositories in the world are planned to be in stable rock formations (e.g. granite) well below the water table (anoxic but saturated). The US was studying a repository above the water table (unsaturated but with unrestricted access of oxygen). Copper, titanium, and carbon steels were determined to be suitable materials for the reducing or anoxic repositories. High-end materials such as alloy C-22 and Ti Gr 7 were being characterized for the mostly dry and oxidizing environment of the US repository in Yucca Mountain. Yucca Mountain has been studied for nearly 25 years as the permanent repository for the US and at one time it was the most advanced developed waste repository in the globe. The construction of the Yucca Mountain repository is currently on hold waiting for the resolution of the Blue Ribbon Commission, who should issue a recommendation in early 2012. Even though the Yucca Mountain repository may not materialize, much knowledge has been gained during the last 15-years of intensive corrosion and materials testing.

REFERENCES

[1] International Atomic Energy Agency (iaea.org)

[2] US Nuclear Regulatory Commission (nrc.gov)

[3] Nuclear Energy Institute (nei.org)

[4] R. B. Rebak and R. D. McCright, "Corrosion of Containment Materials for Radioactive Waste Isolation," ASM Handbook, Volume 13C Corrosion: Environments and Industry, ASM International, 2006, pp. 421-437.

[5] D. W. Shoesmith, Corrosion, 62, 703, 2006

[6] L. O. Werme, "Fabrication and Testing of Copper Canister for Long Term Isolation of Spent Nuclear Fuel," Vol. 608, p. 77 (Materials Research Society, 2000: Warrendale, PA)

[7] P. N. Swift, K. Knowles, J. McNeish, C. W. Hansen, R. L. Howard, R. MacKinnon, and S. D. Sevougian, "Long-Term Performance of the Proposed Yucca Mountain Repository, USA," in the proceedings from the XXXII Symposium on Scientific Basis for Nuclear Waste Management, Volume 1124, Materials Research Society, 2009, pp. 3-14

[8] ASTM C1174-07 "Standard Practice for Prediction of the Long-Term Behavior of Materials, Including Waste Forms, Used in Engineered Barrier Systems (EBS) for Geological Disposal of High-Level Radioactive Waste," (ASTM International, West Conshohocken, PA, 2007).

[9] The Yucca Mountain Project (www.ocrwm.doe.gov)

[10] Nuclear Waste Management in Finland (www.posiva.fi)

[11] SKB, Technical Report TR-07-12, September 2007 (www.skb.se)

[12] Nuclear Waste Management Organization of Japan – NUMO (www.numo.or.jp)

[13] National Radioactive Waste Management Agency France (www.andra.fr)

[14] F. Plas and J. Wendling, "The Geological Research in France – The Dossier 2005 Argile," in the proceedings from the XXX Symposium on Scientific Basis for Nuclear Waste Management, Volume 985, Materials Research Society, 2007, pp. 493-504

[15] D. J. Duquette, R. M Latanision, C. A. W. Di Bella, and B. E. Kirstein, "Corrosion Issues Related to Disposal of High-Level Nuclear Waste in the Yucca Mountain Repository," in the proceedings from the XXXII Symposium on Scientific Basis for Nuclear Waste Management, Volume 1124, Materials Research Society, 2009, pp. 15-28

[16] N. J. Zacha, "Yucca Mountain: Dumped and Wasted?," Radwaste Solutions, The American Nuclear Society, July/August 2009, pp. 12-18

[17] J.C. Estill, K.J. King, D.V. Fix, D.G. Spurlock, G.A. Hust, S.R. Gordon, R.D. McCright, and R.B. Rebak, "Susceptibility of Alloy 22 to Environmentally Assisted Cracking in Yucca Mountain Relevant Environments," Paper No. 02535, Corrosion/2002, (NACE International, Houston, Texas)

[18] K.T. Chiang, D.S. Dunn, and G.A. Cragnolino, "The Combined Effect of Bicarbonate and Chloride Ions on the Stress Corrosion Cracking Susceptibility of Alloy 22," Paper 06506, Corrosion/2006 Conference and Exposition (NACE International, Houston, TX)

[19] R. B. Rebak, "Corrosion Testing of Nickel and Titanium Alloys for Nuclear Waste Disposition," Corrosion, Vol. 65, No 4, 252-271 (2009).

[20] R. B. Rebak, "Factors Affecting the Crevice Corrosion Susceptibility of Alloy 22," Paper 05610, Corrosion/2005 Conference and Exposition (NACE International, Houston, TX)

[21] K. J. Evans and R. B. Rebak, "Measuring the Repassivation Potential of Alloy 22 Using the Potentiodynamic-Galvanostatic-Potentiostatic Method," Journal of ASTM International, Vol. 4, N° 9, Paper ID. JAI101230 (2007)

[22] R. B. Rebak, "Mechanisms of Inhibition of Crevice Corrosion in Alloy 22", Material Research Society, 2007, Warrendale, PA.

[23] A. K. Mishra and G. S. Frankel, "Crevice Corrosion Repassivation of Alloy 22 in Aggressive Environments," Corrosion, Vol. 64, No. 11, 836-844 (2008)

[24] D. S. Dunn, L. Yang, C. Wu, and G. A. Cragnolino. "Effect of Inhibiting Oxyanions on the Localized Corrosion Susceptibility of Waste Package Container Materials." Scientific Basis for Nuclear Waste Management XXVIII, J.M. Hanchar, S. Stroes-Gascoyne, and L. Browning eds, Warrendale, PA: Materials Research Society Symposium Proceedings, Vol. 824, pp. 33-38, (2004)

[25] R. M. Carranza, M. A. Rodriguez, and R. B. Rebak, "Effect of Fluoride Ions on Crevice Corrosion and Passive Behavior of Alloy 22 in Hot Chloride Solutions," Corrosion, Vol. 63, No. 5, 480-490 (2007)

[26] M. Miyagusuku, R. M. Carranza, and R. B. Rebak, "The Effect of Phosphate ions on the Corrosion Behavior of Alloy 22," Paper 10238, Corrosion/2010 Conference and Exposition (NACE International, Houston TX)

[27] T. Lian, G. E. Gdowski, P. D. Hailey, and R. B. Rebak, "Crevice Corrosion Resistance of Alloy 22 in High-Nitrate, High-Temperature Dust Deliquescence Environments," Corrosion,

Vol. 64, No. 7, 613-623 (2008)

[28] K. G. Mon, G. M. Gordon, and R. B. Rebak, "Stifling of Crevice Corrosion in Alloy 22," in Proceedings of the 12th International Conference on Environmental Degradation of Materials in Nuclear Power System – Water Reactors – Edited by T.R. Allen, P.J. King, and L. Nelson, pp. 1431-1438 (TMS The Minerals, Metals & Materials Society, 2005: Warrendale, PA)

[29] K. G. Mon, P. Pasupathi, A. Yilmaz, and R. B. Rebak, "Stifling of Crevice Corrosion in Alloy 22 During Constant Potential Tests," paper PVP2005-71174 in Proceedings of the 2005 ASME Pressure Vessels and Piping Division Conference, July 17-21 2005, Denver, CO, Vol. 7, pp. 493-502 (American Society of Mechanical Engineers, 2005: New York, NY)

[30] D. V. Fix, J. C. Estill, G. A. Hust, L. L. Wong and R. B. Rebak, "Environmentally Assisted Cracking Behavior of Nickel Alloys in Simulated Acidic and Alkaline Waters Using U-bend Specimens," Corrosion/2004, Paper 04549 (NACE International, 2004: Houston, TX)

[31] K. J. King, L. L. Wong, J. C. Estill and R. B. Rebak, "Slow Strain Rate Testing of Alloy 22 in Simulated Concentrated Ground Waters," Corrosion/2004, Paper 04548, (NACE International, 2004: Houston, TX)

[32] R. W. Schutz, "Platinum Group Metal Additions to Titanium: A Highly Effective Strategy for Enhancing Corrosion Resistance," Corrosion, 59, 1043 (2003)

[33] C. S. Brossia and G. A. Cragnolino, "Effects of Environmental and Metallurgical Conditions on the Passive and Localized Dissolution of Ti-0.15%Pd," Corrosion, 57, 768 (2001)

[34] L. L. Wong, J. C. Estill, D. V. Fix and R. B. Rebak, "Corrosion Characteristics of Titanium Alloys in Multi-Ionic Environments," PVP-Vol. 467, 63 (ASME, 2003: New York, NY)

[35] D. V. Fix, J. C Estill, L. L. Wong and R. B. Rebak, "Susceptibility of Welded and Non-Welded Titanium Alloys to Environmentally Assisted Cracking in Simulated Concentrated Ground Waters," Corrosion/2004, Paper 04551 (NACE International, 2004: Houston, TX)

CREEP STUDIES OF MODIFIED 9Cr-1Mo STEEL FOR VERY HIGH TEMPERATURE REACTOR PRESSURE VESSEL APPLICATIONS

Triratna Shrestha[1], Mehdi Basirat[2], Indrajit Charit[1], Gabriel Potirniche[2], Karl Rink[2]
[1]Materials Science and Engineering, University of Idaho, Moscow, ID 83844-3024, USA
[2]Mechanical Engineering, University of Idaho, Moscow, ID 83844-0902, USA

ABSTRACT

Advanced nuclear power reactors are envisioned to meet the increasing energy needs of the world. Structural materials to be used in these advanced reactors need to sustain elevated temperature, extremely corrosive environments and high radiation doses over a long periods of time. Modified 9Cr-1Mo steels, a.k.a. Grade 91, are being considered as a candidate material for the reactor pressure vessels (RPVs) of the Very High Temperature Reactor (VHTR) because of their good creep resistance, high temperature strength and adequate corrosion resistance. This study presents the results of the tensile creep tests conducted in the temperature range of 600-700°C and at stress levels of 100-200 MPa. Analysis of creep results yielded stress exponent values of 10-13. The high stress exponents are likely to be due to the threshold stress imparted by the fine precipitates present in the microstructure. Monkman-Grant and Larson-Miller parameters obtained from the creep tests are also discussed.

INTRODUCTION

The Next Generation Nuclear Plant (NGNP) is expected to address the growing US demand for energy and mitigate greenhouse gas emissions in the US by co-producing hydrogen from the process-heat. The Very High Temperature Reactor (VHTR) is the reactor system (Gen-IV type) at the heart of the NGNP. VHTRs are being designed to operate at temperatures much higher than those of the currently operating reactors. Moreover, they will be designed for service for a much longer period of time (60 years or more) compared to the current operating reactors. Depending on the VHTR design - Prismatic Modular Reactor (PMR) or Pebble Bed Modular Reactor (PBMR), the operating temperature of the reactor pressure vessel (RPV) will vary in the range of 300-650°C. Further, the size of the RPV in the VHTR needs to be more than twice the size of an RPV of a typical Light Water Reactor (LWR). The fabrication of the VHTR vessels remains a challenge because of the need for large ring forgings and the difficulty of welding the heavy-section steels [1,2].

Low alloy ferritic steels (e.g. SA533) are used as the RPV material for the current LWRs. However, because the RPV in the VHTR is expected to experience higher temperatures for a longer time, the material should have good resistance against time-dependent plastic deformation or creep. The current RPV steels are not likely to have adequate creep strength required for the service conditions envisioned for the VHTR pressure vessels. A potential candidate material for the VHTR pressure vessel has been the modified 9Cr-1Mo ferritic/martensitic (F-M) steel. It contains alloying elements like Nb and V, forming very fine and stable particles, which can help improve its creep strength. Other micro and substructural factors are also beneficial to creep resistance. Creep in steels depends on temperature, applied stress, grain size and many other intrinsic material properties. However, there have not been studies elucidating the creep mechanisms occurring in the grade 91 steel. Hence, it is important to understand the creep mechanisms to better predict the creep behavior of the steel under service conditions. At a later stage of creep deformation (tertiary stage), various failure mechanisms (the formation of voids and/or cracks) begin to dominate, thus resulting in the failure of the material. This paper presents some preliminary results on the microstructural characteristics and creep properties of Grade 91 steel.

EXPERIMENTAL PROCEDURE

The chemical composition of ASTM A387 Grade 91 CL2 steel investigated is: Fe - 8.50Cr - 0.90Mo - 0.21V - 0.08Nb - 0.10C - 0.51Mn - 0.20Cu - 0.30Si - 0.035N - 0.15Ni - 0.01P - 0.005S - 0.002Ti - 0.007Al- 0.001Zr (in wt.%). The hot rolled Grade 91 plates were received in a normalized {at 1311 K (1038°C) for 240 minutes} and tempered {at 1061 K (788°C) for 43 minutes} condition in the dimension of 4.1" (10.4 cm) × 4.1" (10.4 cm) × 0.5" (1.27 cm). Optical and scanning electron microscopies were used to characterize the microstructural characteristics of the steel. Creep specimens with a gauge length of 1" (2.54 cm) and gauge diameter of 0.25" (0.635 cm) were machined from the steel plates. According to standard data sheet, Grade 91 steel has yield strength of 533 MPa, an ultimate tensile strength of 683 MPa and percentage elongation of 26% at room temperature. Creep tests were performed at a few combinations of temperature (600-700°C, i.e. 873-973 K) and stress (100-200 MPa) using an ATS lever arm (20:1) creep tester. As-received materials were characterized using optical and scanning electron microscopies (SEM). SEM studies were performed to understand the grain and precipitate morphology of the material.

RESULTS AND DISCUSSION

Microstructural Characteristics

Different surfaces (Figure 1a) of the as-received Grade 91 steels were studied using an optical microscope (RD – Rolling Direction, TD – Transverse Direction, and ND – Normal Direction). Optical micrographs obtained are shown in Figure 1(b), (c) and (d). Figure 1(b) shows a finely dispersed microstructure of the as-received steel in the TD-ND orientation. More detailed microstructural features such as prior austenite grain boundaries, retained austenite and martensitic lath structures are evident in Figure 1 (c) and (d). In almost all normalized and tempered Grade 91 steels, there exists a small amount of retained austenite [3]. This occurs because of the M_f (martensite finish) temperature is lower than the room temperature. Thus, the retained austenite in the normalized structure does not transform to any other phase during the tempering stage. On the other hand, the martensite structure created by normalizing evolves throughout the tempering stage. However, the prior austenite grain boundaries delineating the colonies of lath martensitic structures could still be observed.

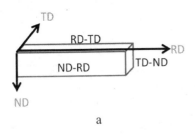

a

b   ~10 µm

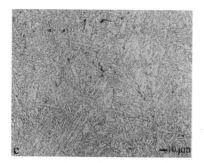
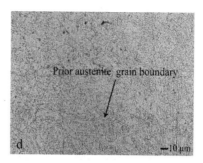

Figure 1 (a). A schematic of the orthogonal reference frame of the rolled Grade 91 steel. Optical micrographs of the as-received Grade 91 steel in different orientations: (b) fine microstructure in the TD-ND orientation, (c) martensite laths in the TD-ND orientation, and (d) prior austenite grain boundary in the RD-TD orientation.

In addition to optical microcopy, SEM using secondary electron imaging mode was used to study the microstructure in detail. Energy dispersive spectroscopy (EDS) was used for elemental analysis and precipitate characterization. Grade 91 steel has various precipitates present, as observed in Figure 2. The prior austenite grain boundaries and evolving new grain boundaries are found in Figure 2b. Larger precipitates are mainly located on the grain boundary while finer precipitates are located inside the grain (Figure 2a). Fine and thermally stable precipitates are crucial for enhanced creep strength in Grade 91 steel as they impede the movement of dislocations.

Coarser $M_{23}C_6$ type precipitates are seen on the grain boundaries (Figure 2c). The size of the $M_{23}C_6$ type precipitates were measured, and it was found that the elongated precipitate has a length of 717.9 nm and width of 113.7 nm (Figure 2 c), while the smaller precipitate has a length of 402.9 nm and width of 235.0 nm (Figure 2d). It is important to note that there are also $M_{23}C_6$ particles with various size distributions. The EDS study of a $M_{23}C_6$ type precipitate shows increased Cr concentration while decrease in the Fe content occurs along the EDS line scan of the $M_{23}C_6$ precipitate (Fig. 3). More precise measurements can be done using EDS in a TEM. A typical composition of $M_{23}C_6$ is $(Cr_{16}Fe_6Mo)C_6$, $(Cr_4Fe_{12}Mo_4Si_2WV)C_6$. A typical chemical composition of MX type precipitates is NbC, NbN, VN, (CrV)N, Nb(CN) and (NbV)C, and they are mainly found at the martensitic lath boundaries [4]. The SEM study of MX type precipitates revealed near-spherical type of precipitates with diameters of ~4-6 nm (image not shown here), which are very small compared to those of the $M_{23}C_6$ precipitates. These types of precipitates are expected to play a critical role for the improved creep resistance of the steel. Transmission electron microscopy (TEM) and other tools will be used to better characterize the finer microstructural details of the Grade 91 steel before and after the creep deformation.

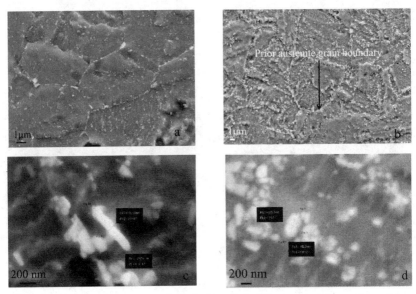

Figure 2. SEM images of as-received Grade 91 steel in different orientations: (a) secondary electron image of different types of precipitates in RD-TD orientation, (b) prior austenite grain boundary in TD-ND orientation, (c) precipitate in TD-ND orientation has length 717.9 nm and thickness 113.7 nm, (d) precipitate in ND-TD orientation has length 402.9 nm and thickness 235.0 nm.

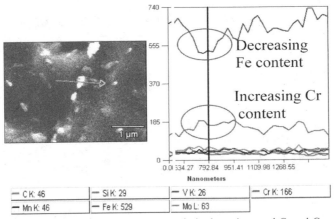

| $-$ C K: 46 | $-$ Si K: 29 | $-$ V K: 26 | $-$ Cr K: 166 |
| $-$ Mn K: 46 | $-$ Fe K: 529 | $-$ Mo L: 63 | |

Figure 3. Energy dispersive spectroscopy analysis shows increased Cr and C content, and decreased Fe content along the line scan of $M_{23}C_6$ precipitate located at a grain boundary.

Creep Properties

At temperatures of $0.5T_m$ (T_m is the melting temperature of the material in K) and above, materials undergo creep deformation leading to significant changes in microstructure and properties. The Grade 91 steel as the VHTR pressure vessel is likely to reach the creep temperature regime even under normal operating conditions. Creep tests were carried out at different temperatures and stresses. Fig. 5a shows a representative creep curve obtained at 650°C and 150 MPa for the as-received Grade 91 steel. The creep rates as a function of time corresponding to the creep curve is shown in Fig. 5b. Different creep regimes (primary, steady state or secondary and tertiary stages) are identified in Fig. 5. The creep rate of Grade 91 steel initially decreases in the primary regime, then becomes constant in the steady state regime, and finally accelerates in the tertiary regime.

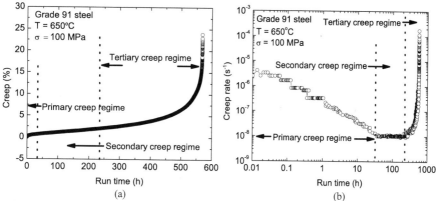

Figure 5. a) A creep curve and b) the corresponding creep rate versus time plot at 650°C and 100 MPa for the as-received Grade 91 steel.

The steady state creep rate increased as a function of stress at a constant temperature as seen in Fig. 6. Among the creep tests completed so far, the highest creep rate (6.4×10^{-5} s^{-1}) was recorded for the creep test (C5, see Table I) carried out at 150 MPa and 700°C. And the creep test (C8) at 650°C and 100 MPa shows the best creep strength ($\dot{\varepsilon}_{ss} = 7.9 \times 10^{-9}$ s^{-1}). Table I summarizes all the creep results obtained to date. The rupture time for C8 sample was 568 hr comprising 31 hr of primary regime, 203 hr of secondary regime and 334 hr tertiary regime. The duration of steady state creep regime was 15% and 36% of the rupture time for C5 and C8, respectively. The duration of the tertiary creep regime in C5 was ~ 61% of rupture time, while only about 24% of the rupture time was spent in the secondary creep regime. Similar results were observed for the creep test (C2) carried out at 200 MPa and 650°C. The early onset of the tertiary regime is possibly due to the coarsening of carbides and formation of creep cavities, but the presence of fine microstructure extends the tertiary regime by hindering the growth of the creep cavities. For a creep test (700°C and 200 MPa) which is not listed in Table I, the creep specimen ruptured before complete loading. The creep sample C9 (600°C and 100 MPa) has been running for more than 3500 hr with a total accumulated strain of 0.73%.

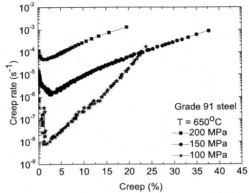

Figure 6. Creep rate versus strain plots for the as-received Grade 91 steel at various stresses at a constant temperature of 650°C.

Table I. Summary of creep test results – times spent in the primary, secondary and tertiary regimes, steady state creep rates, and rupture times along with the respective test temperatures and applied stresses

Creep test	Stress (MPa)	Temp. (°C)	Rupture time (t_R)	Duration of primary creep	Duration of steady state creep	Duration of tertiary creep	Steady state creep rate (s^{-1})
C2	200	650	21 min	2 min	5 min	14 min	4.7×10^{-5}
C3	200	600	35 hr	8 hr	13 hr	14 hr	3.3×10^{-7}
C5	150	700	18 min	3 min	4 min	11 min	6.4×10^{-5}
C6	100	700	10 hr	1 hr	4 hr	5 hr	1.0×10^{-6}
C7	150	650	10 hr	2 hr	3 hr	5 hr	1.7×10^{-6}
C8	100	650	568 hr	31 hr	203 hr	334 hr	7.9×10^{-9}
C9	100	600	-	-	-	-	Ongoing

Steady state creep rate as a function of applied stress can be described by Norton's equation:

$$\dot{\varepsilon}_{ss} = K\sigma^n \tag{1}$$

where $\dot{\varepsilon}_{ss}$ is the steady state creep rate, σ the applied stress, K a constant based on material characteristics and test conditions, and n the stress exponent. Following Eq. 1, the steady state creep rates were plotted against the corresponding applied stresses in a double logarithmic scale in Fig. 7. The slopes of the fitted straight lines gave the relevant stress exponent values. The high stress exponents obtained (10-13) in the present study are close to the stress exponents obtained in other creep studies at higher stress regimes on similar types of steels [5-7]. The stress exponent values may also be influenced by the evolution of the precipitates (such as coarsening) during the creep deformation [8], and such effect needs to be studied in Grade 91 steel. Kloc et al. [7] reported stress exponents ranging from 1 to 10 and higher for a modified 9Cr-1Mo steel crept in the temperature range of 600-650°C and stress range of 1-350 MPa. Notably, the stress exponent was ~ 1 for stress below 100 MPa. Stress exponents of 15 have also been reported for the modified 9Cr-1Mo steel in the stress range of 160-225 MPa at 600°C [7]. In the present study, it is apparent that creep stress exponents tend to have an inverse relation with temperature, i.e., 'n' decreases with increasing temperature (Fig. 7). Spigarelli et al. [7] reported similar stress exponent for weld and HAZ sample of modified 9Cr-1Mo. These changes in stress exponent directs to the change in deformation mechanisms happening during the creep test. The higher stress data cannot be simply extrapolated to lower stress regime because new creep mechanism can set in and one can run the risk of 'blind extrapolation' [9].

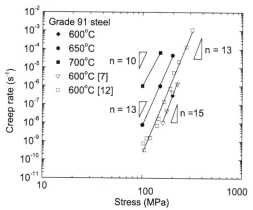

Figure 7. Stress dependence of steady state creep rate for modified 9Cr-1Mo steels. That the data represented by filled symbols are from the current study.

Creep is a thermally activated, diffusion-assisted deformation process. The Bird-Mukherjee-Dorn (BMD) equation best describes the creep mechanism in this regime:

$$\frac{\dot{\varepsilon}_{ss}kT}{DEb} = A \left(\frac{\sigma}{E}\right)^n \left(\frac{b}{d}\right)^p \tag{2}$$

where A is the microstructural mechanical constant, E the elastic modulus, b the Burgers vector, k the Boltzmann constant, T the absolute temperature, d the grain diameter, p the inverse grain size exponent and D the diffusivity. D is described by the following relation:

$$D = D_o exp \left(\frac{-Q}{RT}\right) \tag{3}$$

where D_o is the frequency factor, Q the appropriate activation energy, and R the universal gas constant. The BMD equation is an expanded form of the basic Norton's equation. The creep parameters such as n, p, and Q are the main parameters determining the operating creep mechanism. Table II summarizes various creep mechanisms generally found in the creep deformation of conventional materials. It is important to note that no creep mechanism shown in

Table II shows stress exponents as high as those obtained in Grade 91 steel. The high apparent stress exponents obtained in Fig. 7 make it imperative that threshold stresses be considered in order to fully understand the creep mechanisms in Grade 91 steel. The modified BMD equation incorporating threshold stress can be described by the following form:

$$\frac{\dot{\varepsilon}_{ss}kT}{DEb} = A \left(\frac{\sigma - \sigma_{th}}{E}\right)^n \left(\frac{b}{d}\right)^p \tag{4}$$

where σ_{th} is the threshold stress. The threshold stress is basically the stress below which no creep deformation takes place. Thus, the effective stress, $(\sigma - \sigma_{th})$, actually responsible for the creep deformation. This type of behavior has been found to be prevalent in many particle-containing alloys [9]. However, more creep data needs to be generated to perform a full-scale threshold stress analysis. Further, microstructural evolution, such as particle coarsening process occurring through Ostwald ripening, during the creep deformation will have implications for the creep behavior of the material. More in-depth analysis will be attempted in later studies with the collection of more microstructural and creep data.

Table II. Parametric dependencies of various creep mechanisms [9].

Mechanism	D	n	p	A
High temperature climb	D_L	5	0	~6×10^7
Viscous glide	D_S	3	0	6
Low temperature climb	D_C	7	0	2×10^8
Harper-Dorn	D_L	1	0	3×10^{-10}
Nabarro-Herring	D_L	1	2	12
Coble	D_B	1	3	100

Note. D_L: lattice diffusivity, D_S: solute diffusivity, D_C: dislocation core diffusivity, and D_B: grain boundary diffusivity

Larson-Miller parameter (LMP) is often used to predict the creep-rupture life of materials. LMP can be expressed in terms of rupture time and temperature:

$$\text{LMP} = T (\log t_R + C) \tag{5}$$

where T is the test temperature (in K), t_R the rupture-time (hr), and C the Larson-Miller constant, typically taken as 20. LMP values of Grade 91 steel decreases with increased stress, and the data obtained in the present study remain within the ambit of the data band obtained from other studies (Fig. 8a). For different steel types, Klueh noted that the LMP plots for creep-rupture tests showed greater data scatter at low temperature and high stress than for the high temperature and low stress [5].

Relationship between the secondary creep rate and the rupture time is given by the Monkman-Grant equation:

$$(\dot{\varepsilon}_{ss})^m t_R = \text{Constant.} \tag{6}$$

Using Eq. 6, the obtained fit in Fig. 8b yields data in a logarithmic form:

$$\log \dot{\varepsilon}_{ss} = -1.217 \log t_R - 2.92 \tag{7}$$

In the experimental range of temperatures and stresses, the Monkman-Grant relation seems to follow a straight line (Fig. 8b). These approaches are empirically based, so their use could lead to the overestimation of the creep-rupture life [10]. For better accuracy of Eq. 6, m has to be less than 1 [11]. The value of 'm' was estimated to be ~0.82 for the plot shown in Fig. 8b. Even though the Monkman-Grant relation and LMP can be used to predict rupture life, it does not provide us with insight of the creep failure mechanisms. At present, detailed fractographic SEM studies of the creep-ruptured specimens are being planned to better elucidate the failure mechanisms involved.

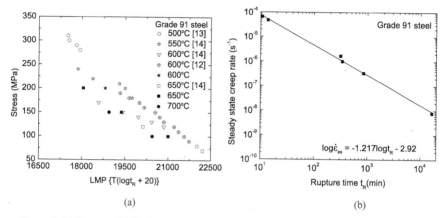

Figure 8. (a) Stress vs. LMP plots (along with the data from literature; the data from the present study are represented by the closed symbols), and (b) the application of the Monkman-Grant relation for Grade 91 steel.

CONCLUDING REMARKS

Even though there has been a significant interest in using the modified 9Cr-1Mo steel for the NGNP and other advanced reactor applications, there is not sufficient data to qualify this type of steel for nuclear reactor applications. Preliminary studies have shown some promising results in the ability of the modified 9Cr-1Mo steel to operate at around 700°C and for long service life. Heat treatment of modified 9Cr-1Mo steel is critical in obtaining the desired microstructure, which determines the service life of the reactor. Due to the mechanistic transition, the creep life cannot be extrapolated based on short-term creep tests. More data is needed to fully describe and understand the creep behavior of Grade 91 steels, which will be useful in the qualification efforts of these steels as a potential VHTR pressure vessel material.

ACKNOWLEDGMENTS

The authors gratefully acknowledge the Nuclear Energy University Program (NE-UP) of the US Department of Energy for supporting this research. The authors would also like to thank Zachary Wuthrich and Peter Wells for their assistance during the course of this study.

REFERENCES
[1]I. Charit and K.L. Murty, JOM, **62**, 67 (2010).
[2]T. Burchell, Bratton R. L., Wright, R.N. and Wright, J., *Next Generation Nuclear Plant Materials Research and Development Program Plan*, Idaho National Laboratory, Idaho Falls, Sep. 2007.
[3]P.J. Ennis, A. Zielinksa-Lipiec, O. Wachter, and A. Czyrska-Filemonowicz, *Acta Metall.*, **45**, 4901 (1997).
[4]R.L. Klueh and D. R. Harris, *High-Chromium Ferritic and Martensitic Steels for Nuclear Applications*, ASTM, Bridgeport, NJ, 2001, p. 39.
[5]Klueh, R. L., *Inter. Mater. Rev.*, **50**, 287 (2005).
[6]L. Kloc, V. Sklenicka, A. Dlouhy and K. Kucharova, *Microstructural Stability of Creep Resistant Alloys for High Temperature Plant Applications*, eds. A. Strang, J. Cawley, and G. W. Greenwood, London Institute of Materials, London, 1998, p. 445.
[7]S. Spigarelli and E. Quadrini, *Materials and Design*, **23**, 547 (2002).
[8]F. Ricardo and G. Gonzalez-Donel, *Acta Mater.*, **56**, 2549 (2008).
[9]I. Charit, K. L. Murty, and C. C. Koch: *Proc. 5th Int. Conf. on Advances in Materials Technology for Fossil Power Plants*, Marco Islands, FL, Oct. 3-5 2007, 281-292.
[10]V. Lupinc, in: Creep and Fatigue in High Temperature Alloys, eds. J. Bressers et al., Applied Science, London, 1981, p. 7.
[11]F.R.N. Nabarro and H. L. de Villiers, The Physics of Creep, Taylor & Francis, London, 1995, p. 22.
[12]P.J. Ennis, A. Zielinska-Lipiec, and A. Czyrska-Filemonowicz, in: Microstructural stability of creep resistant alloys for high temperature plant applications, eds. A. Strang et al., Institute of materials, London, 1998, p. 135.
[13]F. Vivier, J. Besson, A. F. Gourgues, Y. Lejeail, Y. de Carlan, and S. Dubiez, in: Creep behavior and life prediction of ASME Gr. 91 steel welded joints for nuclear power plants, eds. Le Goff et al., Mines-Paris Tech, France, 2008, p. 1.
[14]K. Laha, K.S. Chandravathi, P. Parameswaran, K. Bhanu Sankar Rao, and S.L. Mannan, *Metall. Mater. Trans. A*, **58**, 38A (2007).

DEVELOPING THE PLUTONIUM DISPOSITION OPTION: CERAMIC PROCESSING CONCERNS

[1,2]Jonathan Squire, [1]Ewan R Maddrell,
[1]The National Nuclear Laboratory
Sellafield, Seascale, Cumbria, CA20 1PG, UK

[2]Neil C Hyatt, [2]Martin C Stennett
[2]Department of Material Science and Engineering, The University of Sheffield
Mappin Street, Sheffield, S1 3JD, UK

ABSTRACT

A significant stockpile of plutonium has been accumulated in the UK that is currently assumed to be stored in the long term as a zero value asset. Amongst disposition options, disposal by ceramic is promising but requires further study. Previous work established zirconolite and zirconia as preferred actinide wasteforms. Homogenisation and size reduction of ceramic wasteforms precursors during processing can be provided by attrition milling. This work presents the results of an investigation into the effect of mill operating parameters (tip speed, duration, and lubricant) on the final wasteform quality.

The zirconolite samples, sintered by HIP, failed to reach the target wasteform density (>95% of theoretical) and their densities failed to show any correlation with milling parameters. However, results with zirconia samples, produced by cold press and sinter, did show a correlation between increasingly aggressive milling and increasing ceramic sintered density.

Four lubricants were investigated on their ability to prevent 'foot' formation and aid powder discharge. Oleic acid, polyethylene glycol and zinc stearate (ranked in order) performed well, while Ceri Dust™ failed to prevent foot formation.

INTRODUCTION

The UK has a significant stockpile of separated plutonium. The requirement to reprocess Magnox power station fuel, together with reprocessing operations in THORP, have led to approximately 100 tonnes Heavy Metal of PuO_2 powder accumulating. In the sixty years of nuclear operations in the UK the political direction has changed a number of times. The initial intent was to recover plutonium from spent fuel for recycling in a future Fast Reactor programme. The design philosophy for the Sellafield reprocessing plant was based on this intent; plutonium was considered an asset and not to be wasted. The economics of the nuclear fuel cycle and the maturity of advanced reactor options have changed over several decades and the strategy of continued safe and secure storage is currently implemented.

The establishment of The Nuclear Decommissioning Authority (NDA) in 2005 gave an impetus to define a strategy for the final end state of PuO_2. The NDA Plutonium Strategy Report[1] outlined four generic disposition options: disposal as waste, recycle, store for a limited time or some combination of these. Amongst the conclusions of the report was the need for further research and development of the disposal option.

The disposal option analysed a number of methods including vitrification, encapsulation in cement, and immobilisation as a ceramic. The ceramic wasteform was not explicitly defined. However, a preference for the use of a hot isostatic press (HIP) for sintering was shown. This body of work was commissioned to inform the NDA about the processing of ceramic wasteforms.

241

AIMS AND OBJECTIVES

Inactive research was carried out with the intention that any process developed could be easily implemented into a highly alpha-active operating environment. Within all ceramic wasteform processing routes, a size reduction and homogenisation step is required. Attrition mills have been successfully operated in the manufacture of MOX and can provide this step. To minimise secondary effluents the attrition mill must be operated in the dry mode, ie without a carrier fluid.

Sintering can be achieved by either the cold press and sinter route, or by HIP. Cold press and sinter is used in the manufacture of fuel pellets and a significant level of operating experience has been accumulated. HIP has not yet been fully implemented into an active environment but the process ensures active material is completely contained within a processing can. These cans are sealed and clean on the outside, meaning the HIP can be operated outside of the active environment. An Activity Containment Over Pack (ACOP) provides protection against contamination in the event of a can failure. The design choices, Figure 1, show the attrition mill to be the only significant process to be in the active area.

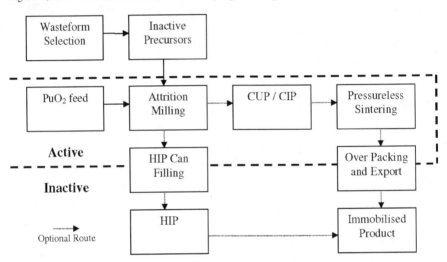

Figure 1. Design philosophy demonstrating the processes that would operate within an active environment.

The wasteform should reliably render the plutonium passively safe and suitable for long-term storage and be able to prevent plutonium dispersal into the environment and illicit recovery[2]. Two candidate wasteforms, zirconolite and zirconia, have been selected on the basis of a large body of work[3-6] on their suitability and of previous working experience[7]. Zirconolite is prototypically $CaZrTi_2O_7$ for the 2M polytype, and zirconia is $Zr_{1-x}Ca_xO_{2-x}$ for the calcium stabilised cubic form.

Demonstrated by Figure 1, the necessary precursors will be selected on the basis of the final wasteform required. The precursors will be prepared to the required specification before entering the process. The attrition mill homogenises these powders with the plutonium feed. A homogenous and correctly sized powder blend will improve sintering characteristics and produce a superior quality wasteform. The powders are then either packed into a can for HIP or cold pressed before sintering.

Previous experiments have shown that the powders in the attrition mill can form a 'foot', which is a compaction of powders in the base of the mill that typically has not been milled and will not be easily discharged. Lubricants, such as hydrocarbon waxes, have shown promise as 'anti-foot formers' and their performance was judged in this work.

In summary, the objectives of the work were to investigate the relationship between attrition mill operating parameters (tip speed and duration) and final wasteform quality and to investigate the ability of selected lubricants to prevent foot formation during milling.

EXPERIMENTAL

There is a preference to sinter by HIP yet it was not clear whether zirconia could be fully consolidated at HIP operating temperatures. Thus, zirconolite was sintered by HIP while the zirconia was prepared by cold isostatic press (CIP) or cold uniaxially press (CUP) and then sintered. Although the two candidate wasteforms require two differing consolidating techniques, the effect of the common milling operation on the wasteform can still be compared.

The attrition mill (Union Process 1SD) was operated in the dry mode. The mill has a gross tank volume of 5.68 l (1.5 US Gallons) and a working capacity of 3.03 l (0.8 US Gallons). The mill was charged with 16.00 kg of 9.52 mm (3/8") diameter stainless steel balls, the quantity chosen according to the manufacturer's recommendation. The milling tank has a water jacket that can be used for heating or cooling. The mill was heated to 65-70 °C, as this has been found to reduce the tendency of foot formation. The precursor mix was added to the top of the mill whilst operating at a tip speed of 1.73 m s^{-1} (200 RPM) and remained in the mill for the necessary duration. The powders were discharged at the same speed.

Operation of the mill was restricted to a maximum speed of 4.32 m s^{-1} (500 RPM) due to acoustic limitations. This was selected as the maximum speed for the work, while 2.16 m s^{-1} (250 RPM) was selected as the lower limit for milling. Durations of 30 and 60 minutes were chosen to produce an equivalent number of rotations.

Four lubricants were selected based on previous experience: zinc stearate, polyethylene glycol (6000 molecular weight), Ceri Dust™ 3620 (a polyethylene wax) and oleic acid. For each lubricant, except oleic acid, 1.5 wt% was added to each of the precursor blends before milling. For the oleic acid 1 wt% was used, as this had previously been shown to be sufficient.

In this inactive work, Ce, added as CeO_2, was used as a surrogate for Pu. The crystal chemistry of Ce^{3+}/Ce^{4+} and Pu^{3+}/Pu^{4+} is similar. This is reflected for 8-fold co-ordination the ionic radii of Ce^{4+} and Pu^{4+} are 1.11 Å and 1.10 Å, respectively[8,9].

Zirconolite produced by HIP

In the zirconolite wasteform, Pu can be substituted on both the Ca and Zr sites with suitable charge balancing on the Ti site if required. In the experiments 0.25 formula units of Ce was targeted onto the A-cation site (Ca in the prototype) with mixed Fe^{2+}/Fe^{3+} (from Fe_3O_4) providing the necessary charge balancing. Fe_3O_4 was chosen for its mixed oxide state and would produce an electrically conductive zirconolite sample enabling scanning electron microscope (SEM) analysis without additional coating. The target formula was constant throughout all permutations of lubricant, milling speed and duration as $Ca_{0.75}Ce_{0.25}ZrTi_{1.625}Fe_{0.375}O_7$.

Batches of 800 g (1:2.5 ratio of volume of material to working volume) of stoichiometric quantities of the metal oxide precursors were weighed out with an additional 1 wt % of ZrO_2 and TiO_2 as a buffer against perovskite ($CaTiO_3$) formation. A significant portion of the precursor was taken from a premixed batch, which was being used for another aspect of wasteform development work. The precursor was charactersised by mass fractions of 21.3% CaO, 8.7% ZrO_2 and 70.0% TiO_2. This material had been calcined for 10 hours at 1200 °C in order to stabilise the CaO content as $CaTiO_3$. The remaining precursors were supplied as metal oxides.

After milling, the powders were packed into stainless steel cans (60 mm tall, 35 mm outside diameter). Two cans were packed from each of the samples along with a control sample of powders that had not been milled. Once the powders were placed into the cans, approximately 30 MPa of uniaxial pressure was applied. More loose powders were added and compressed. This was repeated until the can was full with typically 160-170 g of powder packed into each can. Lids, complete with evacuation tube, were then welded in place. The evacuation tube was connected to a vacuum line and the can placed into a furnace. The lubricant was removed by heating the can at 600 °C for 4 hours.

The cans underwent HIP at 100 MPa at both a high (1320 °C) and a low (1000 °C) temperature. The high temperature run was chosen to ensure full phase formation of the zirconolite. The low temperature run was chosen to form a dense monolith suitable for analysis but that had not undergone full zirconolite phase formation and would capture the as-milled microstructure. Both temperature runs had a 2 hour dwell.

Zirconia produced by cold press and sintering

Zirconia was chosen as an alternative candidate wasteform. Its prototypical composition is $Zr_{1-x}M_xO_{2-x}$, where M can be Ca, Y, lanthanides or actinides. In this work Ca was used to stabilise ZrO_2 in the cubic phase as it is cheaper and more abundant than the commonly used Y. Zirconia requires sintering at approximately 1600–1700 °C possibly precluding the use of HIP since HIP cans are made from stainless steel which has a maximum working temperature of 1350 °C. Cans made from noble metals (eg Nb, Ta, and Mo) would enable HIP at higher temperatures, but they would be prohibitively expensive.

We chose 0.0625 formula units of Ce to be substituted into the cubic zirconia. The target formulation was $0.0625(CeO_2):0.9375(0.85 ZrO_2:0.15 CaO)$. Batches of 1200 g were prepared of the metal oxides with $CaCO_3$ providing the CaO. The precursors were milled according to the previous experimental matrix excluding Ceri Dust™, due to its poor performance, as a lubricant with zirconolite.

The milled powders were then either CUP or CIP. The powders that underwent CUP were placed into a 10 mm diameter die and 125 MPa (1 tonne gauge reading) was applied and held for approximately one minute. The CIP powders were first uniaxially pressed at 27.7 MPa (2 tonnes) in a 30 mm die. The green pellets were CIP at approximately 207 MPa (30,000 PSI). Both sets of pellets were then sintered at 1650 °C for 4 hours in air. The temperature was increased at a rate of 5 °C min^{-1} with a slower rate of 1 °C min^{-1} around the lubricant burn out temperatures (350–400 °C) and the $CaCO_3$ decomposition temperature (800–900 °C).

RESULTS

Lubricant performance

The lubricant performance was judged on whether it allowed a 'foot' to form and at what rate the powders were discharged from the mill. For both sample sets, the lubricants performed comparably. Zinc stearate, polyethylene glycol and oleic acid prevented 'foot' formation while Ceri Dust™ did not. Ceri Dust™ performed so poorly that it was not used in the zirconia work. A crude measurement of discharge rate was used to attempt to rank the remaining lubricants. The powders were timed until approximately 99% of the powder was discharged from the mill. Oleic acid showed the best lubricating performance, typically taking 2 minutes to fully discharge. Powders milled with polyethylene glycol and zinc stearate took between 4 and 10 minutes.

Zirconolite

Figure 2 shows SEM analysis of the samples, which confirmed that the high temperature HIP run fully formed the single-host-phase ceramic. The low temperature run also produced a consolidated sample. More aggressive milling reduced the number of unmilled ceria grains (bright white spots in image B); however, the volume fraction of ceria grains was less than expected. This is most probably due to damage-induced grain pull-out during sample preparation. Future pycnometry studies will confirm the source of the holes shown in the images; whether closed pores introduced during processing or damage during sample preparation. There appears to be more dark grey features in the more aggressively milled high temperature sample; these are yet to be identified. It should also be noted that the high temperature samples appear to be softer and more prone to damage than their low temperature counterparts.

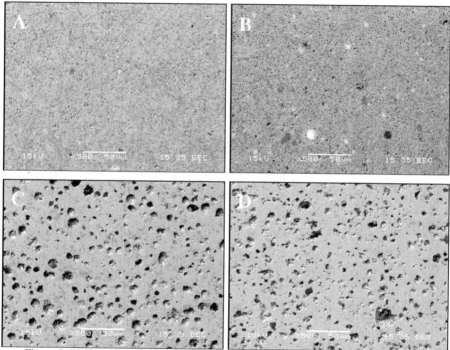

Figure 2. SEM images of four samples, all taken in the backscattering mode at x500 magnification. All samples milled with oleic acid. A: Low temperature HIP run sample milled at 4.32 m s^{-1} for 30 minutes. B: Low temperature sample milled at 2.16 m s^{-1} for 30 minutes. C: High temperature HIP run sample milled at 4.32 m s^{-1} for 30 minutes. D: High temperature sample milled at 2.16 m s^{-1} for 30 minutes.

The sample densities were measured using the Archimedes method with water as the immersion fluid. The target for a zirconolite sample is >95% of theoretical density, calculated as 4.78 g cm^{-3}. Estimated error is ± 0.01 g or ±0.2% of theoretical. Table I gives the densities of the high temperature samples sorted by increasing density.

Table I. High temperature HIP zirconolite samples, sorted by increasing density..

Tip Speed (m s^{-1})	Duration (minutes)	Lubricant	Density (g cm^{-3})	% Theoretical
2.16	30	Ceri Dust	3.80	79.47%
2.16	60	polyethylene glycol	3.90	81.52%
4.32	30	oleic acid	3.98	83.23%
2.16	30	polyethylene glycol	4.06	84.86%
2.16	60	oleic acid	4.06	84.90%
2.16	30	zinc stearate	4.07	85.19%
4.32	30	Ceri Dust	4.10	85.77%
2.16	60	Ceri Dust	4.16	86.98%
4.32	30	polyethylene glycol	4.16	87.05%
2.16	60	zinc stearate	4.18	87.46%
2.16	30	oleic acid	4.26	89.05%
4.32	30	zinc stearate	4.26	89.21%
0	0	None	4.72	98.80%

The most striking result is that the control sample, that had not been milled, is the densest. There is no correlation between how aggressively the sample was milled and its density. Table II gives the densities of the low temperature HIP run, again sorted by increasing density.

Table II. Low temperature HIP zirconolite samples sorted by increasing density

Tip Speed (m s^{-1})	Duration (minutes)	Lubricant	Density (g cm^{-3})	% Theoretical
2.16	60	zinc stearate	3.80	79.47%
2.16	60	Ceri Dust	3.82	79.87%
2.16	60	oleic acid	3.87	80.98%
2.16	30	Ceri Dust	4.05	84.72%
4.32	30	zinc stearate	4.06	84.84%
4.32	30	polyethylene glycol	4.06	84.93%
2.16	30	polyethylene glycol	4.09	85.59%
2.16	60	polyethylene glycol	4.21	87.99%
4.32	30	oleic acid	4.23	88.47%
4.32	30	Ceri Dust	4.25	88.96%
2.16	30	oleic acid	4.25	88.97%
2.16	30	zinc stearate	4.36	91.22%
0	0	None	4.43	92.70%

Once more, the unmilled control sample showed the highest density. Again, there was no correlation between milling and sample density. There was no statistical difference between the high and low temperature runs. Only the unmilled samples approached the 95% benchmark.

Zirconia

The zirconia sample densities were measured by the same technique as the zirconolite with an estimated error of ± 0.01 g or ±0.2% of theoretical. The zirconia samples, again, had a target of >95% of theoretical density; calculated as 5.59 g cm^{-3}. Table III gives the densities of the CIP samples ordered by increasing density.

Table III. CIP zirconia samples sorted by increasing density

Tip Speed (m s^{-1})	Duration (minutes)	Lubricant	Density (g cm^{-3})	% Theoretical
2.16	30	polyethylene glycol	4.86	87.01%
2.16	30	oleic acid	4.95	88.55%
2.16	30	zinc stearate	4.97	88.96%
2.16	60	polyethylene glycol	5.05	90.43%
2.16	60	zinc stearate	5.18	92.61%
2.16	60	oleic acid	5.19	92.84%
4.32	30	zinc stearate	5.27	94.21%
4.32	30	oleic acid	5.29	94.55%
4.32	30	polyethylene glycol	5.30	94.74%

Table III shows a progression of increasing density with increasingly aggressive milling. The lubricant does not appear to directly affect the final sample density. The CUP samples are given as Table IV, again ordered by increasing density.

Table IV. CUP zirconia samples sorted by increasing density

Tip Speed (m s^{-1})	Duration (minutes)	Lubricant	Density (g cm^{-3})	% Theoretical
2.16	30	polyethylene glycol	4.78	85.57%
2.16	30	zinc stearate	4.90	87.60%
2.16	30	oleic acid	4.91	87.84%
2.16	60	polyethylene glycol	4.96	88.69%
2.16	60	zinc stearate	5.04	90.18%
2.16	60	oleic acid	5.11	91.34%
4.32	30	oleic acid	5.16	92.37%
4.32	30	zinc stearate	5.20	93.03%
4.32	30	polyethylene glycol	5.22	93.30%

The progression of increasing density with increased milling intensity is again shown. CIP samples are, on average, 1.70 % greater in density than their CUP counterparts.

DISCUSSION

The initial aims and objectives were to study the effect of the attrition milling conditions on the wasteform characteristics as well as to judge lubricant performance. The zirconia work confirmed the expectation that an increasingly aggressive milling environment would lead to denser ceramics. It has also confirmed that an increased tip speed has a greater milling effect than just the total number of milling revolutions as shown by Table III and Table IV. Comparisons between CUP and CIP show only marginally denser ceramics are produced by CIP, in agreement with the work of Stennett et al[7]. Their work used a lab scale method with a 100g basis and produced zirconia samples densified to 94–96% of theoretical density.

In the zirconolite work two issues were encountered: the lack of a correlation between milling and wasteform density and the failure to reach the >95% of theoretical density target. Two possible explanations are proposed: either the precursors were too coarse or the can bake-out operation is inadequate for this method. The precursors were a blend of metal oxides that had been calcined at 1200 °C with the intention of producing $CaTiO_3$. This would prevent the precursors, specifically CaO, from absorbing CO_2 or water. At the calcining temperature, coarsening of the powders is to be expected, and some agglomerates may form. The agglomerates may be resistant to fracture in the mill, thus the mill will not be milling the precursors sufficiently. It could be that the most aggressive milling conditions are not aggressive enough. The zirconia samples were made from different precursors.

The second possibility is that the lubricant removal step (can bake-out) failed to completely remove the organic material. The bake-out process involved the simultaneous heating and evacuation of the HIP can. Throughout the practical work the bake-out step proved to be problematic as the evacuation system would often block. It was not possible to verify if the lubricant had been fully removed when the cans were sealed. We believe the lubricants may have been removed with varying success; in particular some carbonaceous residues may persist. Any residual carbon may form CO during the HIP run and impede the densification process. The variability in lubricant removal may contribute to the variation in the densities observed in the results. As a conformational study, the same milled powders with polyethylene glycol as the lubricant were CUP at 97 MPa (7 tonnes on 30 mm diameter pellets) and sintered at 1320 °C in air. Similarly to the zirconia samples the three pellets show the same progression of increasing density with increasingly aggressive milling. However, Table V indicates that they still fall far short of the 95% target. In contrast, with the zirconia samples the lubricants were burnt off in an open, oxidising environment with no impediment to their removal.

Table V. Zirconolite samples sorted by increasing density. All samples milled with polyethylene glycol.

Tip Speed (m s^{-1})	Duration (minutes)	Density (g cm^{-3})	% Theoretical
2.16	30	3.59	75.1%
2.16	60	4.09	85.6%
4.32	30	4.16	87.0%

Unmilled samples showed the highest density, which is explained by analysing the powder feed. CeO_2 has a density of 7.65 g cm^{-3}, thus if there are large unmilled ceria grains in the ceramic (see Figure 3) the density will be greater than the fully phase-developed ceramic.

Figure 3 - SEM of unmilled sample. SEM operated in the backscattering mode at x500 magnification.

CONCLUSIONS

A correlation between attrition milling parameters and the density of sintered zirconia samples has been shown. Milling with a tip speed of 4.32 m s^{-1} for 30 minutes produces denser zirconia ceramics than milling at 2.16 m s^{-1} for 60 minutes and 2.16 m s^{-1} for 30 minutes.

The same correlation has not been shown with zirconolite samples produced by HIP. Two possible explanations have been offered; the precursors were too coarse or a fault with the lubricant burn-out procedure. Samples sintered in air have shown the same progression with increasingly aggressive milling as the zirconia samples. The samples are still far below the target 95% of theoretical density. Future work will focus both on using precursors with a far smaller particle size distribution and investigating ways to negate the need for a lubricant in the mill.

In order of discharge performance, oleic acid, polyethylene glycol and zinc stearate were adequate as lubricants as no foots were formed. Ceri Dust™ failed to prevent foot formation.

ACKNOWLEDGEMENTS

The authors gratefully acknowledge EPSRC and The NDA for funding this work.

REFERENCES

1. NDA, *NDA Plutonium Topic Strategy: Credible Technical Analysis*, NDA, 2009.
2. N. C. Hyatt, M. C. Stennett, E. R. Maddrell and W. E. Lee, *Advances in Science and Technology*, 2006, **45**, 2004-2011.
3. F. W. Clinard, D. L. Rohr and R. B. Roof, *Nucl. Instrum. Methods Phys. Res. Sect. B-Beam Interact. Mater. Atoms*, 1984, **1**, 581-586.
4. I. W. Donald, B. L. Metcalfe and R. N. J. Taylor, *J. Mater. Sci.*, 1997, **32**, 5851-5887.
5. G. R. Lumpkin, *Elements*, 2006, **2**, 365-372.
6. A. Ringwood, S. Kesson, K. Reeve, D. Levins and E. Ramm, *Radioactive waste forms for the future*, North-Holland, Amsterdam, 1988.
7. M. C. Stennett, N. C. Hyatt, W. E. Lee and E. R. Maddrell, Environmental Issues and Waste Management Technologies in the Ceramic and Nuclear Industries, Baltimore, Maryland, 2006.
8. R. D. Shannon and C. T. Prewitt, *Acta Crystallographica Section B*, 1969, **25**, 925-946.
9. P. A. Bingham, R. J. Hand, M. C. Stennett, N. C. Hyatt and M. T. Harrison, *Mater. Res. Soc. Symp. Pro,* 2008, **1107**, 421

PORE STRUCTURE ANALYSIS OF NUCLEAR GRAPHITES IG-110 AND NBG-18

G. Q. Zheng*, P. Xu, K. Sridharan, and T. R. Allen

Department of Engineering Physics, University of Wisconsin-Madison
1500 Engineering Drive, Madison, WI 53706, U.S.A.
* gzheng@wisc.edu

ABSTRACT

Nuclear graphite is a major structural component for fuel elements and reactor cores. Some nuclear graphites have been considered as potential candidate structural materials for very high temperature reactor (VHTR) cores. The pore structure in nuclear graphite plays an important role in irradiation performance and oxidation resistance during operation. In this work, the pore structure of nuclear graphite grades IG-110 and NBG-18 was analyzed and compared through calculation and visualized images characterization. Apparent density and porosity were calculated using a geometrical method. A Brunauer-Emmett-Teller (BET) measurement was applied to evaluate the specific surface area, the pore volume and the average pore size of open pores based on the nitrogen adsorption and desorption isotherms at 77.35K. The morphology of the pores within both nuclear graphites was characterized using scanning electron microscopy (SEM). The visualized two dimensional (2-D) and three dimensional (3-D) images were presented to directly reveal the pore structure of nuclear graphite. Cracks of different sizes on the inside surface of the pores were observed for NBG-18. The binder phase and the filler phase were characterized using SEM of the fractured surface of both graphites.

INTRODUCTION

High purity nuclear graphites have attracted much attention as the candidate material for structural components in very high temperature reactor (VHTR) cores because of their attractive high temperature properties. These properties include high strength, low neutron absorption, and high thermal conductivity [1-7]. Nuclear graphite is classified as many different grades depending on the source of the coke, the manufacturing process and density [4]. In the past few decades, the role of radiation on volumetric swelling, creep, changes in elastic modulus and coefficient of thermal expansion have been extensively studied for nuclear graphites [3-9]. In these studies the microstructural change, such as grain size and pore geometry, was determined to be the primary factor that caused mechanical strength, diffusivity, thermal conductivity and oxidation resistance to change in the extreme environment [5-10]. This outcome has led to the development of many methods to characterize the structure of nuclear graphite[8-10]. As early as the 1960's, F. Gianni and F. Potenza developed a spectrographic analysis method to determine the low concentrations of impurities within nuclear graphite [11]. In the 1980's, E. Hoinkis published several reports on the effects of surface area and porosity changes on corrosion-, irradiation-, and oxidation-induced graphitic matrices A3-3 and A3-27 using Brunauer-Emmett-Teller (BET) and the α $_s$-method [8,9]. The fission product transport properties of nuclear graphite were also discussed in his reports. Moreover, the BET method was used to study the change of the specific surface area and pore-radius distribution of the exfoliated graphites synthesized under various conditions and was reported by O. N. Shornikova and his co-workers in 2009 [12]. This report by Shornikova et al. shows that BET could be an acceptable mthod for comparative estimates of the pore structure of different graphite samples. In all their studies, the internal pore structure was characterized using combined approaches of theoretical calculations and experimental visualization of samples cross-sections.

In this paper, the apparent density and porosity of nuclear graphites IG-110 and NBG-18 were measured using a geometric method. The specific surface area, the pore volume, and the average pore size were calculated based on the BET method. The pore structures including open and close pores within nuclear graphites IG-110 and NBG-18 are presented and compared through 3-D micrographs.

EXPERIMENTAL METHOD

Two grades of nuclear graphite, NBG-18 and IG-110, were selected for porosity characterization and analysis. Samples were sectioned into cubes of an approximate size of 10mm×10mm×10mm using a low speed diamond saw. All surfaces were ground and polished down to 1200 grit paper. The polished samples were subsequently cleaned in an ultrasonic bath then rinsed in reverse osmosis water. Length, width and height of the dried cubes were measured multiple times using a digital caliper, and the mean values were chosen to calculate the total volume, denoted as V. The dried weight in air (W_1) and the suspension weight in water (W_2) were obtained by multiple weighings of the sample on a digital balance with a gravity kit. The theoretical density of single crystal graphite, ρ_g=2.265g/cm^3, and water density, ρ_w=1.0g/cm^3, were used in the following calculations.

The apparent density, ρ, was calculated from the average sample weight (W_1) and the total volume (V). Total porosity was determined from the difference between apparent density and the density of graphite crystal. If the closed pores (inaccessible to water) and open pores (accessible to water) are denoted as V_i and V_a, respectively, then the volumes of the closed and open pores plus graphite solid volume (V_g) equals to the total volume. The difference of the weight measured in air and in water results from the buoyancy of graphite plus closed pores within graphite, was calculated according to the relationship, W_1-W_2= ρ_w(V_g+V_i). Finally, the open porosity and the closed porosity were obtained from the ratio of the pore volume to the total volume.

To measure the specific surface area and pore volume, specimens used in the BET measurements were sectioned into slender strips, cleaned and dried, and cut into cubes with dimensions of about 5mm×5mm×5mm. The cubes were dried in a vacuum desiccator for several hours and then put into pellet cell. The pellet cell was degassed at

50° C for 1 hour before adsorbing nitrogen. Based on the assumption of cylindrical pore geometry, the specific surface area, pore volume, average pore diameter and pore size distribution were calculated using computer software (AS1Win Version 1.5X, Quantachrome Instruments Company).

To reveal the structure of the open and closed pores within the samples, 3-D micrographs were obtained by reconstructing the SEM images collected from three orthogonal faces. These faces were polished to a 0.5μm finish using diamond lapping film (DLF) on a Multiprep Polisher (MultiPrep™ System, Allied High Tech.). The SEM images were taken at an acceleration voltage of 3kV using a secondary electron detector (LEO DSM 1530, LEO Electron Microscopy Ltd.). Using the software ImageJ, V1.44h, an area of 1000μm×1000μm was selected from each SEM image and labeled as a topview, sideview and frontview. Each pair of faces share an edge. The 3-D micrographs were constructed by rotating the SEM images and filling them into a 3-D cube as the top-view, side-view and front-view faces using Adobe Photoshop CS4. To compare the morphology of the filler particles, samples were cut into strips and fractured, and the fractured surfaces were examined using SEM.

RESULTS AND DISCUSSION

Porosity, especially open porosity, plays an important role in the properties of nuclear graphite, such as strength and oxidation behavior. The apparent densities and porosities were calculated based on the calculations described in the experimental method section. The results are shown in table 1. The apparent density of nuclear graphites IG-110 and NBG-18 was 1.76 ± 0.02 g/cm^3 and 1.87 ± 0.02 g/cm^3, respectively. Both values are significantly lower than the density of graphite crystal (2.265 g/cm^3) due to internal porosity [6]. The density of graphite becomes lower as the pore volume increases, because the gases introduced during manufacturing occupied a portion of the graphite. Graphite IG-110 has higher porosity than NBG-18 according to the difference in apparent density. This difference is consistent with the values calculated by using the Archimedes' method. In an experiment, the accurate porosity in the nuclear graphite is very difficult to measure due to the complicated geometry of the pores. The open porosity of both samples was roughly estimated by measuring the weight change of open pores occupied by water. As for the closed porosity, it can be obtained numerically by subtracting the open porosity from the total porosity. The results show that IG-110 has about twice the open porosity of NBG-18. However, this open porosity just reflects the open pores that have a big enough size to let water naturally flow in without external force. If the diameter of pores is sufficiently small, it's necessary to add a force to pressurize the liquid or gas into the small pores. So the BET method was used for comparative estimates of small size pores for pore structure analysis.

Table 1. Density and porosity of nuclear graphites IG-110 and NBG-18, measured by geometrical method.

Grade	Apparent Density (g/cm^3)	Total Porosity (%)	Open Porosity (%)	Closed Porosity (%)
IG-110	1.76(2)	22.26	13.59	8.67
NBG-18	1.87(2)	17.27	6.77	10.50

BET is the most widely used procedure to measure the specific surface area and pore volume, as well as pore size distribution in porous materials [8,10,13,14]. In the BET procedure, the data of the volume of adsorbed and desorbed nitrogen at relative pressure, P/P$_0$, was collected to plot an isotherm. The shape of a pore can be deduced from the hysteresis loop of the nitrogen adsorption and desorption isotherms at 77.35K as shown in Figure 1. Based on the hypothesis that (a) the nitrogen gas molecules are physically adsorbed by graphite in layers infinitely and (b) no interaction occurs between each layer, the specific surface area (S) can be calculated from the total surface area (S$_t$) and sample weight (W), according to the following equations:

$$\frac{1}{W((^{P_0}/_p)-1)} = \frac{1}{W_m C} + \frac{C-1}{W_m C}\left(\frac{P}{P_0}\right) \tag{1}$$

$$S = \frac{W_m N A_{CS}}{MW} \tag{2}$$

where W is the weight of gas adsorbed at a relative pressure, P/P$_0$, W$_m$ is the weight of adsorbate constituting a monolayer of surface coverage, C is a BET constant, N is Avogadro's number, M is the molecular weight of the adsorbate, and A$_{cs}$ is the molecular cross-sectional area of nitrogen, 16.2Å^2 at 77K.

The specific surface area calculated using the multipoint BET method was 4.775×10^{-1} m^2/g for IG-110 and 3.601×10^{-1} m^2/g for NBG-18. In Figure 1, the isotherms of both samples formed similar hysteresis loops. According to the classification of de Boer[15], these loops were pertained to type B, which refers to slit shaped pores. The results agree well with the published results[12]. Assuming that all open pores were filled with liquid adsorbate at a relative pressure close to unity, the total open pore volume can be derived from the amount of vapor adsorbed. This value was determined to be 4.615×10^{-3} cc/g for IG-110 and 1.303×10^{-3} cc/g for NBG-18. Therefore, IG-110 has higher open porosity than NBG-18, which is consistent with the result measured using the geometric method. The average open pore diameter was calculated to be 3.866×10^2Å for IG-110 and 1.447×10^2Å for NBG-18 based on the assumption that all pores have cylindrical pore geometry. All these results just represent the open pores with diameter < 350nm for IG-110 and <410nm for NBG-18. The pores with larger diameter than these values were considered as the geometric surface of samples because they are filled out instantly by nitrogen gas at P_0, with no adsorbate change with varying of relative pressure. Based on above data, the exact pore structure in both samples is still not clear, thus 3-D micrographs were constructed using the aforementioned method to further investigate.

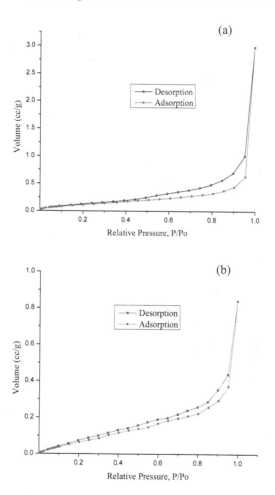

Figure 1. Nitrogen adsorption and desorption isotherms at 77.35K for nuclear graphites (a) IG-110 and (b) NBG-18.

Figure 2 shows the 3-D micrographs of the pore structure of IG-110 and NBG-18. Each 3-D micrograph is composed of three SEM images and each image in top-view, side-view and front-view has the same area of 1mm². 3-D micrographs in Figure 2 clearly show the difference of pore structure in both samples. The pore size of IG-110 varies from a few nanometers to tens of microns and NBG-18 has the pore size ranging from a few nanometers to hundreds of microns. In addition, from the number of pores per unit area, it can be inferred that the pore distribution in IG-110 is more homogeneous than that in NBG-18 even though the anisotropy of the pore structure cannot be quantified. In Figure 2b, a pore with a worm-like shape and a diameter of about 100μm in the side face connects a pore in top face with a pore in front face. For these types of interconnected pores and pore networks, nitrogen or other gases and liquids can pass through without capillary force. When measuring the specific surface area using the BET method, the inside surface of these interconnected pores were counted as a part of geometric surface because the filled nitrogen gas does not change with relative pressure. Nevertheless, such a pore network will act as a passage for gas molecules and thus increase the oxidation rate of nuclear graphite as a result of an increase in exposure area.

(a)   (b)

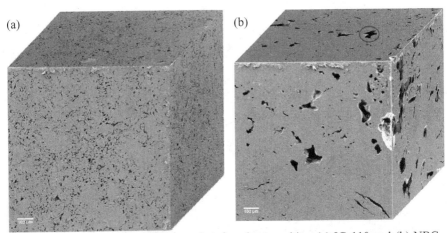

Figure 2. Three dimensional micrographs of nuclear graphites (a) IG-110 and (b) NBG-18. Scale bar is shown on the lower left corner in each image.

One open pore (marked by a blue circle in figure 2b) on the top face of NBG-18 was selected for further examination of the morphology of the pore inside walls. Figure 3a shows a rough surface of the pore interior with many cracks formed randomly on the surface. The cracks have various sizes ranging from a few nanometers to a few microns. The arrows in Figure 3b point to some thin films inside the cracks, which separate a crack into several small cells. These cracks are assumed to have formed during the manufacturing processes [16-19].

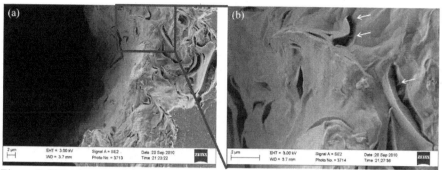

Figure 3. SEM images of the morphology of the inside surface a pore in nuclear graphite NBG-18 (marked with a blue circle on the top face of NBG-18 micrograph in figure 2).

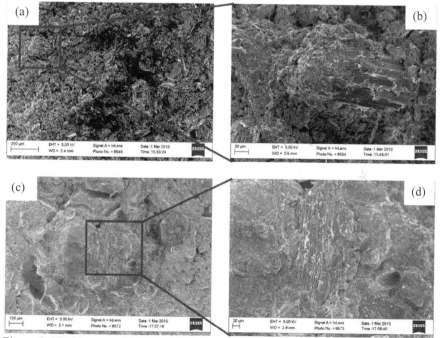

Figure 4. SEM images of fractured surfaces of nuclear graphites (a) and (b) IG-110 and (c) and (d) NBG-18.

The morphology of the filler particles in nuclear graphite plays an important role in the mechanical properties. Figure 4 shows the morphology of the fractured surface of the nuclear graphites IG-110 and NBG-18. Figure 4a shows a graphite filler particle with a length of ~160μm and width of ~100μm in IG-110. The region marked with a blue rectangle in Figure 4a was magnified and shown in Figure 4b. It shows a filler particle composed of many overlaid graphite sheets. Its appearance might be the cross-section of one integrated filler particle, which formed during the breaking of strip samples. Figure 4c and d show the fractured surface of nuclear graphite NBG-18. The shapes of entrapped gas bubbles were revealed from the holes in the fractured surface. Their diameters were up to about 150μm. These gas bubbles were likely formed during manufacturing [16]. In Figure 4d, the filler particles in NBG-18 have a similar appearance as IG-110 even though they were made from different source coke [4]. The difference in the morphology of the fractured surface mainly results from the different forming methods, the iso-static molded method for IG-110 and the vibration molded method for NBG-18 [4]. Comparing the fractured surface of both samples, it seems that the vibration molded method is a preferred way to make more compact nuclear graphite than the iso-static molded method but it produces lots of gas bubbles. This could be the main reason that the porosity of nuclear graphite IG-110 is higher than NBG-18 and that the pore diameter for NBG-18 is much higher than for IG-110.

CONCLUSIONS

Porosity in graphite used in high temperature nuclear reactors can strongly influence their mechanical behavior, oxidation resistance and irradiation response during service. The porosity in two primary grades of nuclear graphites IG-110 and NBG-18 has been investigated.

Using the geometric method, the open porosity of IG-110 was determined to be 13.59%, which is twice high as that of NBG-18 which is 6.77%. Nitrogen adsorption and desorption isotherms in the BET measurements showed that IG-110 had higher specific surface area than NBG-18 and a structure consisting of slit-shaped micro-pores. 3-D micrographs revealed the differences of the pore structures between the two grades of graphites. The pore size varied from a few nanometers to tens of microns in IG-110 and from a few nanometers to hundreds of microns in NBG-18. SEM images of fractured surfaces showed that the filler particles in NBG-18 have a similar appearance as IG-110. A typical filler particle size is up to about 160μm in length and 100μm in width.

ACKNOWLEDGEMENTS

This work is supported by the Nuclear Energy University Programs (NEUP) through the US Department of Energy (DOE) under the contract 00089350. The authors thank Dr. Lance L. Snead from Oak Ridge National Laboratory for providing the graphites NBG-18 and IG-110.

REFERENCES

[1] K. Fukuda, T. Sawai, and K. Ikawa, Diffusion of Fission Products in Matrix Graphite for VHTR Fuel Compacts, *J. Nucl. Sci. Tech.*, **21**[2], 126-32 (1984).

[2] T. Tanabe, K. Niwase, N. Tsukuda, and E. Kuramoto, On the Characterization of Graphite, *J. Nucl. Mater.*, **191-194**, 330-34 (1992).

[3] E. S. Kim, H. C. No, B. J. Kim, and C. H. Oh, Estimation of Graphite Density and Mechanical Strength Variation of VHTR During Air-Ingress Accident, *Nucl. Eng. Des.*, **238**, 837-47 (2008).

[4] S. H. Chi, and G. C. Kim, Comparison of the Oxidation Rate and Degree of Graphitization of Selected IG and NBG Nuclear Graphite Grades, *J. Nucl. Mater.*, **381**, 9-14(2008).

[5] V. Bernardet, S. Gomes, S. Delpeux, M. Dubois, K. Guerin, D. Avignant, G. Renaudin, and L. Duclaux, Protection of Nuclear Graphite Toward Fluoride Molten Salt by Glassy Carbon Deposit, *J. Nucl. Mater.*, **384**, 292-302(2009).

[6] C. E. Vaudey, N. Toulhoat, N. Moncoffre, N. Bererd, L. Raimbault, P. Sainsot, J. N. Rouzaud, and A. P. Mabilon, Thermal Behaviour of Chlorine in Nuclear Graphite at a Microscopic Scale, *J. Nucl. Mater.*, **395**, 62-68(2009).

[7] N. N. Nemeth, and R. L. Bratton, Overview of Statistical Models of Fracture for Nonirradiated Nuclear-Graphite Components, *Nucl. Eng. Des.*, **240**, 1-29(2010).

[8] E. Hoinkis, and E. Robens, Surface Area and Porosity of Unmodified Graphitic Matrices A3-27 and A3-3(1950) and Oxidized Matrix A3-3(1950), *Carbon*, **27**, 157-68(1989).

[9] E. Hoinkis, W. P. Eatherly, P. Krautwasser, and E. Robens, Corrosion- and Irradiation-Induced Porosity Changes of a Nuclear Graphitic Materials, *J. Nucl. Mater.*, **141-143**, 87-95(1986).

[10] Z. Spitzer, V. Biba, F. Bohac, and E. Malkova, Micropore Structure Analysis of Coal from Adsorption Isotherms of Methanol, *Fuel*, **56**, 313-18(1977)

[11] F. Gianni, and F. Potenza, Spectrographic Determination of Boron in Nuclear Graphite, *Anal. Chim. Acta*, **25**, 90-94(1961).

[12] O. N. Shornikova, E. V. Kogan, N. E. Sorokina, and V. V. Avdeev, The Specific Surface Area and Porous Structure of Graphite Materials, *Russia Journal of Physical Chemistry A*, **83**[6], 1022-25(2009).

[13] A. Celzard, J. F. Mareche, and G. Furdin, Surface Area of Compressed Expanded Graphite, *Carbon*, **40**, 2713-18(2002).

[14] J. J. Thomas, J. Hsieh, and H. M. Jennings, Effect of Carbonation on the Nitrogen BET Surface Area of Hardened Portland Cement Paste, *Advn. Cem. Bas. Mat.*, **3**, 76-80(1996).

[15] J. H. de Boer, Symposium, "The Structure and Properties of Porous Materials", Butterworths, London, 68(1958).

[16] H. Kakui, and T. Oku, Crack growth properties of nuclear graphite under cyclic loading conditions, *J. Nucl. Mater.*, **137**, 124-129 (1986).

[17] M. R. Joyce, T. J. Marrow, P. Mummery, B. J. Marsden, Observation of microstructure deformation and damage in nuclear graphite, *Engng. Fract. Mech.*, **75**, 3633-45(2008).

[18] A. N. Jones, G. N. Hall, M. Joyce, A. Hodgkins, K. Wen, T. J. Marrow, B. J. Marsden, Microstructural characterization of nuclear grade graphite, *J. Nucl. Mater.*, **381**, 152-157(2008).

[19] T. D. Burchell, I. M. Pickup, B. Mcenaney and R. G. Cooke, The relationship between microstructure and reduction of elastic modulus in thermally and radiolytically corroded nuclear graphites, *Carbon*, **24**[5], 545-49(1986).

Green Technologies for Materials Manufacturing and Processing

MODIFIED POWDER PROCESSING AS A GREEN METHOD FOR FERRITE SYNTHESIS

Audrey Vecoven and Allen W. Apblett
Department of Chemistry, Oklahoma State University,
Stillwater, OK, USA 74078

ABSTRACT

A new economical method, the modified powder process, was developed for the synthesis of nickel ferrite at temperatures as low as 850°C. The method is based on the pyrolysis of a precursor consisting of a fine iron oxide powder combined with a metal-organic compound, $Ni(L)_2$ (L=acetate, acetylacetonate, gluconate). Upon heating, the decomposition of $Ni(L)_2$ leads to the coating of the iron oxide particles with nickel oxide, yielding a precursor powder in which the two component oxides are more homogeneously mixed than that which can be achieved by simple mixing of the metal oxides according to conventional powder processing. The iron oxide phase was shown to have a slight influence on the final sintering temperature necessary for complete formation of $NiFe_2O_4$. For α and γ iron(III) oxide powders of similar particle size, the ferrite nucleation was found to occur at lower temperature (875°C) for the precursor $Ni(acetate)_2$-γ-Fe_2O_3 while 1000°C was required to generate pure ferrite from $Ni(acetate)_2$-α-Fe_2O_3. The use of nanoparticulate iron-oxide only lowered the formation of nickel ferrite by a small increment. The lowest temperature preparation of nickel ferrite was realized using nickel acetylacetonate and γ–FeOOH. The morphologies and surface area of the products were found to depend strongly on the nickel oxide precursor, high surface area porous solids were obtained when using nickel acetate, large-grained $NiFe_2O_4$ was produced when using nickel acetylacetonate, while the nickel gluconate yielded a tortuous web-like material with large pores.

INTRODUCTION

The conventional route for industrial production of metal oxide ceramic materials is through powder processing that is based on high temperature processing of inexpensive raw materials. The methods consist of mixing stoichiometric amounts of the corresponding oxide powders usually the metal oxides or carbonates (e.g. NiO and Fe_2O_3 for $NiFe_2O_4$ synthesis), after which the mixtures are ground. The powders are then calcined, sometimes after compaction; sintering can be repeated several times with intermediate grinding stages to optimize surface contact between particles. The phase boundary reaction occurs at the points of contact between the two components and later by counter diffusion of the metal cations through the product phase. The relative simplicity of this process and the use of cost effective components such as oxides and carbonates constitute the two major advantages of the powder processing method, also known as the "ceramic method". However, the ceramic method suffers from several disadvantages including loss of volatile oxides during high temperature treatments and the slowing of the reaction rates as the reaction progresses as the diffusion paths of the constituents through the product phase become longer. The intermittent grinding stages between heating cycles help, to some extent, optimize the surface contact between unreacted metal oxide particles. Due to a lack of monitoring ability, the ideal experimental conditions that will lead the reaction to completion are determined by trial error. Therefore, the desired product is often mixed with small quantities of impurities such as reactants or intermediate phases.

Various "pre-ceramic" routes have been developed to overcome the limitations of the ceramic method and yield higher quality products with the desired shapes, forms, and purities for targeted applications [1]. Most have the objective of decreasing the diffusion path lengths of metal ions by bringing down the particle size to a few hundred angstroms, and thus effect a more intimate mixing of the reactants. Metal-organic deposition MOD [2,3] is one such method that provides an alternative method for synthesizing oxides. It is a non-vacuum, solution-based method for depositing thin films. In the MOD process, a suitable metal-organic precursor dissolved in an appropriate solvent is coated on a substrate by spin-coating, screen printing, or spray- or dip-coating. The soft metal-organic film is then pyrolyzed in air, oxygen, nitrogen or other suitable atmosphere to convert the precursors to their constituent elements, oxides, or other compounds. Shrinkage generally occurs only in the vertical dimension so conformal coverage of a substrate may be realized. Metal carboxylates with long slightly-branched alkyl chains (e.g. 2-ethylhexanoate or neodecanoate) are often used as precursors for ceramic oxides since they are usually air-stable, soluble in organic solvents, and decompose readily to the metal oxides

Combining the advantages of the powder process with those of the MOD process yields a hybrid method that we have named modified powder processing. This hybrid method combines the economical aspects of the conventional powder process by using an inexpensive raw material as one reactant (such as Fe_2O_3 or Al_2O_3), and the strengths of the MOD methods by coating the finely ground raw material particles with a second metal oxide phase. This approach is illustrated in Figure 1 (B) in comparison to conventional powder processing (A). It provides a precursor mixture that has homogeneity that is intermediate between that of pure chemical routes and conventional powder processing, and since the ions have less distance to travel, the solid-state reactions are complete at lower temperatures and in less time.

   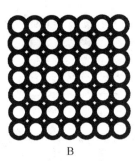

A   B

Figure 1. Comparison of metal oxide distribution in (A) conventional powder processing and (B) modified powder processing:

The first application of the modified powder processing method involved the use of a finely ground ferric oxide and a liquid nickel carboxylate for the synthesis of nickel ferrite [4]. The liquid metal carboxylate, Ni[2-(2-(2-methoxyethoxy)ethoxy)acetate]•$0.5H_2O$ (Ni(MEEA)$_2$•$0.5H_2O$) was employed to uniformly coat a hematite powder with a continuous film of NiO. Pyrolysis of the precursor to 500°C produced a powder with microscopic features similar to those of the starting iron oxide, suggesting a very homogeneous coating of Fe_2O_3 particles with NiO. This was attributed to a possible in-situ CVD reaction during which $Ni(MEEA)_2$ is partially volatilized at its decomposition point and uniformly deposited onto the Fe_2O_3 particles where it undergoes thermal decomposition to form NiO. Upon heating to 800°C the material was completely converted to small porous particles of trevorite, $NiFe_2O_4$ (crystallite size less than 18 nm). The striking difference of morphologies between the initial chunk-like particles and the puffed trevorite particles is accounted for by the diffusion of Fe_2O_3 to the NiO interface, through the product phase as illustrated in Figure 2.

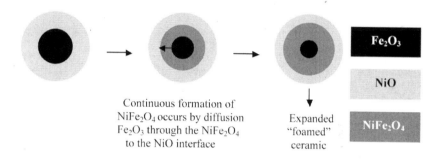

Continuous formation of $NiFe_2O_4$ occurs by diffusion Fe_2O_3 through the $NiFe_2O_4$ to the NiO interface

Expanded "foamed" ceramic

Fe_2O_3

NiO

$NiFe_2O_4$

Figure 2. Mechanism of ferrite formation by nickel oxide coated ferric oxide powders.

Other MEEA salts were also successfully applied to ceramics preparation using the modified powder processing described above. For example $MgAl_2O_4$ was obtained using $Mg(MEEA)_2 \cdot 2H_2O$ and aluminum oxide. However, the morphology of the final product did not exhibit the same features as $NiFe_2O_4$ described above [5]. Despite the proven usefulness of the liquid MEEA for the preparation of spinels, its high cost renders any potential industrial application unlikely. Therefore the research reported herein focuses on the application of the modified powder processing method for low temperature preparation of nickel ferrite by utilizing more economical precursors. In this study, a variety of nickel carboxylates were applied to the modified powder processing method, and the nature of the metal organic compound correlated with the morphology (microscopy, surface area, crystallite size) of the desired final product and the processing temperature necessary for completion of the solid state reaction. In addition to the influence of the metal carboxylate characteristics on the final product, the nature of the starting iron oxide source was also investigated. The precursors that were utilized in this investigation were nickel acetate, nickel gluconate, and nickel acetylacetonate.

EXPERIMENTAL

Hematite nanoparticles [6] and lepidocrocite (γ–FeOOH) [7] were synthesized according to procedures described in the literature. The lepicrodite isolated was a hydrate with 0.60 molar equivalents of water per formula unit. High purity α iron oxide (hematite) was puratronic grade from Alfa Aesar. All other reagents were commercial products (ACS reagent grade or higher) and were used without further purification. Water was purified by reverse osmosis and was deionized before use. X-ray powder diffraction (XRD) patterns were recorded on a Bruker AXS D-8 Advance X-ray powder diffractometer using copper K_α radiation. The diffraction patterns were recorded in the 2θ range of 17-70° with a step size of 0.02° and a counting time of 18 seconds per step. Crystalline phases were identified using a search/match program and the PDF-2 database of the International Centre for Diffraction Data [8]. For infrared spectroscopic measurements, roughly 10 mg of the sample was mixed with approximately 100 mg FTIR-grade potassium bromide and the blend was finely ground. Spectra in the range of 4000 to 400 cm^{-1} were collected by diffuse reflectance of the ground powder with a Nicolet Magna-IR 750 spectrometer. Typically, 128 scans were recorded and averaged for each sample (4.0 cm^{-1} resolution) and the background was automatically subtracted. Thermogravimetric studies were performed on a Seiko EXSTAR 6000 TG/DTA 6200 instrument under a 100ml/min flow of nitrogen. Bulk pyrolyses at various temperatures were performed in ambient air in a temperature programmable muffle furnace using a temperature ramp of 1°C/min and a hold time of 14 hours.. Surface area measurements were performed on a Quantachrome Nova 1200 instrument by nitrogen adsorption and the BET six-point method. Samples were prepared prior to surface analysis by degassing under vacuum at 100°C. Scanning electron micrographs were recorded on a JEOL JXM 6400 Scanning Electron Microscope.

Preparation of Nickel Gluconate Hydrate
1.86 grams (0.020 moles) of $Ni(OH)_2$ and 17.44 grams of a 45-50% solution (0.040 moles) of D-gluconic acid were added to 250 ml of distilled water. The solution was heated under reflux with magnetic stirring for two days. The resulting green solution was concentrated by evaporating 2/3 of the solvent under reduced pressure in a water bath at a temperature range of 50-60°C. Nickel gluconate was isolated by precipitation from the dark green solution after addition of 700 ml of methanol. The resulting pale green solid was collected by filtration using a medium porosity fritted glass filter, then washed twice with 200 ml of methanol, and was then dried under vacuum overnight at ambient temperature. The reaction yielded 8.11 g (83.69 %) of hydrated nickel gluconate, $Ni(C_6H_{11}O_7) \cdot 2.2H_2O$ (484.69 g/mol) as a pale green solid. IR (cm^{-1})(KBr) 3498(s,br), 2886(s), 2823(m), 1642(s), 1553(m), 1479(m), 1295(m), 1138(m), 1103(m), 885(m), 749(s).

General Procedure for Precursor Preparation
The experiments listed in Table 1 consisted of the pyrolysis of 5 mmol of $Ni(L)_2$ with 5 mmol of an iron(III) oxide powder or 10 mmol of FeOOH. In these reactions, L represents acetate, gluconate, or acetylacetonate ligands. Due to the small amount of available nanoparticulate Fe_2O_3, experiment #3 was performed with only 2.2 mmoles of $Ni(acetate)_2$ and 2.2 mmoles of Fe_2O_3. A predetermined volume of the appropriate solvent was added to the metal-organic reactant and the mixture was gently heated (about 50°C) to obtain a clear homogeneous solution. The metal oxide powder was then rapidly added to the warm solution and the resulting mixture was slowly stirred it cooled down and thickened to a paste. The precursors were subsequently heated to 450°C with a 1°C/min ramp rate in a muffle furnace. The resulting brown solids were weighed and then ground in a mortar and pestle. A fraction was conserved in a vial for X-ray and microscopy studies while the remainder was pyrolyzed to 800°C. Reaction completion temperatures were determined by collecting the X-ray diffraction pattern of the sintered sample, starting at 800°C, and working by increments of 50°C until disappearance of the starting metal oxide peaks. Metal gluconates and acetates are highly water-soluble metal-organic compounds so 2.5 ml of water was sufficient to dissolve the metal precursor as described above. Exp. #3 only required 1ml. On the other hand the solubility of metal acetylacetonates in water

was not sufficiently high to reach complete dissolution of the metal precursor, even under heating. Therefore, N,N-Dimethylformamide (DMF) was used instead and 2 mL of it was sufficient to appropriately dissolve the precursor. In order to estimate the importance of the solvent on the solid-state reaction completion, All pyrolyzed samples were compact powders with the exception of experiments that included nickel(II) or iron (II) gluconate, whose thermal decompositions were accompanied by a "foaming" process.

Table I. Nickel Ferrite Precursor Compositions

#	Nickel source	Weight (g)	Iron source*	Weight (g)	Solvent	Yield g (%)
1	Nickel gluconate	2.43	α-Fe_2O_3	0.799	H_2O	1.10 (94.1)
2	Nickel acetate	1.25	α-Fe_2O_3	0.799	H_2O	1.16 (98.9)
3	Nickel acetate	0.55	α-Fe_2O_3 NP	0.351	H_2O	0.501 (97.2)
4	Nickel acetyl acetonate	1.46	α-Fe_2O_3	0.800	DMF	1.17 (100)
5	Nickel gluconate	2.42	γ-Fe_2O_3	0.799	H_2O	1.12 (95.5)
6	Nickel acetate	1.25	γ-Fe_2O_3	0.800	H_2O	1.13 (96.2)
7	Nickel acetyl acetonate	1.44	γ-Fe_2O_3	0.798	DMF	1.12 (95.6)
8	Nickel gluconate	2.42	γ-FeOOH	0.994	H_2O	1.16 (99.1%)
9	Nickel acetate	1.24	γ FeOOH	0.993	H_2O	1.16 (99.2%)
10	Nickel acetyl acetonate	1.44	γ-FeOOH	0.995	DMF	1.16 (99.0%)

NP: Nanoparticulate oxide

RESULTS AND DISCUSSION

The yield of the metal oxides (Table I) from all of the experiments were somewhat lower than the theoretical amounts calculated from the metal content of the reagents. Nevertheless, unreacted metal oxide peaks were not observable in the X-ray powder patterns once the ferrites had crystallized. Therefore, the stoichiometry for ferrite formation seems to have been conserved. Yields were closest to total conversion for samples containing gamma iron oxyhydroxide. The commercially available iron oxide powders were used without further purification or drying, and, therefore, may have contained adsorbed molecules unaccounted for in the molecular weight. In the case of iron oxyhydroxide, which had to be prepared, a small amount of iron reagent was pyrolyzed to 450°C to determine the exact water content and molecular weight. Thus, the yields of nickel ferrite from γ-FeOOH were closest to unity (Figure 2).

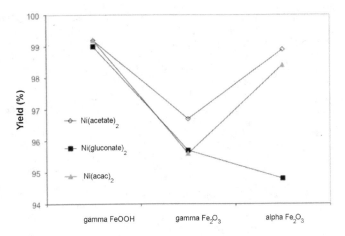

Figure 2. Nickel Ferrite Yields

In addition to molecules adsorbed on oxide powders, partial evaporation of the metal organic precursor during thermal decomposition can also account for the lowered yields. Depending on the ligand and the metal atom, a small fraction of the compound is evolved and undergoes pyrolysis in the gas phase. The extent of this effect was evaluated for each precursor by heating a pre-weighed amount of the metal organic compound to 450°C. The resulting powder was considered to be pure metal oxide, since 450°C is well above the decomposition temperatures of the metal-organic compounds utilized. The metal oxide yields reported in Table 3 illustrates the degree of volatility of each reagent. Acetylacetonate metal salts are common reagents for Chemical Vapor Deposition and have high volatility. It is suspected that a fraction of Ni(acac)$_2$ sublimed during thermal treatment with the majority depositing nickel oxide onto the iron oxide surface, but a small amount was lost from the reaction mixture. It should be noted that the trends observed in Table II match those in Figure 3: precursors containing gluconate metal salts produced ferrites in markedly higher yields than did acetate and acetylacetonate precursors.

Table II. Metal Oxide Yield (%) from Pyrolysis of Ni(II)(L)$_2$ at 450°C

Ni(gluconate)$_2$•2H$_2$O	98.1
Ni(acac)$_2$•H$_2$O	97.2
Ni(acetate)$_2$•2H$_2$O	99.9

The plots of reaction completion temperatures of nickel ferrite and zinc ferrite versus iron oxide form are shown in Figure 3. Reported temperatures for nickel ferrite preparation using the conventional method range from 1100 to 1200°C. The modified powder process allows one to obtain nickel ferrites at temperatures between 860 and 1100°C. It was found that the required processing temperatures were lower when using γ-Fe$_2$O$_3$ rather than α-Fe$_2$O$_3$ as the iron source and this effect was most pronounced when using Ni(acetate)$_2$ as the nickel oxide source. Two aspects of the reaction must be considered to account for this trend: first, the thermal behavior of γ-Fe_2O_3 and, second, the microstructural properties of each starting metal oxide powder. The α structure is the thermally stable form of iron oxide, and exposure of the metastable γ-phase to heat results in its irreversible transformation into α-Fe_2O_3, usually at temperatures ranging from 400 to 550 °C, depending on the microstructural properties [9]. The starting γ-phase possesses an inverse spinel structure with lattice parameter a=8.352 Å, and consists of face centered cubic (fcc) stacking of O^{2-} and random distribution of Fe^{3+} in both octahedral and tetrahedral sites along with a high concentration of cation vacancies [10]. Investigation of the nature of the γ to α phase conversion revealed the transformation takes place along the [110] direction of the spinel lattice in a topotatic fashion and is accompanied by the restacking of O^{2-} from fcc to hcp ionic arrangement (a=5.035 Å, c=13.75 Å) [11]. Phase transformations are characterized by an increase in reactivity and are therefore more favorable to chemical reactions with surrounding metal oxide particles, which results in early ferrite nucleation [12]. Previous studies on the comparison of the preparation of ferrites with α and γ iron oxides reported lower sintering temperatures and increased rates of "ferritization" when the γ form was employed [13, 14, 15].

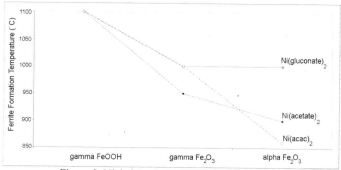

Figure 3. Nickel Ferrite Processing Temperatures

Microstructure is an essential factor in the improvement of reactivity of the starting metal oxide powders. For example, smaller particles have high surface energy, and are characterized by a low particle volume:particle surface ratio, they spontaneously attempt to minimize the latter by reacting with surrounding material and thereby increase their volume:surface ratio. Surface area measurements allow one to estimate the specimen grain size, hence the reactivity, and the surface available for NiO deposition around the iron oxide particles. The surface areas of the iron oxide powders investigated are listed in Table 4. The surface area measurements are in good agreement with the trends observed in Figure 3 where the processing temperature drops more or less sharply for each nickel precursor in the series α-Fe_2O_3, γ-Fe_2O_3, γ-FeOOH. The values reported in Table II suggest that the particle size of the gamma iron oxide is considerably smaller than that of the alpha iron oxide powder. In order to estimate the effect of the structure of Fe_2O_3 on the formation temperature, $NiFe_2O_4$ was prepared with a combination of α-Fe_2O_3 nanoparticles with Ni(acetate)$_2$, and a precursor that combined α-FeOOH with Ni(acetate)$_2$. Both precursors yielded pure nickel ferrite at 850°C, which is slightly lower than the 900 and 950°C required for trevorite preparation from the corresponding γ-Fe_2O_3-Ni(acetate)$_2$ and γ-

FeOOH-Ni(acetate)$_2$ precursors It can be concluded than the choice of the iron oxide phase influences the completion temperature for formation of nickel ferrite but the effect is a small one.

Table II. Iron Oxide Powder Surface Area Measurements:

Iron Oxide powder	Surface area (m^2/g)
γ-Fe$_2$O$_3$	83.2
α-Fe$_2$O$_3$	15.4
γ-FeOOH	183.3

Figure 4 shows the X-ray diffraction patterns of the products obtained by heating the Ni(acac)$_2$-γ-Fe$_2$O$_3$ precursor at three different temperatures: 300°C (a), 600°C (b), 1000°C (c). The removal of the acetylacetonate ligands occurs gradually over a wide temperature range, as shown in the TGA diagram Figure 5, and is fairly complete by 350°C. The initial weight loss is attributed to the evaporation of the water of hydration and of N,N dimethylformamide. The TG/DTA study of Ni(acac)$_2$ showed the decomposition was complete by 338°C and was then followed by a slight weight gain of 1.9% attributed to nickel metal oxidation to metal oxide. This was confirmed by the precursor's X-ray powder pattern at 300°C where the characteristic peaks of nickel oxide and nickel metal (2θ = 44.5 and 51.9) were identified. At 600°C sharp peaks for hematite (α-Fe$_2$O$_3$) and nickel oxide were observed while at 1000°C, trevorite, NiFe$_2$O$_4$ had formed.

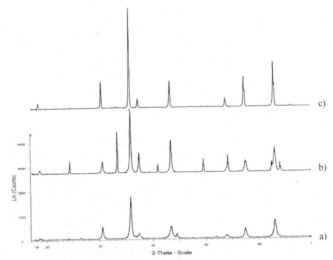

Figure 4. XRD of Ni(acac)$_2$-γ-Fe$_2$O$_3$ Pyrolyzed at 300°C (a), 600°C (b) and 1000°C (c)

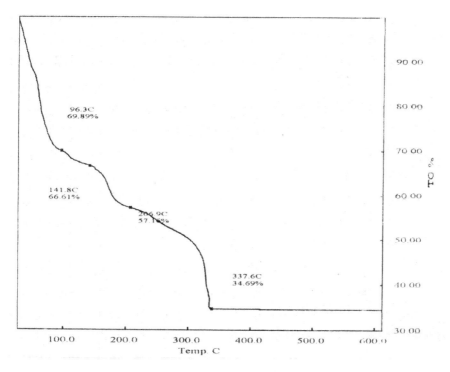

Figure 5. TGA of NiFe$_2$O$_4$ Precursor Ni(acac)$_2$-γ-Fe$_2$O$_3$

    The SEM micrographs of Ni(acetate)$_2$-γ-Fe$_2$O$_3$ at 450°C (a) and 850°C (b) shown in Figure 6 provide a good illustration of the modified powder process. The starting γ iron oxide is a fine powder with particle size in the 80-90 nm as determined by SEM microscopy. After Ni(acetate)$_2$ is coated onto the iron powder and decomposed at 450°C, a solid is created that is composed of what appears to be smooth nickel oxide films with embedded iron oxide grains (Figure 6 (a)). When the mixture is heated to 850°C and it converts to nickel ferrite, the resulting solid maintains a sheet-like appearance but has become significantly more porous. There are large micron-sized pores similar to those present at 450°C but there are also numerous smaller pores that approximate the size of the original iron particles. As a result of the macroporous nature of the nickel ferrite product, it still maintains a significant surface area of 14.3 m^2/g. In general sintering would be expected to lead to densification and collapse of porous networks. In fact this has occurred since the total surface area has dropped but this appears to be due to the loss of small pores in the web-like strands of the final product. The opening of larger pores is the result of the mechanism of solid-state reaction between NiO and Fe$_2$O$_3$. In this reaction, NiFe$_2$O$_4$ forms at the interface between the nickel and iron oxide. Then subsequent formation of nickel ferrite occurs by diffusion of iron oxide through the nickel ferrite barrier film to the point where it can react with nickel oxide. Since nickel oxide does not diffuse inward, the result is an outward migration of iron. In the case of this precursor, the iron flows into the sheet-like networks of nickel oxide, leaving pores behind and generating a web-like morphology for the final product. The observed behavior is different than that of Ni(MEEA)$_2$ in which uniformly coated particles puffed into broccoli flower-like collection of small particles [4]. Thus, it has been demonstrated that the morphology at the deposited nickel oxide, as controlled by the choice of nickel precursor, has a tremendous influence on the morphology of the final product.

As expected, Ni(acac)$_2$ produced a more homogeneous coating of nickel oxide than did the acetate. The SEM images of Ni(acac)$_2$-γ-Fe$_2$O$_3$ pyrolyzed at 130°C (a), 450°C (b), and 1000°C (c) are shown in Figure 7. At the lower temperature, crystals of dehydrated Ni(acac)$_2$ are clearly visible (Figure 7(a)) and cover much of the iron oxide particles so that the latter may be described as embedded within the crystals. The intervention of an intermediate phase such as the Ni(acac)$_2$ melt at 230°C, resulted in a very homogeneous solid with highly uniform flat surfaces. Subsequent thermal treatment to 450°C caused the structure to densify, and produce a very compact material mostly consisting of NiO, and γ-Fe$_2$O$_3$ as evidenced by the X-ray pattern. Presumably the temperature gap between the melting point and the decomposition temperature allowed for Ni(acac)$_2$ to uniformly diffuse and deposit on the iron particles before decomposing to nickel oxide. Sintering to 1000°C resulted in a very fine powder with crystallite and particle sizes of 48.9 nm and 250 nm, respectively, and lower porosity than the nickel ferrite produced by Ni(acetate)$_2$-γ-Fe$_2$O$_3$. However, the morphology of the final ferrite was completely different than that from the Ni(MEEA)$_2$ experiments [4] since this product consists of micron-sized grains of nickel ferrite (see Figure 7 (b)). The difference between the two procedures may arise from a difference in the extent of nucleation of nickel ferrite. In the case of Ni(MEEA)$_2$ no intermediate formation of NiFe$_2$O$_4$ was observed before a very abrupt transition from a NiO-α-Fe$_2$O$_3$ mixture to NiFe$_2$O$_4$. On the other hand, the XRD pattern of the Ni(acac)$_2$ precursor at 600°C contained reflections from all three crystalline phases. The cause of this difference might be attributed to two separate phenomena, one physical and one chemical. As previously demonstrated, the phase transition from γ to α-Fe$_2$O$_3$ can influence nucleation of NiFe$_2$O$_4$ if nickel ions migrated into lattice positions that were being vacated by ferric ions. If this were the case, the Ni(MEEA)$_2$ experiments which were performed with α-Fe$_2$O$_3$ would not have benefited from this phase transition. Secondly, Baron et al. have demonstrated the metal acetylacetonates can undergo metal ion exchange with aluminum-oxygen polymers, producing Al(acac)$_3$ and substituting a new metal with the metal oxide backbone [16]. A similar phenomenon in the Ni(acac)$_2$-γ-Fe$_2$O$_3$ mixture would lead to significant mixing of iron and nickel and the surface of the iron oxide and promote nucleation of NiFe$_2$O$_4$.

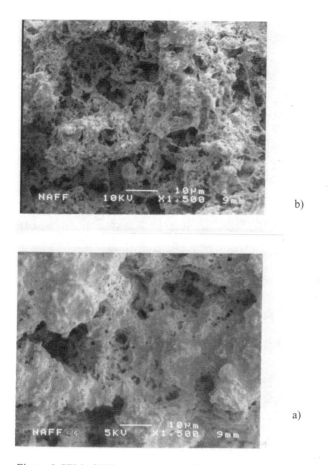

Figure 6. SEM of Ni(acetate)$_2$-γ-Fe$_2$O$_3$ Pyrolyzed at 450°C (a) and 800°C (b)

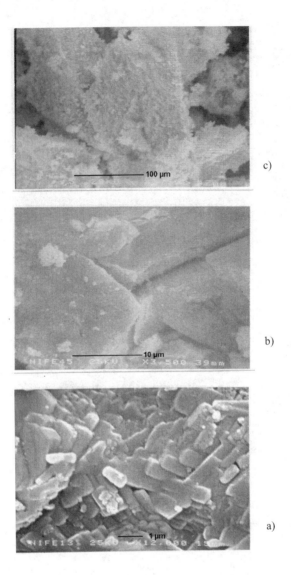

Figure 7. SEM of Ni(acac)$_2$-γ-Fe$_2$O$_3$ Precursor Pyrolyzed at 130°C (a), 450°C (b), and 1000°C (c).

The use of metal gluconates as the oxide coating source led to yet another morphology. Gluconate salts tend to caramelize upon heating in a similar fashion to glucose from which it is derived. Dehydration and combustion of this sticky intermediate leads to an expanded foamy solid. The morphology of the resulting ferrites (see Figure 8) is a tortuous web-like material with large pores. The walls of the metal oxide network appeared to be non porous and as a result, the overall surface area is low.

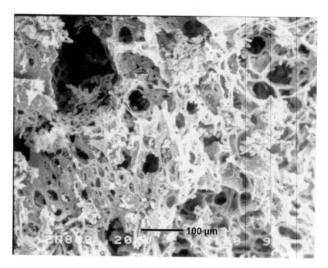

Figure 8. SEM of Ni(gluconate)$_2$-γ-Fe$_2$O$_3$ Precursor Pyrolyzed at 850°C

Precursors containing α or γ iron oxyhydroxide did not exhibit morphological features particularly different from the samples containing iron oxide. Although the particle and crystallite sizes did change with the nature of the iron source, the morphology seemed to be more dependent on the metal ligand type.

CONCLUSIONS

A new economical method, the modified powder process, was developed for the synthesis of nickel ferrite at temperatures as low as 875°C. The method is based on the pyrolysis of a precursor consisting of a fine iron oxide powder combined with a metallo-oranic compound Ni(L)$_2$. Upon heating, the decomposition of Ni(L)$_2$ leads to the coating of the iron oxide particles with the corresponding metal(II) oxide, resulting in a more homogeneous mixture of Fe$_2$O$_3$ and NiO than is typical of the conventional powder processing procedure. Based on previous studies and the final ferrite powder morphologies, it was concluded that the solid-state mechanism following NiFe$_2$O$_4$ nucleation takes place through diffusion of iron oxide through the ferrite phase to react with the outer layer of metal oxide. The iron oxide phase was shown to have a slight influence on the final sintering temperature. For α and γ iron oxide powders of similar particle size the ferrite nucleation was found to occur at lower temperatures for the precursor Ni(acetate)$_2$-γ-Fe$_2$O$_3$ and resulted in reaction completion at 900°C, while pure ferrite was obtained at 1000°C from Ni(acetate)$_2$-α-Fe$_2$O$_3$. Presumably, this effect is caused by an increase in reactivity during the phase conversion of gamma iron oxide to the alpha structure throughout the 450 to 550°C temperature range. The microstructure of the ferrite powder was not affected by the nature of the iron oxide powder phase. The nature of the organic ligand did not affect the final processing temperature to a great extent, however it was demonstrated it has a profound effect on the morphology of the ferrite powder. For example, pyrolysis of the $NiFe_2O_4$ precursor $Ni(acetate)_2$-γ-Fe_2O_3 produced a web-like material with large pores and high surface area, whereas the thermal decomposition of $Ni(acac)_2$-γ-Fe_2O_3 resulted in a pore-free very fine material. The new modified powder process allows one to prepare ferrites at a lower cost than most chemical techniques. Furthermore, the ferrite powder morphological characteristics can be tailored by using a specific metal-organic compound for precursor preparation. Further study of the modified powder process with a wider variety of coordination compounds would most likely result in the production of a range of metal oxide ceramics with diverse morphologies.

REFERENCES

[1] C.D. Chandler, C. Roger, M.J. Hampden-Smith, Chemical Aspects of Solution Routes to Perovskite-Phase Mixed-Metal Oxides from Metal-Organic Precursors, *Chem. Rev.* **93**, 1205-1241 (1993).

[2] J. V. Mantese, A. L. Micheli, A. H. Hamdi, and R. W. Vest, "Metal Organic Deposition" *Materials Research Society Bulletin.*, **XIV**, 1173 (1989).

[3] R. W. Vest, "Electronic Thin Films from Metal-organic Precursors"; pp. 303-347 in *Ceramics Films and Coatings*, Edited by J. B. Wachtman and R. A. Haber, Noyes Publications, Park Ridge, N.J., 1993.

[4] E. H. Walker, Jr., M. L. Breen, and A. W. Apblett, Preparation of Nickel Ferrite Using Liquid Metal Carboxylates, *Chem. Mater.*, **10**[5] 1265-1269 (1998).

[5] A. W. Apblett, L. E. Reinhardt, and E. H. Walker, Jr., Liquid Metal Carboxylates as Precursors for Aluminum-Containing Ceramics," *Comments Inorg. Chem.*, **20**[2-3], 83-99 (1998).

[6] M. Ozaki, S. Kratohvil and E. Matijevic Formation of Monodispersed Spindle-Type Hematite Particles, *Journal of Colloid and Interface Science*, **102** [1], 146-151 (1984).

[7] G. Bauer, Handbook of Preparative Inorganic Chemistry, **Vol. 2**, (Academic Press, New York, 1963) pp 1500-1501.

[8] "Powder Diffraction File (PDF-2)" (International Centre for Diffraction Data, Newtown Square, PA).

[9] E. Karmazsin, P. Satre and P. Vergnon, Simultaneous Thermomagnetic and Dilatometric Measurements in a Study of the Fe_2O_3 Metastable Transformation, *J. Thermal Anal.*, **28**, pp 279-284 (1983).

[10] G. Hagg, The Crystal Structure of Magnetic Ferric Oxide, Gamma-Fe_2O_3, *Z. Phys. Chem.*, **29**, 95-103 (1935)

[11] S. Kachi, N. Nakanishi, K. Kosuge, H. Hiramatsu, and M. Kiyama, Ferrites: Proceedings of the International Conference, 141-143 (1971).

[12] V.V. Boldyrev, M. Bulens, and B. Delmon, The Control of the Reactivity of Solids, Elsevier Scientific Publishing Company, 1979, pp 32-38.

[13] K.S. Rane, V.M.S. Verenkar, and P.Y. Sawant, Hydrazine method of synthesis of γ-Fe_2O_3 useful in ferrites preparation. Part IV. Preparation and characterization of magnesium ferrite, $MgFe_2O_4$, from γ-Fe_2O_3 obtained from hydrazinated iron oxyhydroxides and iron (II) carboxylatohydrazinates. *J. Mat. Science, Mat. Elect.*, **10**, 133-140 (1999).

[14] H.J. Huhn, Influence of α-Iron Sesquioxide on Ferrite Formation, *Zeitschrift fuer Chemie* **27**, 334-335 (1987).

[15] J.J. Shrotri, a.g. Bagul, s.d. Kulkarni, C.E. Deshpande, and S.K. Date, in "Proc. VI Int. Conf. On Ferrites" Tokyo Japan, 1992, pp 404.

[16] A.Kareiva, J. Harlan, D. MacQueen, D. Brent, R. Cook And A.R. Barron, Carboxylate-Substituted Alumoxanes as Processable Precursors to Transition Metal-Aluminum and Lanthanide-Aluminum Mixed-Metal Oxides: Atomic Scale Mixing via a New Transmetalation Reaction, *Chem. of Mater.*, **8(9)**, 2331-2340 (1996).

NOVEL METHOD FOR WASTE ANALYSIS USING A HIGHLY LUMINESCENT DIPLATINUM (II) OCTAPHOSPHITE COMPLEX AS A HEAVY METAL DETECTOR

Nisa T. Satumtira[†], Ali Mahdy[‡], Mohamed Chehbouni[†*], Oussama ElBjeirami[¥], and Mohammad A. Omary[†*]

[†] Department of Chemistry, University of North Texas,
Denton, Texas USA
[‡] Department of Chemistry, Computer, and Physical Sciences, Southeastern Oklahoma State
University, Durant, Oklahoma USA
[¥] Department of Chemistry, King Fahd University of Petroleum & Minerals
Dhahran, Kingdom of Saudi Arabia

* Corresponding authors

ABSTRACT

The platinum phosphite compound potassium tetrakis(μ-diphosphito)-diplatinate (II) dihydrate, $K_4[Pt_2(POP)_4] \cdot 2H_2O$ ("PtPOP"), is a well-known phosphor that absorbs light in the visible region of the spectrum, resulting in an intense green luminescence both in the solid state and in aqueous solutions. We have previously presented our oxygen-sensing capabilities for this complex and the utilization thereof in bioimaging (Satumtira et al., *PMSE Preprints* **2009**, *101*, 1203). Herein, we report that the interaction of PtPOP with metal cations of lead, thallium, silver, and tin, results in quenching of the phosphorescence, shifts in emission energy, and/or precipitation by cation exchange products characterized by XRD. These findings have significant consequences for real-time detection of heavy metals from aqueous solutions since PtPOP is highly soluble in water. This provides a novel method for waste water analysis in the metal and ceramic industry.

INTRODUCTION

$K_4[Pt_2(P_2O_5H_2)] \cdot 2H_2O$, referred to as "Platinum POP" (or PtPOP), is a well studied dimeric Pt(II) complex, with an interesting molecular structure (Figure 1). The original synthesis first published by Sadler, et al.[1] eluded to a highly emissive, water soluble phosphor whose bright green emission at 515 nm was elicited by excitation at 368 nm. Though originally thought to be a complex synthesized much earlier[2], the published crystal structure [1] proved otherwise. The emissive nature of PtPOP was attributed to the intramolecular interactions between the two platinum atoms. Absorbance spectra showed two bands at 367 and 452 nm (much weaker in intensity than the band at 367 nm), which were assigned to the $^1A_{2u} \leftarrow {}^1A_{1g}$ and $^3A_{2u} \leftarrow {}^1A_{1g}$ transitions, respectively[3]. Upon excitation of an aqueous solution at these wavelengths, a corresponding photoluminescent emission centered roughly at 515 nm was observed. Chemical quenching of the emission of PtPOP has also been observed by Peterson, et al. [4]. Though these studies were mostly limited to salts with controlled solution acidity and ionic strength, as quenchers, it has also been found that complexation of PtPOP with certain metals will extinguish the bright emission. The biological importance of PtPOP has also being investigated[5] due to the oxygen sensitivity and water solubility of PtPOP. Such characteristics make it a potentially successful candidate for use in biological imaging.

Studies by the Nagle group indicated that heavy metal ions in aqueous solutions, such as Tl^+, Ag^+, Pb^{2+}, and Sn^{2+}, can complex with PtPOP changing the emission profile[6; 7]. Structurally,

279

the Pt(II) atoms in PtPOP possess a square planar orientation. The feasibility for another metal ion to complex with PtPOP is facile, as there is an open axial position available for bond formation. In the case of Tl^+ (as reported by Nagle, et al. [8]), the changes are attributed to the formation of luminescence exciplexes between the metal ion and the lowest energy triplet excited state of Pt_2* resulting from metal-metal bonding between the reactants. The half-filled σ* $(5d_z^2)$ and σ $(6p_z)$ orbitals of Pt_2* interact at the axial site with the filled 6s and empty $6p_z$ orbitals of Tl^+ respectively forming a covalent bond between Pt_2*and Tl^+.[9] The reaction of silver ion with PtPOP is very similar to that of thallium ion.[10] Lead and tin have also been documented in their interactions with PtPOP. The reaction of Pb^{2+} and Sn^{2+} with PtPOP has resulted in the disappearance of the 368 nm band of Pt_2 and the appearance of broad absorption bands located at 393 nm and 550 nm. This behavior can be explained by the metal to metal charge transfer (MMCT) from Pt_2 to Pb^{2+} and Sn^{2+}.[11] Unlike Tl^+, which forms exciplexes and interacts only weakly with Pt_2 in the ground state, tin and lead divalent cations interact strongly with Pt_2 in the ground state and lead to quenching of the *Pt_2 phosphorescence.[11] The difference in the behavior of the metal ions can also be explained by their difference in reduction potentials. Tl^+/Tl^0 has a value of -1.94 ± 0.05 V (versus NHE) in comparison to Pb^{2+}/Pb^+ (-1.0 ± 0.1 V versus NHE). The quenching of $Pt_2^+/$*Pt_2 (-1.6 ± 0.2 V) is exothermic with Pb^{2+} but endothermic by Tl^+. This explanation demonstrates that the Tl^+ will not directly quench the emission from PtPOP; rather, it will interact directly with the excited state.[5]

The quenching PtPOP by the compounds provides for convenient on-site heavy metal sensing using a portable UV lamp. PtPOP is soluble in water which makes this application simple and realistic. While the reaction with these four metals has been well documented, it provides a limited scope of PtPOP's properties. The reactions of PtPOP with several multivalent metals, that have not been tested as of yet, such as gadolinium, mercury, iron, copper and cobalt, will be investigated. This utilization of the properties of PtPOP can be especially valuable in fields dealing with heavy metal contaminated waste, such as the metal and ceramic industries.

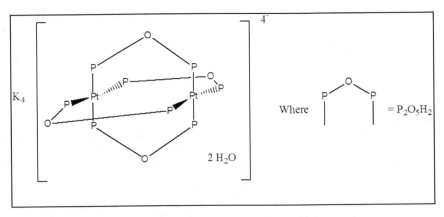

Fig.1. Structural representation of the PtPOP ion.

EXPERIMENTAL METHODS

All chemical reagents were purchased via commercial resources. Potassium tetrachloroplatinate was purchased from Pressure Chemical Co. and hypophosphorous acid was purchased from Sigma Aldrich. $Pb(NO_3)_3$ and $Gd(NO_3)_3$ were purchased from Sigma Aldrich. $Gd(C_2H_3O_2)_3$ and $Cu(NO_3)_3$ were purchased from Alfa Aesar. $Co(NO_3)_3$ was purchased from Mallinckrodt Baker. $Hg(C_2H_3O_2)$ was purchased from MCB Reagents. Millipore © water was used throughout the experiment.

Synthesis of PtPOP

A novel synthetic approach was developed with shortened reaction time and no loss in yield[5]. Modifications of previously published syntheses were utilized[1; 3]. 0.4 grams of K_2PtCl_4 and 1.5 grams of H_3PO_3 were combined into a 10 mL microwave reactor flask. Approximately 5 mL of Millipore water was added, and the flask containing the dark red slurry was sealed. The reaction was completed in 30 minutes at 90°C using a CEM Discover S-Class Microwave Reactor. The light brown solution was allowed to cool then moved to a watch glass which was placed into a 100°C oven for the water to slowly evaporate. The yellow and green solids (it was found that the bubbling the solution with Argon before reacting helped uniform the solids to yellow, however this did not improve the yields) were washed first with ethanol then acetone in a fritted filter flask. The powder was allowed to dry under vacuum overnight. Yields were approximately 30-40 %. Light yellow crystals were grown by dissolving solid in a small amount of water and layering first methanol then acetone over the solution.

Absorption studies

Absorption experiments were carried out on a Dual-beam Perkin Elmer Lambda 900 UV-VIS-NIR system using 1 cm quartz cuvettes in Millipore © water. Spectra were measured before and after addition of aqueous heavy metal ions (Pb^{2+}: 2.4 x 10^{-2} M, Gd^{3+}: 2.36 x 10^{-2} M, Fe^{2+}: 2.34 x 10^{-4} M, Cu^{2+}: 2.78 x 10^{-2} M, Co^{2+}: 1.75 x 10^{-2} M) to aqueous solutions of PtPOP (2.82 x 10^{-5} M). The ratios were determined by relating the absorbance of each heavy metal ion to PtPOP and diluted accordingly to achieve comparable values. The solutions were first purged in nitrogen gas for 20 minutes before testing.

Photoluminescence studies

Steady-state luminescence spectra were acquired with a PTI QuantaMaster Model QM-4 scanning spectrofluorometer equipped with a 75-watt xenon lamp, emission and excitation monochromators, an excitation correction unit, and a PMT detector. The emission spectra were corrected for the detector wavelength dependent response. The excitation spectra were also corrected for the wavelength dependent lamp intensity, but the correction was carried out at $\lambda >$ 240 nm due to the unreliability of the correction at shorter wavelengths at which the samples here absorb and the xenon lamp output is rather low. Long-pass filters were used to exclude light scattering due to the excitation source from reaching the detector. Lifetime data were acquired using a nitrogen laser interfaced with a tunable dye laser and a frequency doubler, as part of fluorescence and phosphorescence sub-system add-ons to the PTI instrument. The 337.1 nm line of the N_2 laser was used as the source of excitation.

Quenching studies

Aqueous heavy metal ions ($\sim 8 \times 10^{-5}$ M) were carefully titrated in 5μL aliquots, reaching a total of 1 mL, into aqueous solutions of PtPOP ($\sim 5 \times 10^{-6}$ M, 2.5 mL). Emission, excitation, and lifetime values (described in detail previously) were recorded after the addition of each aliquot. Using the Stern-Volmer equation (see Equation 1), quenching constants were calculated and evaluated.

$$\frac{F_o}{F} = 1 + k_q \tau_o[Q] = 1 + K_D[Q]$$

Equation 1.[12]

Where F_o and F = fluorescence intensities in the absence and presence of quencher, k_q = bimolecular quenching constant, τ_0 = lifetime of the fluorophore in the absence of quencher, and Q = concentration of quencher. The Stern-Volmer quenching constant is given by $K_D = k_q \tau_o$. Plots of F_o/F vs. [Q] were generated.

RESULTS AND DISCUSSION

The absorbance profile of PtPOP changes after the addition of the various heavy metals. Before addition of any heavy metal, PtPOP has a maximum absorbance at 368 nm (attributed to the allowed transition $^1A_{1g} \rightarrow ^1A_{2u}$ [13]). Upon the addition of the cations, this peak either disappears and is replaced by another peak, or shifts slightly (Figure 2). In the case of Gd^{3+}, Cu^{2+}, and Co^{2+}, absorbance at 368 nm appears to be weaker in intensity. The interactions of PtPOP with Ag^+ resulted in formation of a precipitate with corresponding disappearance in emission. Precipitation is a strong indicator of agglomeration and formation of metallic silver. Our observations indicate that this may be a possibility. The addition of Pb^{2+} to PtPOP resulted in the disappearance of the absorption at 368 nm and the formation of a peak at 380 nm and broad shoulder centered at 550 nm. Similar behavior has also been observed by Nagle, et al.[11] In their study, the drastic change in absorbance has been attributed to strong ground state interactions between the metal ion and PtPOP.[6] Similarly, in the addition of Fe^{2+} to PtPOP, one can observe a broad peak at 360 nm. The quenching of the Pt_2 peak and the formation of new peaks can be attributed to ground state metal-metal interactions of the metal ions with Pt at the axial site of Pt_2.[6; 7]

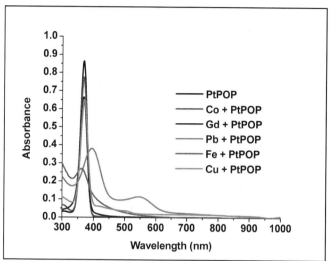

Fig. 2. Absorption changes of PtPOP upon addition of aqueous heavy metal ions

Photoluminescence spectroscopy revealed that the emission of PtPOP is easily quenched upon addition of Cu^{2+}, Gd^{3+}, or Fe^{3+}. Emission quenching of PtPOP by Cu^{2+} was plotted (Figure 3).

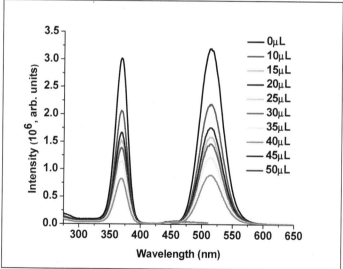

Fig 3. Photoluminescence excitation (left) and emission (right) spectral changes of PtPOP upon quenching by Cu^{2+}(λ_{exc} = 370 nm; λ_{em} = 514 nm) in 5 μL aliquots.

The above graph showed continuous drop in the emission and excitation intensity of Pt_2 upon sub-molar addition of Cu^{2+} ions. To determine the type of quenching, lifetime measurements were conducted with different quencher concentrations. A plot of τ_o/τ versus different copper II concentrations showed a straight line implying a dynamic quenching (Figure 4).[14] The life time for PtPOP, τ_o was 2.892 μs.

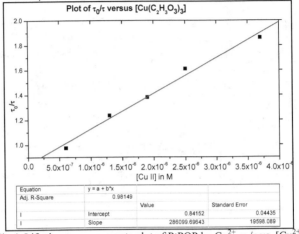

Fig 4. Life time measurements plot of PtPOP by Cu^{2+}, τ_o/τ vs. $[Cu^{2+}]$

In order to determine the quenching rate constant k_q, the Stern-Volmer equation of the reaction of PtPOP with the copper (II) acetate was plotted (Figure 5)

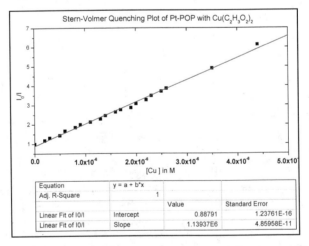

Fig 5. Stern-Volmer quenching plot of PtPOP by Cu^{2+}, I_o/I vs. $[Cu^{2+}]$

The slope of the Stern- Volmer constant was $K = \tau_o . k_q = 1.139 \ 10^6 \ M^{-1}$. The quenching rate constant k_q for Cu^{2+} ions was then determined to be: $k_q = \dfrac{K}{\tau_O} = 3.94 \ 10^{11} \ M^{-1}s^{-1}$.

Quenching of PtPOP by Fe(III) was treated in a similar fashion (see Figure 6).

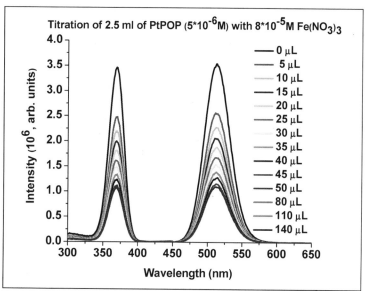

Fig 6. Photoluminescence spectra of PtPOP quenching by Fe^{3+}
($\lambda_{exc} = 370$ nm; $\lambda_{em} = 514$ nm) in 5 µL aliquots.

The above graph showed that the intensity of the Pt_2 peak decreased by 70 % upon addition of 140 µL of $Fe(NO_3)_3$. Life time measurements of PtPOP were experimentally determined to be $\tau_o = 3.00 \ 10^{-6}$ s. The slope of the Stern-Volmer plot of the PtPOP with Iron (III) nitrate (Figure 7) was $K = 1.094 \ 10^6 \ M^{-1}$. Therefore, the quenching rate constant was determined to be:

$$k_q = \frac{K}{\tau_o} = \frac{1.094 \times 10^6 \ M^{-1}}{3.00 \times 10^{-6} \ s} = 1.64 \times 10^{11} M^{-1} s^{-1}$$

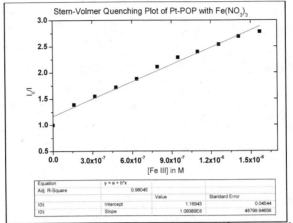

Fig 7. Stern-Volmer quenching plot of PtPOP by Fe^{3+}, I_0/I vs. $[Fe^{3+}]$

The quenching rate constant of Iron III ions was similar to that of Copper II ions. The fact that the quenching rate constant of iron and copper ions was higher than $10^{10}\,M^{-1}s^{-1}$ indicates some type of binding interaction between the fluorophores and the quencher.[14]

In a similar fashion, aliquot volumes of gadolinium (III) acetate were titrated with PtPOP (Figure 8)

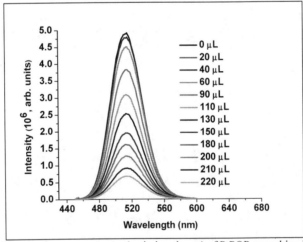

Fig 8. Photoluminescence spectra (emission shown) of PtPOP quenching by Gd^{3+} (λ_{exc} = 368 nm; λ_{em} = 514 nm) in 10 µL aliquots.

The quenching rate constant of gadolinium acetate with PtPOP was determined to be $k_q = 8.20 \ 10^{11} \ M^{-1}s^{-1}$. The results are in line with the other metal ions tested.

CONCLUSION

The addition of various heavy metal ions to aqueous solutions of PtPOP provided a rainbow of results, both in quenching values and physical characteristics observed. It can be said, however, that any addition of heavy metals resulted in the quenching of emission from PtPOP, and the disappearance of the bright green phosphor of Pt_2. While the investigation had shed the light on the use of PtPOP as possible heavy metal detector in aqueous solutions, further studies on the quenching mechanism is yet to be determined.

ACKNOWLEDGMENTS

Acknowledgment is made to the Oklahoma NASA EPSCoR, Oklahoma State Regents for Higher Education for studentand equipment supports of this research.

The authors also thank the Welch Foundation, the University of North Texas Departments of Chemistry, King Fahd University of Petroleum and Minerals Department of Chemistry for valuable discussion, and the Department of Chemistry, Computer and Physical Sciences of Southeastern Oklahoma State University for supporting this research.

REFERENCES

[1]M. A. F. D. R. Pinto, P. J. Sadler, S. Neidle, M. R. Sanderson, A. Subbiah and R. Kuroda. A novel di-platinum(II) octaphosphite complex showing metal-metal bonding and intense luminescence; a potential probe for basic proteins. X-Ray crystal and molecular structure. *Journal of the Chemical Society, Chemical Communications*, 13-15.(1980).

[2]A. P. Zipp. The behavior of the tetra-[upsilon]-pyrophosphito-diplatinum(II) ion Pt2(P2O5H2)4-4 and related species. *Coordination Chemistry Reviews*, **84**, 47-83.(1988).

[3]C. M. Che, L. G. Butler and H. B. Gray. Spectroscopic properties and redox chemistry of the phosphorescent excited state of octahydrotetrakis(phosphorus pentoxide)diplatinate(4-) ion (Pt2(P2O5)4H84-). *Journal of the American Chemical Society*, **103**, 7796-97.(1981).

[4]J. R. Peterson and K. Kalyanasundaram. Energy- and electron-transfer processes of the lowest triplet excited state of tetrakis(diphosphito)diplatinate(II). *The Journal of Physical Chemistry*, **89**, 2486-92.(1985).

[5]N. T. Satumtira, S. Marpu, O. Elbjeirami, P. Neogi, T. Lon, V. Nesterov and M. A. Omary. Platinum POP Revisited: Improved Microwave-Assisted Synthesis, Structure, Concentration-Sensitized Triplet Excitation, and Gas-Sensitive Luminescence in Aqueous and Biological Media. *Inorganic Chemistry Communications*.(2010 (Submitted)).

[6]D. P. Bender and J. K. Nagle. Metal-metal interactions of aqueous lead(II) and tin(II) ions with the ground and excited states of tetrakis([mu]-diphosphito)diplatinate(II). *Inorganica Chimica Acta,* **225**, 201-05.(1994).

[7]S. A. Clodfelter, T. M. Doede, B. A. Brennan, J. K. Nagle, D. P. Bender, W. A. Turner and P. M. LaPunzina. Luminescent metal-metal-bonded exciplexes involving tetrakis(.mu.-diphosphito)diplatinate(II) and thallium(I). *Journal of the American Chemical Society,* **116**, 11379-86.(1994).

[8]J. K. Nagle and B. A. Brennan. Luminescent exciplex formation involving tetrakis(Î¼-diphosphito)diplatinate(II) and -thallium(I) in aqueous solution. *Journal of the American Chemical Society,* **110**, 5931-2.(1988).

[9]J. K. Nagle, A. L. Balch and M. M. Olmstead. Tl2Pt(CN)4: a non-columnar, luminescent form of Pt(CN)42- containing platinum-thallium bonds. *Journal of the American Chemical Society,* **110**, 319-21.(1988).

[10]M. Christensen, K. Haldrup, S. Kjaer Kasper, M. Cammarata, M. Wulff, K. Bechgaard, H. Weihe, H. Harrit Niels and M. Nielsen Martin. Structure of a short-lived excited state trinuclear Ag-Pt-Pt complex in aqueous solution by time resolved X-ray scattering. *Physical chemistry chemical physics : PCCP,* **12**, 6921-3.

[11]D. P. Bender and J. K. Nagle. Metal-metal interactions of aqueous lead(II) and tin(II) ions with the ground and excited states of tetrakis(Î¼-diphosphito)diplatinate(II). *Inorganica Chimica. Acta,* **225**, 201-5.(1994).

[12]J. R. Lakowicz. *Quenching of Fluorescence.* 3rd; Springer. 2006; 'Vol.' 13.

[13]W. A. Fordyce, J. G. Brummer and G. A. Crosby. Electronic spectroscopy of a diplatinum(II) octaphosphite complex. *Journal of the American Chemical Society,* **103**, 7061-64.(1981).

[14]J. R. Lakowicz. *Principles of Fluorescence Spectroscopy.* third; Springer. 2006; 'Vol.' 954.

GEOPOLYMER PRODUCTS FROM JORDAN FOR SUSTAINABILITY OF THE ENVIRONMENT

Hani Khoury*, Yousif Abu Salhah, Islam Al Dabsheh, Faten Slaty, Mazen Alshaaer
University of Jordan, Amman, 11942, Jordan

Hubert Rahier
Research Group of Physical Chemistry and Polymer Science (FYSC), Vrije Universiteit Brussel (VUB)

Muayad.Esaifan, Jan Wastiels
Department of Mechanics of Materials and Constructions (MEMC), Vrije Universiteit Brussel (VUB)

ABSTRACT

Geopolymerization is the process of polymerizing minerals with high silica and alumina at low temperature by the use of alkali solutions. Geopolymers could be a substitute for Portland cement and for advanced composite and ceramic applications. The geopolymer technology would eliminate the need for energy requirement as they may be cured at ambient temperature.

Current research at the University of Jordan concentrates on developing building products (geopolymers) through geopolymerization. The goal is to produce low cost construction materials for green housing. The produced construction materials are characterized by high strength, high heat resistance, low production cost, low energy consumption, and low CO_2 emissions.

The results have confirmed that natural kaolinite satisfy the criteria to be used as a precursor for the production of high quality inexpensive, stable materials. The kaolinite geopolymer specimens from El-Hiswa deposits with compressive strength of 44.4MPa under dry conditions and 21.8 MPa under water immersed conditions were obtained by using around 16 % NaOH at 80°C after 14 hours. At higher temperatures these geopolymers maintained or improved their mechanical and physical performance after heating up to 600 °C. At 400 °C the compressive strength was 52.1 MPa under dry conditions and was 39.1 MPa under water saturated conditions. At 1000°C the mechanical strength was 15.1 MPa under dry conditions and was10.5 MPa under water saturated conditions. The density of the geopolymers has dropped down at dry conditions from 2.02 at 80 °C to 1.93 at 1000 °C. The presence of two refractory phases' sodalite and mullite make them sufficiently refractory at 1000°C. The lower densities at higher temperatures enable their use as an insulating material.

INTRODUCTION

Geopolymer technology has recently attracted researchers because the products are non-combustible, heat-resistant, formed at low temperatures, fire/acid resistant and have environment friendly applications [1, 4, 11, 15] Geopolymers have been proposed as an alternative to traditional Ordinary Portland Cement (OPC) for use in construction applications, due to their excellent mechanical properties [2]. Their physical behavior exceeded that of Portland cement in respect of compressive strength, resistance to fire, heat and acidity, and as a medium for the encapsulation of hazardous or low/intermediate level radioactive waste [3, 4] Chemical polymerization reactions of aluminosilicate minerals could be hardened and transformed into aluminosilicate geopolymers as a result of polycondensation [5]. Clay minerals (calcined clay), mining wastes, and slag are considered as a good source of aluminosilicate precursor. Geopolymers consist of three-dimensional mineral phases resulting from the polymerization of two dimensional sheet-like aluminosilicates in an alkaline solution. The exact mechanism of the geopolymerization is not known precisely until now. The structure maintains electrical neutrality as a result of aluminum substitution for silicon in the tetrahedral layer and the compensation of the negative charge by the available cations such as Na^+. Inexpensive functional fillers like silica sand and zeolitic tuff were used in different proportions to help in stabilizing the produced geopolymers [6].The excellent mechanical properties of the geopolymers have attracted the researchers to focus on the effect and application of different raw materials on the compressive strength, chemical impurities, and the effect of the chemical composition of the alkali activating solutions [6].

Recently, geopolymers produced from kaolinite/NaOH system gave a higher unconfined compressive strength values up to 52 MPa for dry test. This value has been increased to 57 MPa by the immersion of the geopolymer samples in 10% solution of triethylene glycol. The maximum obtained unconfined compressive strength is 90 MPa, this high value was obtained by using rock wool as an additive material during the geopolymerization process and heating to 500 °C for one day [7]. New low temperature functional geopolymeric materials were prepared exhibiting adsorption capacity for pollutants [8, 9]. Only a relatively small number of investigations have specifically studied the effects of high-temperature on geopolymeric gels [10].The thermal properties (dehydration and shrinkage) were only investigated of a geopolymer produced from metakaolin and sodium silicate solution. [11, 12]. The samples exhibited shrinkage of approximately 6% during dehydration, with significant densification observed at approximately 800 °C. The thermal behavior up to 1000°C of geopolymers produced from metakaolin activated with sodium silicate solutions was studied and characterized [6, 11, 12].

The following work will focus on evaluating the influence of high temperature on the mechanical performance and stability of low temperature geopolymers produced from kaolinitic raw material. The thermal behavior and phase transformation of the Na-geopolymers will be characterized to investigate its possible use as a low temperature refractory material (<1000°C).

MATERIALS AND TECHNIQUES
Materials:

Jordanian kaolinitic clay (as a source of aluminum silicate) with a purity of 60% from El-Hiswa deposit was used. The kaolinitic clay deposits are located in the south of Jordan about 45km to the east of Al-Quweira town [8, 13]. Preparation of the Jordanian kaolinite samples involved crushing, grinding and sieving of an oven dried sample (at 105 °C) to obtain a grain size less than 425μm. Then the samples were mixed in 50 L plastic drum for several times for homogenization. The plasticity limit of El-Hiswa kaolinite was measured according to the ASTM D4318 [14] and was found to be 22%.
Silica sand with 99% quartz content was used as a filler to provide high mechanical properties. The sand was washed with distilled water and sieved to obtain a grain size between100 and 400 μm.

NaOH (GCC, 96%) was used as an alkaline activator for the dissolution of aluminoslicate precursor. Water was the reaction medium and the optimum water content was close to the plasticity limit. The optimum curing time at 80°C was determined to be around 14 hours [8, 15]. The optimal ratios of the mixture were determined depending on the best compressive strength, the optimal curing temperature and time for the geopolymer specimens [8, 15]. The composition of the optimized mixture to produce geopolymers at 80°C after 14 hours curing time is given in Table 1.

Table 1. Composition of the geopolymer mixture

Composition	Clay	Sand	NaOH	H2O
%	100	100	16	22

Fabrication of geopolymers' specimens:

The geopolymers' specimens were prepared from from El-Hiswa kaolinitic deposit as a source of aluminum silicate. Silica sand was used as a filler to provide high mechanical properties (SS) and NaOH solution was used as an alkaline solution.

Homogeneous mixtures were prepared (Table 1) using a controlled speed mixer (mixing speed was 107rpm for 2 min followed by 198 rpm for 10min). Good mixing is important to obtain homogeneous and comparable specimens and to avoid the agglomeration of the mixture. Each mixture (series) was divided into specimens (50 g each). The mixture was molded immediately after weighing to avoid drying and decrease of the workability of the mixture. The paste was molded in a stainless steel cylinder (diameter of 25mm and height of 45mm) at a pressure of about 15MPa (Carver hydraulic laboratory press). The molded specimens of each series were cured by placing them in a ventilated oven (Binder-ED115) at 80 °C for 24 h. After curing, the specimens were cooled down at room temperature. The specimens were tested for physical properties and characterized using microscopic and XRD techniques. The compressive strength for the fabricated heated geopolymer specimens (80 - 1000 °C) was measured using CONTROLS testing machine (Model T106 modified to suit with standard testing), where the load was applied on surface area = $(12.5 \times 12.5 \times \pi)$ mm^2 and height = 50 mm, and increased by a displacement rate of 2 mm/min. The density for the heated specimens and water absorption of the immersed heated specimens were measured [8, 15].

Analytical techniques:

The zero measurements (reference) were recorded immediately after curing at 80°C for 24 hours then the geopolymer specimens were heated in the furnace up to 100, 200, 400, 600, 800, and 1000 °C. The required temperature was 24 hours. After heating, the specimens were brought back to ambient temperature before further testing. The properties of interest were measured (e.g. density, strength). An average of at least three specimens was calculated for each test.

Thin sections were made for all the heated geopolymer specimens and were studied using the polarized microscope. X-ray diffraction analysis is used to identify crystalline phases of the materials. Representative portions of the ground heated geopolymers were randomly X-rayed using Philips 2 kW model, Cu Kα radiation (λ= 1.5418 Å nm) with a scan rate of 2°/min. X-ray diffractograms were recorded for powdered geopolymers at 80, 100, 200, 400, 600, 800, and 1000 °C to detect the phase changes.

The SEM/EDX techniques were used for obtaining mineralogical and textural details. The platinum coated geopolymer samples were scanned using high-energy beam of primary electrons in a raster scan pattern using model FEI- INSPECT-F50 of SEM/EDX. ^{29}Si and ^{27}Al spectra were obtained using Bruker AC250 spectrometer that operates at 49.70 MHz and 65.18 MHz for the ^{29}Si and ^{27}Al resonance frequencies, respectively.

RESULTS AND DISCUSSION

Thin sections were prepared for the heated specimens and studied under the polarized microscope. This microscopic study of the prepared thin sections illustrated that the specimens are massive, cohesive, and show no significant voids.

There is a firm and tight adhesion between quartz grains and polymerized kaolinitic matrix in all the studied specimens in the temperature range 100 - 1000 °C (Figure 1).

Fig. 1. A geopolymer specimen heated at 100°C showing polymerized kaolinitic matrix and reaction rims around quartz grains. (XPL, × 100).

In the geopolymer specimens heated up to less than 800°C, no significant change in the texture and reaction rims were observed. Reaction rims, possibly sodalite became significant and clear in specimens heated up to 800°C and sodalite was very clear along fractures in specimens heated up at 1000°C (Figure 2).

Fig. 2. A photomicrograph of reaction rims around and along quartz grains. Isotropic sodalite patches within the matrix. (XPL, × 400).

Sodalite was formed at 80°C [16] and the crystallinity has increased as it was confirmed by the X-ray diffraction results (Fig. 5). Sodalite is an isotropic mineral that forms as a result of the reaction between the decomposed kaolinite (metakaolin) and NaOH in the matrix between quartz grains. The values of the densities of the geopolymer specimens along with compressive strength results are listed in Table 2.

Table 2. The physical properties of the geopolymer specimens (densities and compressive strength results) at different temperatures.

Temperature °C	Compressive Strength Dry (MPa)	Compressive Strength Immersed (MPa)	Density Dry g/cm³	Density Immersed g/cm³
80	44.4	21.8	2.02	2.18
100	42.1	16.9	2.01	2.20
200	44.1	26.2	1.99	2.19
400	52.1	39.1	1.97	2.20
600	42.2	31.9	1.93	2.17
800	10.0	8.5	1.93	2.14
1000	15.1	10.5	1.93	2.07

The effects of the different temperatures on the compressive strength of the geopolymers' specimens from El-Hiswa kaolinite after heating at different temperatures up to 1000°C are illustrated in Figure 3. The effects of the different temperatures on the density of the geopolymer specimens are indicated in Figure 4.

The kaolinite geopolymer specimens from El-Hiswa deposits with compressive strength of 44.4MPa under dry conditions and 21.8 MPa under water immersed conditions were obtained by using around 16 % NaOH at 80°C after 14 hours.

The compressive strength of the geopolymers' specimens from El-Hiswa kaolinite increases with the increase of temperature up to 52.1 MPa under dry conditions and 39.1 MPa under water immersed conditions at 400°C. Afterwards, the compressive strength decreases with increasing the temperature up to 800°C. The compressive strength of the immersed specimens shows the same trend of dry specimens by increasing the temperature up to 1000°C. The comparison of the compressive strength at 80 °C with the tested specimens at higher temperatures has indicated that the geopolymers' specimens exhibit maximum mechanical strength at 400 °C. The specimens have a high mechanical strength (42.2 MPa) at 600 °C slightly below the initial compressive strength (44.4 MPa). The mechanical performance remarkably has decreased at a temperature higher than 600 °C (10 MPa) then it has increased above 800°C (15.1 MPa).

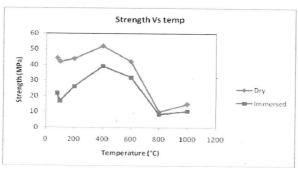

Fig. 3. Compressive strength of the geopolymers at different temperatures under dry and wet conditions.

Figure 4 shows relatively higher densities of the immersed heated specimens. The densities were almost constant above 600 °C for the dried specimens and have decreased for the immersed specimens. The most likely explanation is that the decrease in density is related to the removal of water and breaking of the kaolinite bonds as a result of dehydroxylation at a temperature higher than 400°C, causing the opening of pores and cracks. The almost constant density of the dry specimens from 600-1000°C is attributed to sintering and crystallization of new phases. The decrease in densities of the wet specimens is due to the dissolution of non reactive excessive salts, disappearance of poorly developed Na- Al silicate phases, and to the increase of crystallinity of sodalite, with its channel-like structure.

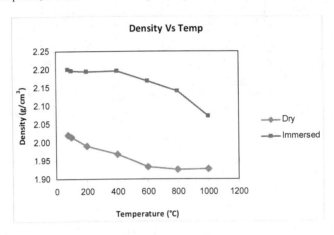

Fig. 4. Densities of the geopolymers at different temperatures under dry and wet conditions.

The XRD traces of all the geopolymers heated up to 100°C, 200°C, 400°C, 600°C, 800°C, and 1000°C showed that the original geopolymer phases cured at 80 °C. As identified by XRD techniques, the mineral phases of the geopolymer cured at 80 °C [15, 16] are mainly composed of kaolinite, muscovite, quartz, Na-Al silicate phases (Na- phillipsite and natrolite structures). Figure 5 shows the phases change with the increase of temperature. Kaolinite major peaks (7 Å and 3.5 Å) diappear at 600°C as a result of dehydroxylation and change into metakaolinite. Figure 6 illustrated the geopolymer at 100 °C where Na-Al silicate matrix encompassing relicts of unreacted kaolinite.

Unstable Na-Al silicate phases (major peaks at 4.7 Å and 3.25 Å) disappear at 800°C. Na-Al silicate phases in general change into sodalite as a result of dehydroxylation and recrystallization. The d-spacings of sodalite at 3.76 Å indicates a better crystallinity at 1000 °C. Figures 7 and 8 illustrate β-quartz and a porous texture with the collapse of kaolinite at 600°C.. Mullite appears at 1000°C. It could be seen that α-quartz exhibits phase change to β-quartz at 600 °C. Quartz inverts to high temperature β-quartz at about 573°C. Kaolinite peaks disappeared at 600°C as a result of dehydroxylation, while Na-Al silicate phases disappeared at 800°C. Figures 7 and 8 illustrate to β-quartz and a porous texture with the collapse of kaolinite at 600°C.

The drop in the compressive strength values at a temperature higher than 600°C is related to the dehydroxylation of kaolinite and the disappearance of unstable Na-Al – silicate phases possibly as a result of melting and/or recrystallization (Figure 9). The increase in the compressive strength values above 800°C is related to the better crystallinity of sodalite and the appearance of mullite. Sodalite is the result of reaction between metakaolinite and NaOH at high temperature. Heating the geopolymer at 1000°C for 24 h did not show a collapse of the texture and this is an empirical indication of refractoriness. The presence of sodalite and mullite refractory phases should at 1000°C made these geopolymers sufficiently refractory for continuous use up to this temperature. Figure 10 illustrates the porous nature of the geopolymer at 1000 °C, with well developed porous laminated texture and sodalite crystals.

A geopolymer made with sand as filler gave a compressive strength of ~ 52 MPa at 400 °C and ~ 42 MPa at 600 °C is sufficiently high. At ~ 1000°C, the compressive strength is much lower (~ 15 MPa.), but as an aluminosilicate geopolymer, it could be used as a thermal insulator. Thermal insulators are used for lining structurally supporting refractory facilities or as mortars in such structures. Hence, a high temperature high strength is not a pre-requisite for their use. The low densities of the geopolymers at higher temperatures, and the presence of zeolites (sieve-like structure) enable their use as insulating materials.

CONCLUSIONS

Specimens with compressive strength of 44.4 MPa under dry conditions and 21.8 MPa under water saturated conditions were obtained. The obtained materials are environmentally friendly and need only a low energy to be produced. Observations confirmed that natural geomaterials, e.g. kaolinite, satisfy the criteria to be used as a precursor for the production of inexpensive and durable construction materials. At higher temperatures these geopolymers maintained or improved their mechanical and physical performance after heating up to 600 °C. At 400 °C the compressive strength was 52.1 MPa under dry conditions and was 39.1 MPa under water saturated conditions. At 1000°C the mechanical strength was 15.1 MPa under dry conditions and was 10.5 MPa under water saturated conditions. The density of the geopolymers has dropped down at dry conditions from 2.02 at 80°C to 1.93 at 1000° C. The geopolymers heated up to 1000°C did not show any major melting. The presence of two refractory phases' sodalite and mullite should make them sufficiently refractory at 1000°C for its continuous use. The lower densities at higher temperatures enable their use as an insulating material. Therefore, they can be used as insulating material and for applications where fire safety is required.

ACKNOWLEDGEMENTS

The financial support of the Deanship of Scientific Research, University of Jordan of the project "Chemical stabilization of natural geomaterials for construction and industrial applications" is highly appreciated. The support of the Flemish (Belgium) Vlaamse Interuniversitaire Raad (VLIR, Contract ZEIN2006PR333) within the "Own Initiatives" program is gratefully acknowledged.

*Corresponding Author: khouryhn@ju.edu.jo

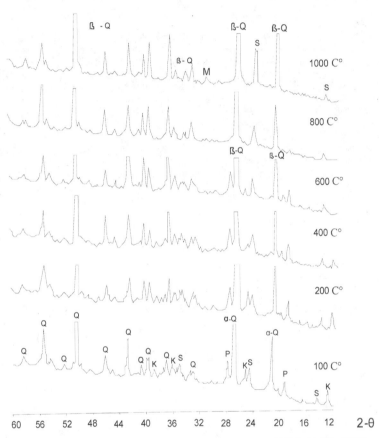

Fig. 5. The XRD traces of geopolymers heated up to 100°C, 200°C, 400°C, 600°C, 800°C, and 1000°C.

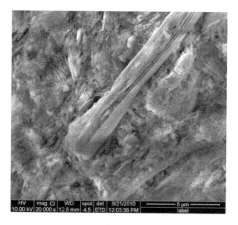

Figure 6: SEM image the heated geopolymer at 100 °C with relicts of kaolinite flakes and Na-Al silicate phases as a matrix .

Figure 7: SEM image the heated geopolymer at 600 °C with beta-quartz and fractured texture.

Figure 8: SEM image the heated geopolymer at 600 °C with collapsed porous texture.

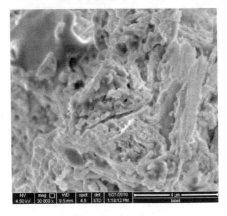

Figure 9: SEM image of heated geopolymer at 800 °C, showing porous and fused glassy texture.

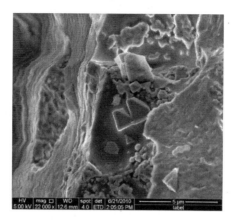

Figure 10: SEM image of heated geopolymer at 1000 °C, showing porous laminated texture and sodalite crystals.

REFERENCES

[1] P.Duxson, A. Fernández-Jiménez, J.L Provis., G.C. Lukey, A. Palomo and J.S.J. van Deventer. Geopolymer technology: the current state of the art. Journal of Materials Science, 42, 2917-2933, (2007).

[2] M. Rowles, B. O'Connor, Chemical optimisation of the compressive strength of aluminosilicate geopolymers synthesised by sodium silicate activation of metakaolinite, J. Mater. Chem. 13 1161–1165, (2003)

[3] J. Davidovits, " Geopolymers: Man-Made Rock Geosynthesis and the Resulting Development of very Early High Strength Cement," Journal Materials Education, 16 [12] 91-139, (1994).

[4] J. Davidovits,"Chemistry of Geopolymeric Systems, Terminology," Geopolymere '99, Geopolymer International Conference, Proceedings, 30 June – 2 July, 1999, pp. 9-39, Saint-Quentin, France. Edited by J. Davidovits, R. Davidovits and C. James, Institute Geopolymere, Saint Quentin, France, (1999).

[5] J. Davidovits, Geopolymers and geopolymeric new material, J. Therm. Anal. 35, 429–441, (1998)

[6] P. Duxson, G. C. Lukey, J. S. J. van Deventer, Physical evolution of Na-geopolymer derived from metakaolin up to 1000 °C. J Mater Sci 42:3044–3054 (2007).

[7] H. Khoury, H. Al Houdali, Y. Mubarak, N. Al Faqir, B. Hanayneh, and M. Esaifan: Mineral Polymerization of Some Industrial Rocks and Minerals in Jordan. Published by the Deanship of Scientific Research, University of Jordan. (2008).

[8] H. Khoury, M. AlShaaer (2009): Production of Building Products through Geopolymerization. GCREEDER. Proceedings Global Conference on Renewable and Energy Efficiency for Desert Regions. P 1-5, (2009)

[9] R. I. Yousef, B. El-Eswed , M. Alshaaer, K. Fawwaz, H. Khoury, The influence of using Jordanian zeolitic tuff on the adsorption, physical, and mechanical properties of geopolymers products, J. Hazardous Materials, 165, Issues 1-3, 379-387, (2009)

[10] D. Perera and R.Trautman, Geopolymers with the Potential for Use as Refractory Castables."Advances in Technology of Materials and Materials Processing", 7[2] 187-190 (2005).

[11] H. Rahier, , B. Van Mele, and J. Wastiels, Low Temperature Synthesised Aluminosilicate Glasses. Part B: Rheological Transformations during Low-Temperature Cure and High- Temperature Properties of a Model Compound, J. Mater. Sc., 31, 80-85, (1996).

[12] V. F. F. Barbosa and K. J. D. MacKenzie, "Synthesis and Thermal Behaviour of Potassium Sialate Geopolymers," Materials Letters, 57 1477-82 (2003).

[13] H. Khoury, Clays and Clay Minerals in Jordan, The University of Jordan, p.116, (2002).

[14]   ASTM D4318 D 4318, Standard Test Methods for Liquid Limit, Plastic Limit, and Plasticity Index of Soils, vol. 04.08, American Society for Testing and Materials, (2003)

[15] F. Slaty "Durability of Geopolymers Product from Jordanian Hiswa Clay" PhD dissertation (2010).

[16] H. Rahier, F. Slaty, I. Al dabsheh, M. Al Shaaer, H. Khoury, M. Esaifan and J. Wastiels. Use of local raw materials for construction purposes (2010).

LEACHING OF CALCIUM ION (Ca^{2+}) FROM CALCIUM SILICATE

Vandana Mehrotra
Graduate Institute of Ferrous Technology
Pohang University of Science and Technology
Pohang 790-784, South Korea

ABSTRACT

A process and mechanism of leaching Ca^{2+} ion from calcium silicate with hydrochloric acid was investigated. The extraction reaction of calcium ions from calcium silicates was induced by using concentrated hydrochloric acid. The process feasibility was evaluated based on the experimental data. It was believed that the rate was controlled by a chemical reaction in laboratory conditions. The ^{29}Si MASS NMR analysis characterized the coordination change of Si atom after the leaching. The NMR results provided an important evidence for process which facilitates calcium leaching and this result can be utilized to predict the chain of reactions during leaching. The experimental results suggested the optimal conditions to obtain the maximum amount of calcium leaching in acidic medium. It was concluded that the entire process of calcium leaching is effected by the silica content which inhibits 100% calcium leaching equilibrium value due to gel formation .The entire process conditions can be applicable to the blast furnace slag and basic oxygen furnace slag which is a valuable secondary resource containing about 46% percent of CaO, which is similar to commercial calcium silicate (CCS). The work in this paper have thus been concerned with enhancing calcium silicate dissolution rates from BF and BOF slag due to their low saleable value and finding the optimal conditions for maximum calcium leaching from them.

1. INTRODUCTION

About 400 kg of slag is being generated in integrated steel plants for producing 1 ton of steel. It comprises 300 kg of iron making slag and 100 kg of steelmaking slag. Almost all the iron making slags have been utilized as raw materials for cement, road building materials, fertilizers and other uses.[1] Although steelmaking slags show complex matrix structure consisting primarily of calcium silicates, aluminosilicates[2] and aluminoferite, they are rich in free lime and iron source which may be utilized as valuable recycling resources. In particular, many efforts have been made to utilize the lime contained in iron making and steelmaking slag in more economic and environmentally-friendly manner.

Leaching may be defined as the extraction of a soluble material from an insoluble solid by dissolution in a suitable solvent and the process of leaching encompasses the physical, chemical and biological reactions that mobilize a chemical species as well as the transport mechanisms that carry it away from the matrix into the surrounding environment. Several researches have already been performed mainly about the reaction of slag cements and concrete formation in alkali medium. In those researches the focus was put on the reaction between calcium-based sorbents and various bases. Along with it several other studies have also been performed of acidic attack on hydrated cement material. Another interesting process route [7] used acetic acid to dissolve calcium from natural calcium silicates. In the first step of the proposed process route, calcium silicate was dissolved in an aqueous solution of acetic acid, generating a solution of calcium acetate and a solid residue of silicon oxide. It has been reported that the calcium extraction yield of 48% was obtained in 250 min at 60 °C and that the rate enhancement was ascribed to the increase of acid concentration[3,5]. Calcium silicate dissolution was incongruent with calcium rates which are 3 times as high as the silica rates at room temperature.[8]. Unfortunately, almost few studies have been performed in acidic media with HCl [3-6].

However, most of the previously proposed methods have not been industrially implemented due to low product quality and cost efficiency, high energy and time consuming, or dubious product quality. And most of them have focused mainly on carbon dioxide sequestration first by leaching out calcium ion from slags and finally its conversion into calcium carbonate. Therefore, it is very essential to develop a novel chemical process which is cost and energy efficient, simultaneously reducing the total CO_2 production in a steel plant with minimum chemical steps and simple process.

The current research aims to enhance a calcium dissolution process to extract most of the calcium from iron making and steelmaking slag with decreased and delayed and subsequent conversion of the calcium ion to the corresponding oxide. This will not only help to reuse CaO in the whole steel production cycle but will also effectively reduce the energy consumption and CO_2 emission generated from the limestone burning process. In the current investigation, all the experiments were performed using chemical-grade calcium silicate for the preliminary examination of calcium leaching because the CaO content in iron making and steelmaking slag are similar to that in commercial calcium silicate.

2. EXPERIMENTAL

The experiments in this study were carried out employing a magnetic stirrer on which a 300-mL beaker was placed. 15 g of calcium silicate (Sigma Aldrich Co.) was dissolved using 150 mL of water in the beaker as shown in Fig. 1. Different amounts of concentrated HCl solution (80% by volume) were added drop by drop to the beaker containing the sample solution to maintain different pH values of 3, 5 and 7. All the reactions took place at room temperature and 1 atm pressure. In case the pH of the resulting solution started to get stabilized at a particular pH value in 20 sec, 10 mL of the solution was subsequently taken out at regular intervals of 5 min from the reacting mixture and filtered. The filtrate was separately kept for ICP analysis of calcium and silicon ion. The residue was collected at the end of reaction (after 45 min) and was kept separately. The total experimental duration was 45 min. The residue obtained at the end of the reaction was dried in a heating oven maintained at 60°C for 24 hours and was subsequently analyzed with X-ray diffraction (XRD) and nuclear magnetic resonance (NMR).

3. RESULTS

3.1. XRD (Residue)

According to XRD results original calcium silicate (2-A) mostly shows strong peaks of wollastonite (CaOSiO$_2$) at 2θ angle of 28 and 32. XRD of pH 3 (2-B) mainly indicates commercial silica gel structure similar to (2-E) with strong peak at 2θ angle of 26°which highlights the silica rich nature of layers present in this residue, whereas XRD for pH 5 show peaks at 20, 28, 32° 2θ angles with slight appearance of initialization of loss of crystallinity as shown in fig (2-C). Sample maintained at pH 7(2-D) mostly shows peaks resembling crystalline nature of sample after hydrolysis reaction.

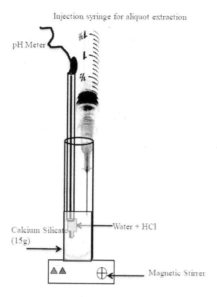

Fig. 1. Experimental apparatus and procedure for chemical extraction.

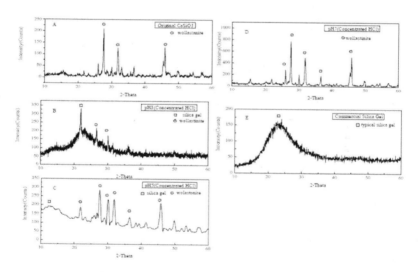

Fig. 2. X-ray diffraction results for different samples.

3.2. NMR: (Residue)

Nuclear magnetic resonance (NMR) spectroscopy can provide composition and structural information about solid materials. In the current investigation, NMR spectroscopy was employed to characterize the site symmetry and bonding environment of silicon atom after hydrolysis reaction. The qualitative NMR analysis commonly referred to as CP/MASS (cross polarization – magic angle sample spinning) can provide the information about the ^{29}Si nuclei near to the protons. The broad peaks in Fig. 3 indicate that numerous protons are surrounding a particular Si atom.

Fig. 4 shows the quantitative NMR MASS analysis results. It provides the information about isotropic chemical shift of a single silicon crystal. In particular, ^{29}Si MASS NMR can be evaluated to be one of the effective methods of characterizing silicates, especially for amorphous substances such as glass or gel due to the limitation of the structural information by X-ray diffraction. The NMR peaks of amorphous materials are broader than those of crystalline materials, but the chemical shift values are almost identical. Thus, ^{29}Si MASS NMR spectroscopy was employed to investigate the local homogeneity of amorphous silicate structure.

In Fig 4, the peaks for the original CaSiO$_3$ were identified near -75.69 and -83.33, which are attributable to Q^1 and Q^2 Si sites, respectively. In the case of pH 7, Q^1 is not visible and the chemical shift of Q^2 is about (-83.64) which corresponds to chemical shift (-) of approximately 0.06 ppm. Q^2 is absent in sample with pH 5 and 3. Instead of it, Q^3 peak is prominent and this indicates the polymerization reaction of silica chain.

3.3. ICP: (Filtrate)

The maximum leaching was obtained at pH 3 maintained by concentrated HCl as shown in Fig. 5. The yield of leached calcium was 78 %. It means this pH was helpful for enhancing calcium in the solution. Samples maintained at pH of 5 and 7 showed Ca yield of 53 and 22 %. Considering the Si amount in the filtrate; Fig 6 shows high amount of silicon for pH 3 which indicates that the increase in calcium amount, results in the relative increase of silicon to balance the Ca/Si ratio of the solid.

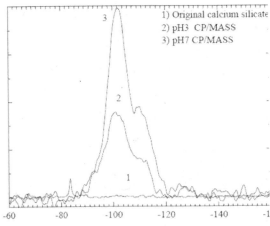

1) Original calcium silicate
2) pH3 CP/MASS
3) pH7 CP/MASS

Fig. 3. CP/MASS NMR result.

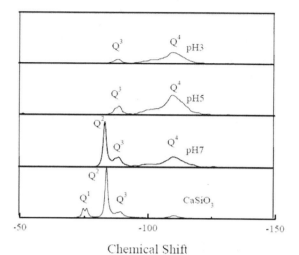

Chemical Shift

Fig. 4. 29Si MASS NMR result.

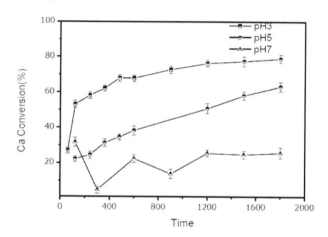

Fig. 5. Optimal conditions for enhancing calcium leaching into aqueous solution.

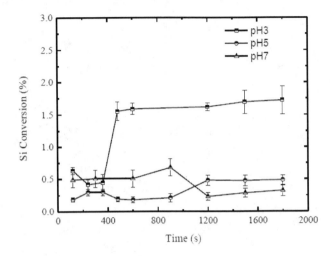

Fig. 6. Comparative graph for silica concentration in the solution with time.

4. DISCUSSION:

4.1. Optimal Condition for Maximum Calcium Leaching

In the current experimental layout it is very important to ensure high amount of calcium in the filtrate. The maximum leaching was obtained at pH 3 maintained with concentrated HCl .The yield of calcium was 78 %. This means that at pH of 3, calcium was easily able to come out in the aqueous solution preferably in form of (Ca-OH) ˙ due to the attack of H₃O˙ ions. This mechanism can be explained by applying the concept of ionization potential. At pH 3, there is abundance of H₃O˙ ions in the aqueous solution, which readily attack calcium silicate structure and as calcium has lower ionization potential (6.09 eV) than Si atom (8.9 eV) and, so calcium readily forms Ca(OH)˙ and comes into the aqueous solution as soluble phase , resulting in increase of calcium ion in the filtrate to a very large extent compared with silicon from calcium silicate . Besides this chemical reaction mechanism, various other diffusion mechanisms also play a major role in calcium leaching. [9].

To counteract the increasing amount of calcium ion in the aqueous solution or high molar ratio in liquid Fig 7, silicon atoms from (Si-OH) bonds balance by relatively varying their concentration by their rearrangement or polymerization in the solid residue and gradually form C-S-H gel with different Ca/Si molar ratios as shown in Table 1. The Ca/Si ratio for different samples was calculated by the mass balance calculations after the quantitative analysis of calcium and silicon ion in the solid samples by Titration method. It appears that polymerization of silicon atoms helps in releasing few Si ions in the aqueous solution and reorienting the remaining Si (OH) bonds. The different amount of silicon atoms going into the solution depends on the degree of polymerization of the silica structure which in turn depends on the amount of calcium leached into the aqueous solution. In future some experiment may be conducted to check the degree of polymerization in the samples. More is the calcium leaching higher is the level of polymerization of the silicon atom which results in larger decrease of Ca/Si molar ratio in solid. This means that incase of higher calcium leaching, there is also larger degree of polymerization of silicon atoms assisting in the outflow of calcium ions in the aqueous solution which is also evident from the NMR results (discussed in detail in Section 4.2) . Fig 8 also shows the comparative outflow of calcium and silicon ions / liter of solution. At pH 3, the amount of calcium ions leached into the solution is 60 with 1.6 moles of silicon. These values are better than those with other pH values. Thus it's very clear that in the case of aqueous solution with pH 3, due to abundant H_3O^- ions available in the solution, they can be utilized for leaching the calcium ion from the calcium silicate structure into the solution, resulting in increasing molar ratio of the solution.

Table I. Ca/Si molar ratio in residual solid after leaching reaction and the leached fraction.

Samples	pH	Ca/Si molar ratio in residual solid	Fraction of Ca Leached (%)
1	3	0.20	78
2	5	0.50	67
3	7	0.70	22

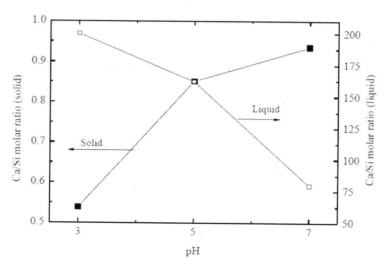

Fig. 7. Change of Ca/Si molar ratio with pH.

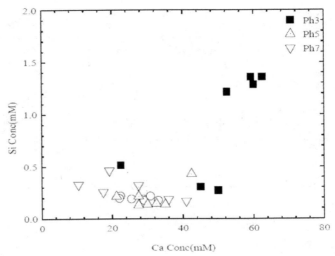

Fig. 8. Solubility results for calcium silicate.0204060800.00.51.01.52.0 Ph3Ph5Ph7Si Conc(mM)Ca Conc(mM)

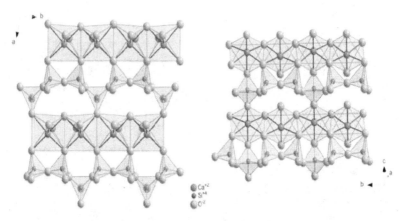

Fig.9. Schematic diagram for Wollastonite crystal structure.
(Available at http://commons.wikimedia.org/wiki/File:Wollastonite.)

4.1. Characterization of Silicon Atom (NMR)

The Q designation is used for structural assignments. The exponent refers to the number of silicon-oxygen attached to the silicon of interest. Q is generally expressed by the following equation:

$$Q^n = Si \bullet (O - Si)_n \bullet (OH)_{4-n} \tag{5}$$

Here Q^0 represents monomer, $[Si(OH)_4]$, and Q^1 is mono-replaced, $[Si-(O-Si)-(OH)_3]$, and so on. Q^1 and Q^2 peaks can be assigned to $CaSiO_3$ crystalline powder. Since the peaks in the XRD data show loss of crystalline phase after hydrolysis reactions, Q^3 and Q^4 were assigned to the amorphous phase, which is considered to be rich in SiO_2. In acid catalyzed reactions, an up field shift (-) of approximately 0.06 ppm has a great influence upon a mono-replaced (Q^1) or di-replaced (Q^2) silicon atom. Extensive ^{29}Si MASS NMR work indicates that the intermediate structures (Q^2 and Q^3) of polymers are a determining factor for the calcium leaching and duration of gel formation with calcium silicate in aqueous medium at different pH values.

In the current study, ^{29}Si MASS NMR experiments were very helpful in reasonably predicting the co-ordination change of Si atom in calcium silicate hydrolyzed at various pH values by using concentrated HCl. The co-ordination change for Si atom clearly indicates that it is possible to leach calcium ion by the ion exchange mechanism as long as large number of OH groups remain attached to any silicon atom. Therefore, if the Si atom orientation changes from Q^2 to Q^4 through Q^3, the leached calcium amount will also be increased.

The co-ordination change for Si atom occurred when H_3O^+ ion from acid acted as a proton donor at the Q^1 Si site $Ca-Si-(O-Si)_1-OH_3$ bond to release a calcium ion with the formation of Q^2 and Q^3 silicon structure for pH7 and mainly Q^3 and some Q^4 silicon structure for pH5, as shown in NMR result. However, for pH3 Q^3 was vulnerable to further attack due to excessive hydronium (H_3O^+) ions in the aqueous solution and finally became further polymerized to Q^4. This interpretation can also be verified by the analysis of Ca/Si ratio in residue which is equal to .5 and 0.7 initially at pH values of 5 and 7, ultimately changed to 0.2 at pH 3 that is the amount of silica in the remaining sample is comparatively higher with lower pH value. The X-ray diffraction pattern also supports similar information.

As the leaching amount of calcium ion increase, the peaks in the spectrum of samples indicated the transformation from Q^2 to Q^4 through Q^3. However, the intensities of Q^2 and Q^3 peaks gradually decreased in the successive samples and Q^2 peak completely disappeared. The obtained results can be understood by the structural homogeneity difference of the sample, as is evident by the XRD results indicating that the crystalline phase change is very fast and rapidly converted to amorphous phase. It is believed that the sample at pH 3 contains lesser OH groups attached to the silicon atom and it appears that OH from most of the $Si-(O-Si)-OH$ bonds has disintegrated due to the addition of HCl. Therefore, it can be considered that the formation of Q^4 Si structures is related to the enhanced calcium leaching, which was also verified by the ICP-AES analysis result. The polymerization results in the formation of calcium silicate hydrate (C-S-H) structure with different Ca/Si ratios. A fundamental explanation of the current results is that polymerization reaction of Si atom in calcium silicate can arise from more than one ionic constitution. The different Ca/Si ratios according to different pH can be obtained according to the following steps:

(a) Bridging silicate tetrahedra
(b) Decrease in Ca balanced by OH (i.e., Ca–OH),

Thus, in the range of Ca/Si ratio below or about 1, a given Ca/Si ratio can theoretically be achieved by either:

(1) Combination of low mean chain length (i.e., many missing bridging tetrahedra) and low Ca–OH content, or

(2) A high mean chain length (i.e., more bridging tetrahedra) and high Ca–OH content

$$(OSi)_2 Ca(OSi)_2 (OH)_3 (s) + HO^+(aq) \rightarrow Si(OSi)_3 (OH)^+ (s) + Ca(OH)_2 (aq) \tag{6}$$
$$(Q^2) \qquad\qquad\qquad\qquad (Q^3)$$

$$(OSi)_3 CaSi(OSi)_3 (OH)(s) \cdot Si(OSi)_3 (s) + Ca(OH)_2 (aq)$$
$$(Q)^3 \qquad\qquad (Q)^4 \tag{7}$$

In most of the current experiments Eqs. (6) and (7) are mostly applicable, resulting in enhancing calcium leaching .

4.2. Proposed Leaching of Ca Ion

The crystal structure of a wollastonite (calcium silicate) is shown in Fig. 9. It is single chain silicate with chain repeat units of three tetrahedra. Each tetrahedron is linked to a column of calcium centered octahedron. Because of the size mismatch of the tetrahedra repeat units and octahedral bands, the tetrahedra are tilted to accommodate their apices to the octahedral apices. In each tetrahedron there is a silicon atom bonded with 4 oxygen atoms. The Si-O bonds in the tetrahedra are partially covalent and partially ionic. They carry a net negative charge and bond with each other and other cations to form silicates. Calcium atom is there to balance the net negative charge in case of calcium silicate. It can be assumed that during the hydrolysis reaction, the hydronium ions (H3O$^+$) generated by acids are transported by diffusion through a boundary layer to the silicon atoms which can be considered as a reaction interface.[10, 11]. Simultaneously, the reaction product, Ca (OH)$_2^+$, diffuses back from the reaction interface into the bulk solution.

The continuous leaching of Ca ion from calcium silicate results in the formation of calcium silicate hydrate gel (C-S-H) due to continuous depolymerisation of silicon atom and allowing the release of Ca ion from the wollastonite structure of calcium silicate into the aqueous solution. The hyphens indicate indefinite stoichiometry of CaO, SiO$_2$ and H$_2$O and the hydrate is sometimes referred to as "silicate gel". C-S-H is produced along with calcium hydroxide Ca(OH)$_2$ in the reaction of the silicate phases..

At pH of 5 and 7 treated with concentrated HCl, there is not much acid resistance once the reaction starts even after 20 min of reaction which is quite clear from the X-ray analysis of residue. The residue has low amount of polymerization of silicon atom, being much similar to original calcium silicate and the Ca/Si ratio is about 0.75. For the residue of pH 3, the Ca/Si ratio is about 0.2 and the calcium leaching amounts to 77% as already discussed in section 4.1 of this paper.

5. CONCLUSION

A calcium leaching process from calcium silicate was investigated by the addition of acid. The low pH values maintained by addition of acid were considered appropriate for effective calcium leaching. The proposed idea for enhanced leaching is easily applicable in slags. Polymerization of the silicon atom appears to be the most common consequence of enhanced calcium leaching and this result can be utilized to predict the chain of reactions during leaching. Acid catalyzed the hydrolysis reaction to a very large extent. It can be predicted in the current study that the polymerization of silica and acidic conditions will be the key factor to enhance leaching in slags also besides the effect of aluminum. From the findings, the following conclusions were obtained:

(1)   Continuous leaching of the calcium ion due to attack of hydronium ion in the aqueous solution resulted in the formation of silica gel commonly referred to as calcium silicate hydrate (C-S-H).

(2)   The solution of pH 3 with concentrated HCl is believed to be the most optimal condition for calcium leaching at room temperature.

(3)   It can be concluded that the formation of Q^1 intermediate structures of Si atom has great influence on enhancing calcium leaching from C-S-H gel.

ACKNOWLEDGEMENT

Financial support from POSCO is gratefully appreciated .The authors are very thankful to Research Facility Centre, GIFT Postech where quantitative analysis of all samples was done by ICP-AES and elemental analysis was done by X-ray diffraction. The authors are also thankful to K.B.S.I, Daegu for ^{29}Si MASS NMR experiments.

REFERENCES

1) H. Okumura, Proc. 1st Int. Conf. on Processing Materials for Properties, ed. by H. henien and T. Oki, The Minerals, Metals and material Society, Hawai,803,(1993).

2) G. Bandt, K. Schaper and D. Hesse, Adsorption of HCl with calcined dolomite: investigation on the kinetics of the formation of the product layer ,Proc. of the 1992 I. Chemical Engineering Research Event, 305-307,(1992).

3) H. T. Karlsson, J. Klingspor and I. Bjerle, Adsorption of hydrochloric acid on solid slaked lime for fuel gas clean up ,*J. of Air Pollution Control Association*, **31**,1177-1180,(1981).

4) G. Mura and A. Lallai, On the kinetics of dry reaction between calcium oxide and gas hydrochloric acid. *Chemical Engineering Science*, **47**, No. 9-11, 2407-2411,(1992).

5) W. Peukert and F. Loeffler, Sorption of and HCl in granular bed filters. In: Clift, R., Seville, J.P.K. (Eds.), Gas Cleaning High Temperatures, Blackie, Glasgow, 604-623, (1993).

6) H. E. McGannon, The Making, Shaping and Treating of Steel, 8th ed., Berne and Pan-American International, Pittsburgh, PA, (1964).

7) S. Uchida, S. Kageyama, M. Nogi, H. Karakida, T. Kakizaki and K. Tsukagoshi, Reaction kinetics of HCl and limestone *J. Chinese Inst. Chem. Eng.*, **10**,45-49, (1979).

8) S. A. Carroll and K. G. Knauss, Experimental Determination of Ca-Silicate Dissolution Rates: A Source of Calcium for Geologic CO_2 Sequestration, Lawrence Livermore National Laboratory – (PhD Thesis).

9) C.H. Rawlins, Geological sequestration of carbon dioxide by hydrous carbonate formation in steel making slag, Ph.D. diss, Missouri University of Science and Technology, 2008

10) K. Fujii, W. Kondo, Heterogeneous equilibria of calcium silicate hydrate in water at 30 °C, J. Chem. Soc. Dalton Trans. **2**, 645– 651, (1981).

11) P. Barret, D. Bertrandie, Fundamental hydration kinetic features of the major cement constituents: Ca3SiO5 and h-Ca2SiO4, J. Chim. Phys.**83 (11/12)** 765– 775, (1983).

GREEN ENERGY AND GREEN MATERIALS PRODUCTION ACTIVITY AND RELATED
PATENTS

J. A. Sekhar and M. C. Connelly
University of Cincinnati, 45221-0012

J. D. Dismukes
University of Toledo, 43606-33

ABSTRACT
    A connection between life patterns of Green Materials and Green Energy is formally
established for the first time. Life-cycles studies provide a comprehensive insight into comparative
innovation behavior and innovation constants. In this article we study Green Materials as being related
to Green Energy for their innovative life pattern behavior.   We note that production and patent
activities may be correlated to such an extent that they may be superimposed to a large degree, for all
growth stages, simply by an origin-shift (OR) in the life cycle. The relative drive-force (defined as the
ratio of the production and patent growth constants, DF) is noted to scale with the origin-shift. The
value of this drive-force determines the amount of production that is influenced by patents. The slope
of drive- force against the origin-shift ratio curve is noted to be a constant across all energy categories.
We find that even early stage production displays an origin-shift. . Energy materials are also studied in
their broader class of general materials where the total usage is considered i.e., exceeding that for
energy production alone.   The life-cycle approach collapses the energy categories/sources into two
groups. Group 1, containing coal, natural gas, wind, renewable, fossil fuel, solar and total energy, is
composed of energy categories/sources whose patent activity could be inferred as driving their
production. On the other hand energy production from biomass, biofuel, geothermal and nuclear
energy are identified in Group 2, in which patent activity is driven by production activity.   A good
situation vis-a-vis life is when the (OR) is slightly less than one and the (DF) less than one as this leads
to a situation where with time, a higher growth stage III in production output may be encountered. A
very low (OR) and a low (DF) on the other hand lead to a transition from growth stage III to a no-
growth (stage IV) with time. We discuss these groups in the Schumpeterian framework of constructive
and destructive innovation.

1.   INTRODUCTION

Life-cycles studies have provided valuable information on innovation enhanced supply
[1,2,8,12,13,15] and comparative innovation activity [3-8, 12-15]. Life-cycles are ubiquitous and are
able to describe the *production amount/year* or *cumulative totals* as a function of normalized time for
the *entirety* of a transforming material, biological cell reproduction-cycle or even the quantity of a
material or energy produced per year [7-15]. Such cycles display a characteristic S-shaped pattern with
a long initial and final tail, sometimes with a final drop-off, depending on whether the total amount
(integrated production) or the amount per year (activity) is being considered. Product life-cycles on the
other hand are typically bell-shaped (or skewed bell shapes) and are associated with the pricing of a
very specific product or software that is short-lived. Sometimes, long life cycles are also approximated
by a bell shaped curve, for example, oil production by Hubbert [8], however, this simplifies the overall
pattern considerably and introduces artificial elements of symmetry. In the case of life-cycles for
materials production, four stages I to IV, are typically noted across the life [1,15]. A final decay stage
V is also sometimes noted in the amount produced per year [12]. Stage I is the stage that begins with
the initial discovery or creation of the object of the activity, and ends when the development leads to

313

some tangible increase in activity. Yerramilli and Sekhar [1] and later Sekhar and Dismukes [13, 15] have indicated that there is a blip that is noted during the next stage which is akin to a bubble or early market acceptance in the activity (i.e., the amount produced per year) prior to a stronger growth period during the stage III life. Such blips occur just prior to a lull before the stage III. The Stage II (containing the blip) begins with a rapid rise in the activity followed by a quick, rapid fall in activity. The Stage II ends at a low point (or time-frame). Next is the Stage III, i.e., the revival and re-growth stage, which is the high growth stage. This stage, if reached, generally includes a significant take-off in the activity (high-growth), with a much higher level of activity than seen even during the maximum point in the stage II. The Stage II blip has some similarities to the short term life–cycle shape. Finally, a Stage IV is seen, also called the survival or low-growth stage where the activity has reached maturity and has leveled off or has begun to die (presumably because of a paucity of supply and/or equalization of technical know how across producers). The various stages and features are now thought to include a final death stage, Stage V [12]. The term "Activity" may be used for other long-term variables, i.e., not only for the *production per year*; it could mean *patents per year* or other continuing features related to the initial discovery. A typical long-term cycle for a metal with plots for the activity, i.e., the production per year is shown in Figure 1.

In the terminology of inventions (ideas that are often patentable) and innovation (the market driven profitable commercialization of a physical product or service) [1,12,13,15], invention-driven activity dominates the stages I and II and market-forces based activity (innovation) is dominant in stages III and beyond. A long-term life-cycle for the *production/year* of a material thus appears to adequately segment the lower activity domains of a life-cycle into the invention stages, and the higher activity domains into the innovation stages [1,15]. A rapidly accelerating activity-stage, namely Stage III, has been shown to be possible by technical drivers alone *only* during the early periods of this stage, but requires expansion of the facilities (production facilities) during the latter stages if a high growth feature is to be maintained [13,15]. For the production of materials' [12] or production of energy a plot of the published data on amount produced per year vs. time often easily reveals the existence of the features that segment in to the stages of their life-cycle (e.g., as seen in Figure 1) simply by looking at the pattern. However, the precise identification of the transitions years between the stages is a non-trivial problem. Normative predictions are difficult to make, especially for the Stage III to Stage IV transition. Regardless, a pattern-based, life-cycle model has been refined over the years to capture the, overall life-cycle pattern [1,12,15] and related groupings. The procedures for establishing the fitness of this model to a large number of sectors has involved the tedious task of fitting the closest possible parameters for the multivariate model with the published data and establishing the confidence of the fit with non-linear confidence parameters [12,17]. We have also carefully studied the correlation between the number of patents per year and the amount produced, particularly in the Stage III regime of the patent activity plots. Although energy distribution and particular usage for materials processing [9-13] is also strongly influenced by patents, we have not included use aspects in this study, limiting it to only the energy production data. The patent data is more encompassing as described below. This has led us to identify transitions between the various stages and relate them to early stage activity behavior, thus yielding clues that may allow better predictions of a particular life-cycle transition from an early to a later stage. For the rest of the article, some particular biases in our statements may be noted. For example we implicitly assume that Stage III offers better value propositions compared to stage IV. We also associate inventions with patents and innovations as including both technical ideas (patents) as well as business processes that are associates with successful innovation. In this article, we focus on the Green terminology by examining materials that are used for Green energy production..

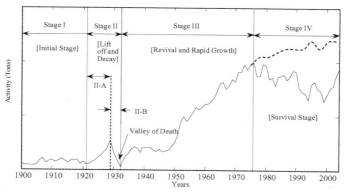

Figure 1. Schematic illustration of a typical Long-term life-cycle for a Material [1, 12, 13, 15]. Stage 1 and II are Invention Stages and Stage III is the major Innovation Stage. In the terminology of inventions (ideas that are often patentable) and innovation (the market driven profitable commercialization of a physical product or service) [1,12,13,15], invention-driven activity dominates the stages I and II and market-forces based activity (innovation) is dominant in stages III and beyond. The overall pattern of the plot indicates the division of the life-cycle into four stages that is common to metals. All metals may not have all four stages depending on the length of time that the metal has been in use. Such life-cycles are also applicable to non-metals [13] and in this article we show similar patterns for various energy categories/sources. The valley of death period could be sharp as shown or spread out [1]. For the rest of this article we group the two early stages as invention driven and the later stages as innovation dominated. The time axis (for the year) spans is the typical time frame during the 21st century during which most growth of metals and energy usage has occurred.

2.   LIFE CYCLE MODEL

The first life-cycle model relating the amount produced per year (y) to the year of production (x) for metals, was proposed by Yerramilli and Sekhar, 2006 [1] who showed that an equation of the form Equation 1 below, was able to fit the production data for five metals:

$$y = x^n \left[\alpha x^2 + \beta x \sin(\omega x)\right] + (exp[(x - \mu) / v] \, exp[-exp[(x -\mu) / v]] \, \delta / v) \qquad (1).$$

Equation 1, from reference [12], is a modified form of the first published form of equation 1, in references [1,15]. It is further noted that n is non-dimensional while α and β have the same units as y. We define $x=(x_r-x_0)$, x_r is the actual year of the data and x_0 is the first year of the data set. Also, $n=n/n_0$ where n_0 always takes the value of 1 (below we note that n_0 appears to be a universal constant[72,73], and other similar plots shown in reference 12). A plot of equation (1) made with a constant best-fit number for α and n was able to describe the four-stages of a long-term life-cycle [2]. α and n were identified [2] as being related to a resource constraint and a multiplier respectively. A constant value of α was akin to a fully elastic demand situation or, alternately, to a condition where there were no constraints (from availability or cost) on the supply; thus leading to a prolonged stage III behavior (e.g., aluminum and copper). If α decreased with time, then a Stage IV was indicated, i.e., the consequence of diminished resource availability was reflected in a decreasingα. An especially low α

was found to describe the conditions for a prolonged exposure to the valley of death. Subsequently, Sekhar and Dismukes, [15] further showed that a negative value of α could lead to the onset of a very early stage IV (low growth) behavior sometimes triggered by a lower cost technology substitution. If alpha is small, the inventive stages are in a sense stretched out, i.e., are longer, and dominate the plot. Conversely, if alpha is large, then the inventive stages are shorter in comparison and the innovative stages are longer and dominate the overall life-cycle plot.

The pattern equation (1) has been modified subsequently [12]. Equation (1) was modified by Connelly and Sekhar, [12,17,18] by raising the α and β parameters to the power of n (thus making n a resource multiplier). This raising of α and β to the power of n resulted in more consistent and reliable fitting results and also a greater dominance of the Stage III activity [17]. Note that regardless of the actual underlying phenomena behind the precise value of n, it is easily made dimensionless by dividing it by constant n_0, with the same units as n and which is numerically equal to one. In many instances, especially those relating to the lack of availability of data in the power sector, there is also no loss in description by introducing an additive constant as shown below. In this article we use the platform equation described in equation (1) in the modified form:

$$y = C1 + x^n [\alpha^n x^2 + \beta^n x \sin(\omega x)] + (exp[(x - \mu) / v] \, exp[-exp[(x - \mu) / v]] \, \delta / v) \qquad (2)$$

where $C1$ is a scaling constant. $C1$ was equal to the first year of production for the specific material and allowed the fitted curve to match up better with the actual production producing an R^2 closer to one. In effect, the addition of $C1$ is a form of scaling similar to that discussed above causing the actual production data to be closer to the generated fitted data.

The first part of equation (2), $x^n [\alpha^n x^2 + \beta^n x \sin(\omega x)]$, is primarily responsible for the shape of Stage III (and Stage IV if a variable α is used) in the fitted curves, while the second part, $[(exp[(x - \mu) / v] \, exp[-exp[(x - \mu) / v]] \, \delta / v)]$ establishes the shape of stages I-II in the fitted curve. Since α, β and x are all raised to the power n, it is likely that n could be a measure of the overall innovation pattern (i.e. an innovation constant).

Table 1. Common Pattern Variables. Variables to be determined in connection with the common pattern equation (1) that describes   life-cycles. Normalized years are the span of years under consideration and are represented by x. Production is y (i.e. activity such as energy produced per year or patents per year). The remaining constants are found through trial and error.

α	Called the "Take-off constant" (especially when not a function of x)[1,12,15,17,18]. α facilitates the rate of take-off after the end of Stage II [2]. The rate of growth of activity is very sensitive to α. By using n_0 the dimension is now independent of n'. *This parameter is a primarily a supply related constant and n/n_0 enhances or decreases this number supply by innovation.*
β	Increases the amplitude (visibility) of the cyclicity. The magnitude of cyclicity increases as β decreases. The dimensions of β are also now independent of n' *because of n_0.*
ω	Called the "wavelength constant". An increase in ω increases cyclicity. Value of ω is expressed in "per year" and equals $(2*\pi)$/wavelength.
μ	Called the "Stage II location constant". The position of the Stage II hump is shifted to the right as the value of μ increases. Dimension is years. Sometimes the difference in correlated data sets may be measured by the difference of μ.
ν	Called the "Stage II scaling constant". As the value of ν increases, the Stage II hump is stretched out. Dimension is years.
δ	As the value of δ increases, the peak (amplitude) of the Stage II hump increases. The units of δ is mass (tons) or number of patents per year.
n	Along with α has a strong influence on the shape of the curve. It is a positive number between 0 and 2. n is dimensionless as it is divided by n_0. The value of n_0 is one.
x	Time in years. Actual year of data (x_r) minus year of origin (x_0). $(x_0$ is determined from the earliest production data available and is commonly 1900. x_0 later than 1900 occur due to lack of data back to 1900, missing data or production beginning after 1900. Patent data before 1900 is very scarce making comparisons with production data meaningless. See reference 2 for data sets where x_0 is not 1900.
y	Production Activity in metric tons per year or the number of patents per year.

These seven parameters, as well as the date of origin, x_0, of the data are entered into a MatLab computer program [15] which generates an actual curve of the data, a fitted curve and an R^2 value, which is an established measure of best-fit, and which needs to be as near one as possible to obtain the best fitted curve [12]. The origin, x_0, is determined from the earliest available production data and is usually around the year 1900 for materials. However, the data for energy is not as comprehensive and we have therefore had to take into account a scaling constant $C1$ shown in equation 2.

3. RESULTS:

The data for the production of energy and materials was gathered from references 60 and 71. Figures 2 and 3 show examples from materials and energy life cycle analysis of various materials and energy sources Calculation details are to be found in references 1, 12, 13, 72 and 73. We define:

Origin-shift ratios (OR)= $(x_0 + (OS))/x_0$

and

Drive-force ratios $(DF) = (\alpha^n)_{activity} / (\alpha^n)_{patent}$.

When plotted against each other three patterns emerge. The first is that when a positive origin-shift is indicated, the drive-force ratio is always above one (the ratio is always less than one and approaches zero when the origin-shift is negative). Second, when n_a is divided by n_p the resulting ratio is always less than one for energy sources that have negative shifts in origin and the resulting ratio is always one or above for energy sources with a positive origin-shift. Lastly, as the origin ratios move away from one in either direction the modified patent R^2, which is generated by patent data being run with the common pattern equation production parameters, generally becomes smaller than one. As indicated in Figures 4 and 5 materials such a W and energy categories/sources such as coal and natural gas energy have origin ratios the farthest below one and lower modified R^2 than do sources with ratios nearer to one. In the same way nuclear energy has the highest origin ratio and has an R^2 less than sources with origin ratios near to one. This could indicate that energy sources may approach Stage IV when they reach origin ratio extremes (especially on the low ratio end) and enter stage IV when modified R^2 values become too small to support an origin-shift or origin ratio. Fig. 4, 5 and 6 show the division of the energy categories/sources into two groups. For energy production the Group 1, containing coal, natural gas, wind, renewable, fossil fuel, solar and total energy, is composed of energy categories/sources whose patent activity could be inferred as driving their production as suggested by the lag in production. Biomass, biofuel, geothermal and nuclear energy are in Group 2, in which patenting is driven by production suggested, conversely, by a lag in patenting.

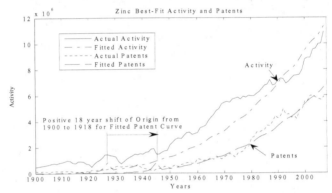

Figure 2. Plot depicting the origin shift of patent and activity best-fit curves for zinc. The shift is positive, indicating patent activity occurring after production activity and thus possibly being driven by the production. All parameters for the pattern equation are identical for the patent and production activity curves except for the origins (the matching results in the positive origin shift).

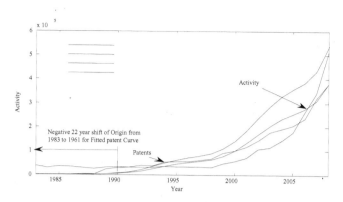

Figure 3. U.S. Wind Energy Best-Fit Activity and Patents. Plot showing the origin-shift(OS) of patent and activity best-fit curves for wind energy production in kJ. The shift is negative, indicating patent activity occurring before production activity and thus possibly driving the production. All parameters for the pattern equation are identical for the patent and production activity curves except for the difference in the origin that result in the negative origin-shift.

The pattern equation in the case of energy sources creates a normalized relationship between α and n that can be evaluated and compared to origin-shifts produced by independent patent and production activity best-fit derivations for a grouping of energy categories. Table 2 shows the materials that have been studied [72]. A plot of DF vs. OR, shown in Figures 4 and 5, indicates a comparison of innovation behavior across categories depending on whether they fall into group 1 or 2. There are some unique issues when considering energy production. Note for example that the activity of fossil fuel energy is possibly being driven the most by its patents since its drive-force ratio is nearest to zero for systems whose activity is driven by patents. A low origin-shift ratio is an indicator that there will be considerable difficulty for patents to prevent stage IV type behavior regardless of the driving force. Fossil fuels on the other hand offer possibilities for employing patents for obtaining robust stage III behavior. Similarly, the patents of nuclear power may be driven the most by its activity because its ratio is farthest from one for energy sources whose patents are driven by activity. Some words of caution are in order before the outlier data points are analyzed in such a manner.

Table 2. Materials studied.

Aluminum	Antimony	Arsenic	Asbestos	Barite
Bauxite/Alumina	Beryllium	Bismuth	Boron	Cadmium
Chromium	Cobalt	Copper	Feldspar	Fluorspar
Gold	Graphite	Gypsum	Helium	Hydraulic Cement
Iodine	Iron	Kyanite	Lead	Lithium
Magnesite	Magnesium	Manganese	Mercury	Molybdenum
Nickel	Niobium	Nitrogen	Phosphate	Platinum
Potash	Rare Earths	Salt	Selenium	Silicon
Silver	Sulfur	Talc	Tantalum	Tin
Titanium	Tungsten	Vanadium	Zinc	Zirconium

Stage IV may be entered along diminishing origin-shift ratio, Path A in Figure 4 and 5, from the constructive mode of innovation. The dashed line in Fig. 5 represents a possible trend line for the remaining energy source data points when coal, natural gas and total energy are removed or move into Stage IV. The energy sources represented along the dashed line are generally composed of sources that are relatively new and not particularly mature in their life-cycles. They might not as yet exhibit a long Stage III if at all and could be in stages I or II. Coal, natural gas and total energy (which may be dominated by coal) are older technologies that have data available only in Stage III. A separate (dashed) slope was generated to illustrate possible differences in the plots of these sources. R^2 values for energy sources are in some cases low.   Note that this can include fossil energy. One explanation for the high slope may be that evaluations were attempted here on data sets of production and patents that may not contain data for all stages of the life-cycles whereas the model presented herein is designed to model a situation with very strong Stage III life. Older energy sources such as coal, natural gas and total energy may have data sets that are missing Stage I or II data since the history of these go further back than 1900. Solar and wind energy are relatively new and may not yet be in a strong Stage III thereby skewing the resulting R^2 values. Nevertheless outlier points like Fossil Fuel allow for a reasonable speculation that the life path has a steep slope (Path B dashed line in Figure 10) in the constructive group that leads it to the destructive grouping and then reverses when the stage III to stage IV transition becomes imminent.

The production activity per patent is high in Group II and low in Group 1. Prolonged slippage in the Origin-ratio results in the drive-force diminishing to such an extent that stage IV is inevitable. This happens if the patenting is not adequately resourced.   Results which indicate a path by which increasing the patent activity leads to good stage III behavior for the production output activity is identified in Figure 5. A good situation is (OR) slightly less than one and (DF) less than one as this leads to a situation where with time a higher growth stage III in production output may be encountered. A very low (OR) and a low (DF) on the other hand lead to a transition from growth stage III to a no-growth (stage IV) with time.

4.   GREEN MATERIALS
Several energy categories/sources are associated with specific materials which we will refer to as energy materials or energy producing materials. Some of these were studied in reference 12, 15, 72. This section contains correlation, best-fit and origin-shift analysis data for the production and patents of energy producing materials Coal, Natural Gas, Uranium, and oil. The production is now studied in tons, barrels or cubic feet, of coal, natural gas, oil and uranium rather than energy (kJ) produced by them, which was done for energy categories/sources above. The production data is from the EIA web site [60] and the patent data is from the EPO site [59]. All data gathering techniques, correlation, best-fit and origin-shift analyses were carried out in an identical manner as for energy sources. It should be pointed out that the data origins for these four energy materials are not the same as their associated energy sources due to availability of data. Generally, the data for the material is available farther back than that for the energy source leading to older origins for the materials. An explanation could be that the material was discovered and used years previous to its development and employment as an energy source. Energy materials behaved in a similar manner to energy sources for life-cycle patterns. These four materials all exhibit Stage III behavior with associated origin-shifts. The patterns and innovation inferences are now different than when studied for energy content alone. Table 2 shows that oil has a positive origin-shift and therefore exists in the destructive mode of innovation while coal, natural gas and uranium have negative shifts and are in the creative mode of the innovation process. Table 2 continues the trends displayed by materials and energy sources of positive origin-shifts producing origin ratios and drive-force ratios over one and negative shifts producing ratios below one. The modified R^2 also tends to be farther from one the larger the shift in origin is. Figure 5 shows that oil is

in group 2 and natural gas, coal and uranium are in group 1. By comparison in Fig.6, nuclear energy is in group 2 while coal and natural gas energy are in Group 1 and oil energy is in Stage IV with no origin-shift. We note that when considering usage broader than for energy use, the groupings change (compare Figure 5 to Figure 6).

4.1   Green Energy and Green Material Connection

In several cases the best-fit and correlation evaluations of energy sources revealed a connection between such energy sources that are predicted to be in Stage III, or highly active and materials that are also highly active, or in Stage III, of their life-cycles. The materials associated with a non carbon source are sometimes called Green Materials. Table 3 gives examples of energy sources and the materials that are both being influence in their production and helping in the production of the corresponding energy production. Vanadium and silicon are Stage III materials in the constructive mode and are commonly used in products related to solar and wind power production [64]. Other materials such as graphite, nickel, cobalt, silver, manganese, the rare earths and lithium, which are in Stage III [12], are being produced largely in support of energy production and storage. Even Stage IV materials (cadmium, selenium, and fluorspar) are being produced in the more limited energy generation field [64]. In three cases active energy methods namely coal energy, nuclear energy and natural gas energy use coal, uranium and natural gas that are also in Stage III for the corresponding fuel. With energy resources declining worldwide and with growing environmental concerns, it seems logical that materials that are being "innovatively" optimized to meet energy requirements in an environmentally friendly manner result in Stage III growth. It appears that where the need for optimizing energy category/sources arises, the materials production that might fill that need rises as well.

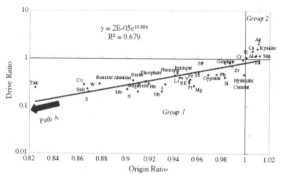

Figure 4. Drive Ratio vs. Origin Ratio [72] displaying relative strength of innovation enhanced supply driving force of either patents or production activity. The driving force of nitrogen is strongest for a material whose production activity is driven by its patents since its ratio is farthest below one. The driving force of silver is strongest for a material whose patents are possibly driven by its production activity since its ratio is farthest above one. Alternately, the difference could be couched in terms of a hypothesis that elements like nitrogen are driven by technical reasons or in other words the innovation is highly technically influenced whereas the influence of new metal technology on innovation diminishes for materials like silver and Cu. Note also that the cross over point occurs at 1 (y-axis). The origin shift is the shift described in section 3 between the best-fit activity and best-fit patent evaluations for each material using the common pattern equation (1). The sensitivity of R^2 to changes in the origin shift allow for an accurate estimation of the optimal origin shift. Note that although the fit is exponential the data-points appear to lie in a straight line for the range of years shown in the figure. Path A may indicate the path by which a material may leave Stage III and enter Stage IV.

Table 3 indicates that in a majority of cases the origin-shift direction of the materials follows that of their particular related energy category. Likewise, the origin and drive-force ratios are higher or lower than one for the material as that of its related energy source. Nuclear energy is the only obvious exception to this pattern, for the corresponding materials life-cycle studies for this category. The degree of the origin-shift ratio (OR) for many materials also seems to have limited relation to that of the energy category/ source, however when they are similar (e.g. vanadium for wind) it is likely that the particular material growth is considerably dominated by that energy source/category. The patent activity in such cases may be a strong driver for rapid growth. Coal and Coal energy display very different drive-force ratios indicating that coal usage as a material is very different from coal usage for energy production. This is a surprising result that will be studied in detail in the future. The ability for patent activity to enable coal energy usage such that it continues in a stage III mode is limited, whereas when considered as a material there is a good possibility for patents to be a driver which retain the material production in stage III.

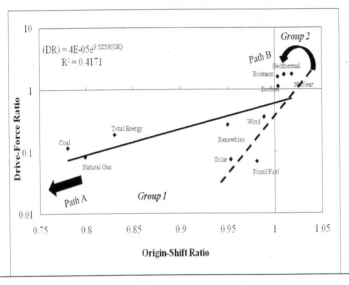

Figure 5. Energy Source Drive-force Ratio (DR) a function of the Origin-shift Ratio (OR) [73]. This graph displays relative strength of driving force of either patents or production activity. Possible path A into Stage IV depicted by arrow. The predicted trend-line of energy source data without coal, natural gas and total energy illustrated by dashed line. R^2 for the dashed line is .6632 with an equation for (DR) = $1E\text{-}19e^{43.483(OR)}$. Path A indicates the path an energy category may leave Stage III and enter Stage IV. Path B shows one possible path trajectory of entering Stage III from the early stages.

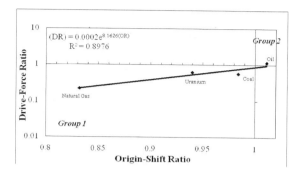

Figure 6. Green Energy Materials Origin-shift Ratio(OR) vs. Drive-Force Ratio(DR). Displays relative strength of driving force of either patents or production activity for energy materials.

Table 3. Green Materials. Comparison of origin-shifts, origin ratios and drive-force ratios of six energy categories and the materials that are related to them [12, 18, 73]. The particular material (s) that is relevant for the energy category is listed below the category [12, 73]. The materials associated with a non carbon source are sometimes called Green Materials

	Energy Category Materials	Origin-shift	Origin-shift Ratio	Drive-force ratio
1	**Solar Energy**	**-93 years**	**0.953**	**0.074**
	Vanadium	-90 years	0.954	0.363
	Silicon	-30 years	0.985	0.435
2	**Wind Energy**	**-22 years**	**0.989**	**0.36**
	Vanadium	-90 years	0.954	0.363
3	**Nuclear energy**	**+36 years**	**1.018**	**1.75**
	Uranium	-130 years	0.94	0.61
	Fluorspar	-113 years	0.941	0.444
4	**Coal Energy**	**-428 years**	**0.780**	**0.115**
	Coal	-30 years	0.984	0.57
5	**Natural Gas Energy**	**-392 years**	**0.799**	**0.082**
	Natural Gas	-326 years	0.832	0.22
6	**Renewable Energy**	**-97 years**	**0.950**	**0.267**
	Graphite	-24 years	0.987	0.813
	Nickel	-69 years	0.964	0.824
	Rare Earths	-101 years	0.947	0.389
	Cobalt	-256 years	0.865	0.284
	Lithium	-106 years	0.945	0.396
	Lead	-41 years	0.978	0.434
	Manganese	+23 years	1.012	1.418
	Silver	+21 years	1.011	2.060

This article summarizes our recent past research relating to long-term production life-cycles in materials' and energy production. The study of life-cycles furthers the understanding of the innovative process in general and for materials and energy sources in particular. The transition from Stage III to

Stage IV in long-term life-cycles is described by the drive-force and origin-shift ratio plot. The main conclusions are:

- A correlation is noted between the production data and related patent activity for materials and energy. The same correlation holds for Green Materials and related Technologies.
- The four life-cycles stages (initial stage, lift-off and decay, revival and rapid growth and survival) that were found to exist for the production of materials are also found to exist for energy categories.
- The driving or driven behavior of patents, represented by an origin-shift or lag, may indicate the dual nature of Schumpeterian innovation concepts by identifying the constructive and destructive modes of the innovative process.
- The production activity per patent is high in Group II. Although the driving force by patents is enhanced by low DF, it also correponds to a low origin shift ratio for Path A thus towards a situation where the patents become ineffectual. Path B is when the patents are stil effective.
- Stage III energy category/sources were often found to be supported by Stage III materials that are active and contribute to the innovation of the energy production.

Acknowledgment. We are grateful for discussions with Dr. A. A. Vissa. Non-financial support from MHI Inc. is also greatefully acknowledged.

REFERENCES
1. Yerramilli, C. and Sekhar, J.A., 2006. A common pattern in long-term metals production. Resources Policy 31, 27-36.
2. Rogers E.M., 2003. *Diffusion of Innovation*, 5th Edition.
3. Levitt, T., 1965. Exploit the product life-cycle. Harvard Business Review, November-December.
4. Freeman. C., (Ed.), 1996. Long Wave Theory. Elgar Publishing Limited, Cheltenham, U.K..
5. Jenner, R. A., 2004. Real wages, business cycles and new production patterns. Small Business Economics 23, 441-452.
6. Keklik, M., 2003. Schumpeter Innovation and Growth: Long Cycle Dynamics in Post WWII American Manufacturing Industries. Ashgate, Vermont.
7. Phillips, K. L. and Wrase, J., 2006. Is Schumpterian 'creative destruction' a plausible source of endogenous real business cycle shocks? Journal of Economic Dynamics & Control 30, 1885-1913.
8. Hubbert, M. King, 1956, The Energy Resources of the Earth. Energy and Power-A Scientific American Book, 1971, 31-39. see also Hubbert, M. K, "Nuclear energy and the fossil fuels." American Petroleum Institute, Drilling and production practices, pp. 7-25
9. Zhang, H., De Nora, V. and Sekhar, J.A., 1993. Materials Used in the Hall-Heoult Cell. TMS Publications.
10. Rajamani, V., Anand, R., Reddy, G.S., Sekhar, J.A. and Jog, M.A., 2005. Heat Transfer Enhancement Using Weakly Ionized Atmospheric Plasma in Metallurgical Applications. Met and Mat. Trans B., Vol. 37B, 565-570.
11. Liu, J., Duruz J.J, de Nora, V. and Sekhar, J.A, 1999. Clean Technologies and Environmental Policy, Vol. 1, No. 3. Springer, Berlin/Heidelberg, pp. 180-186.
12. Connelly, M.C, Dismukes, J.P. and Sekhar, J.A., 2010. New Relationships between Production and Patent Activity during the High Growth Life-cycle Stage for Materials. Technological Forecasting and Social Change. Accepted for publication May 2010.

13. Sekhar J.A., Yeramilli C, and Dismukes J.P.; Linking Productivity Analysis and Innovation For Materials And Energy. A Common Platform Approach. Edited by S. Freiman, American Ceramic Society, John Wiley and Sons, NJ, USA, 2007, pg 143. See also M.G. Laksmikantha and J. A. Sekhar, Metallurgical and Materials Transactions A Volume 24, Number 3, 617-628, DOI: 10.1007/BF02656631

14. Avrami, M., 1939. Kinetics of Phase Change. I. General Theory. Journal of Chemical Physics 7 (12), 1103–1112.

15. Sekhar, J. A., & Dismukes J., 2009. Generic innovation dynamics across the industrial technology life-cycle: Platform equation modeling of invention and innovation activity. Technological Forecasting and Social Change, Vol. 76, Issue 1, 192-203.

16. Branscomb, L., 2004. Duke L. & Tech. Rev.005 web source.

17. Connelly, M.C. and Sekhar, J.A., 2008. Inventions and Innovation: A Case Study in Metals. Key Engineering, Innovation in Materials Science, eds. J. A. Sekhar and J. D. Dismukes, Key Engineering Materials, v.380, 15-39.

18. Connelly, M.C., 2010. An Analysis of Innovation in Materials and Energy. Ph.D. Dissertation, University of Cincinnati and Connelly, M.C., 2007. The Relationship between Patents and technical Innovation: Innovation measurement as applied to metals. M.S. Thesis, University of Cincinnati.

19. Berry, D., 2009. Innovation and the price of wind energy in the US. Energy Policy 37, 4493-4499.

20. Dismukes, J.P., Miller, L.K. and Bers, J.A., 2009. The industrial life-cycle of wind energy electrical power generation ARI methodology of life-cycle dynamics. Technological Forecasting & Social Change 76, 178-191.

21. Inoue, Y. and Miyazaki, K., 2008. Technological innovation and diffusion of wind power in Japan. Technological Forecasting & Social Change 75, 1303-1323.

22. Muylaert de Araújo, M.S. and Vasconcelos de Freitas, M.A., 2008. Acceptance of renewable energy innovation in Brazil-case study of wind energy. Renewable and Sustainable Energy Reviews 12, 548-591.

23. Shikha, Bhatti, T.S. and Kothari, D.P., 2005. New Horizons for Offshore Wind Energy: Shifting Paradigms and Challenges. Energy Sources 27, 349-360.

24. Kobos, P.H., Erickson, J.D. and Drennen, T.E., 2006. Technological learning and renewable energy costs: implications for US renewable energy policy. Energy Policy 34, 1654-1658.

25. Harborne, P. and Hendry, C., 2009. Pathways to commercial wind power in the US, Europe and Japan: The role of demonstration projects and field trials in the innovation process. Energy Policy 37, 3580-3595.

26. Wüstenhagen, R., Wolsink, M. and Bürer, M.J., 2007. Social acceptance of renewable energy innovation: An introduction to the concept. Energy Policy 35, 2683-2691.

27. Negro, S.O. and Hekkert, M.P., 2008. Explaining the success of emerging technologies by innovation system functioning: the case of biomass digestion in Germany. Technology Analysis & Strategic Management 20, 465-482.

28. Louime, C. and Uckelmann, H., 2008 Cellulosic Ethanol: Securing the Planet Future Energy Needs. International Journal of Molecular Sciences 9, 838-841.

29. Kelly-Yong, T.L., Lee, K.T., Mohamed, A.R. and Bhatia, S., 2007. Potential of hydrogen from oil palm biomass as a source of renewable energy worldwide. Energy Policy 35, 5692-5701.

30. van der Laak, W.W.M., Raaven, R.P.J.M. and Verbong, G.P.J., 2007. Strategic niche management for biofuels: Analyzing past experiments for developing new biofuel policies. Energy Policy 35, 3213-3225.

31. Wonglimpiyarat, J., 2010. Technological change of the energy innovation system: From oil-based to bio-based energy. Applied Energy 87, 749-755.

32. Suur, R.A. and Hekkert, M.P., 2009. Cumulative causation in the formation of a technological innovation system: The case of biofuels in the Netherlands. Technological Forecasting and Social Change Vol 76 Issue 8, 1003-1134.

33. Vertès, A.A., Inui, M. and Yukawa, H., 2008. Technological Options for Biological Fuel Ethanol. Journal of Molecular Microbiology and Biotechnology 15, 16-30.

34. Trancik, J.E., 2006. Scale and innovation in the energy sector: a focus on photovoltaics and nuclear fission. Environmental Research Letters 1, 014009 (7 pp.).

35. Lee, T.J., Lee, K.H. and Oh, K.B, 2007. Strategic environments for nuclear energy innovation in the next half century. Progress in Nuclear Energy 49, 397-408.

36. Li, J., 2009. Scaling up concentrating solar thermal technology in China. Renewable and Sustainable Energy Reviews 13, 2051-2060.

37. Faiers, A., Neame, C. and Cook, M., 2007. The adoption of domestic solar-power systems: Do consumers assess product attributes in a stepwise process? Energy Policy 35, 3418-3423.

38. Guidolin, M. and Mortarino, C., 2010. Cross-country diffusion of photovoltaic systems: Modeling choices and forecasts for national adoption patterns. Technological Forecasting and Social Change Vol. 77 Issue 2, 279-296.

39. Faiers, A. and Neame, C., 2006. Consumer attitudes towards domestic solar power systems. Energy Policy 34, 1797-1806.

40. Huang, A.Y.J. and Liu, R.H., 2008. Learning for supply as a motive to be the early adopter of a new energy technology: A study on the adoption of stationary fuel cells. Energy Policy 36, 2143-2153.

41. Suurs, R.A.A., Hekkert, M.P. and Smits, R.E.H.M., 2009. Understanding the build-up of a technological innovation system around hydrogen and fuel cell technologies. International Journal of Hydrogen Energy 34, 9639-9654.

42. Avadikyan, A. and Llerena, P., 2010. A real options reasoning approach to hybrid vehicle investments. Technological Forecasting and Social Change Vol. 77 Issue 4, 649-661.

43. Hellman, H.L. and van der Hoed, R., 2007. Characterizing fuel cell technology: Challenges of the commercialization process. International Journal of Hydrogen Energy 32, 305-315.

44. vanBree, B., Verborg, P.J. and Kramer, G.J., 2010. A multi-level perspective on the introduction of hydrogen and battery-electric vehicles. Technological Forecasting and Social Change Vol. 77 Issue 4, 529-540.

45. Rourke, F.O., Boyle, F. and Reynolds, A., 2010. Tidal energy update 2009. Applied Energy 87, 398-409.

46. Bañales-López, S. and Norberg-Bohm, V., 2002. Public policy for energy technology innovation: A historical analysis of fluidized bed combustion development in the USA. Energy Policy 30, 1173-1180.

47. Tsoutsos, T.D. and Stamboulis, Y.A., 2005. The sustainable diffusion of renewable energy technologies as an example of an innovation-focused policy. Technovation 25, 753-761.

48. de Vries, B.J.M, van Vuuren, D.P. and Hoogwijk, M.M., Renewable energy sources: Their global potential for the first-half of the 21st century at a global level: An integrated approach. Energy Policy 35, 2590-2610.

49. Bergek, A., Jacobsson, S. and Sandén, B.A., 2008. 'Legitimation' and 'development of positive externalities': two key processes in the formation phase of technological innovation systems. Technology Analysis & Strategic Management 20, No. 5, 575-592.

50. Sovacool, B.K., 2009. Resolving the impasse in American energy policy: The case for a transformational R&D strategy at the U.S. Department of energy. Renewable and Sustainable Energy Reviews 13, 346-361.

51.   Bonilla, S.H., Almeida, C.M.V.B., Giannetti, B.F. and Huisingh, D., 2010. The roles of cleaner production in the sustainable development of modern societies: an introduction to this special issue. Journal of Cleaner Production 18, 1-5.
52.   Schmidt, R.C. and Marschinski, R., 2009. A model of technological breakthrough in the renewable energy sector. Ecological Economics 69, 435-444.
53.   Norberg-Bohm, V., 2000. Creating Incentives for Environmentally Enhancing Technological Change: Lessons From 30 Years of U.S. Energy Technology Policy. Technological Forecasting and Social Change 65, 125-148.
54.   Narayanamurti, V., Anadon, L.D. and Sagar, A.D., 2009. Transforming Energy Innovation. Issues in Science & Technology, 57-64.
55.   Bürer, M.J. and Wustenhagen, R., 2009. Which renewable energy policy is a venture capitalist's best friend? Empirical evidence from a survey of international cleantech investors. Energy Policy 37, 4997-5006.
56.   Weiss, M., Junginger, M., Patel, M.K. and Blok, K., 2010. A review of experience curve analyses for energy demand technologies. Technological Forecasting and Social Change Vol. 77 Issue 3, 411-428.
57.   Bonvillian, W.B. and Weiss, C., 2009. Stimulating Innovation in Energy Technology. Issues in Science & Technology, 51-56.
58.   Wang, T.J. and Liu, S.Y., 2010. Shaping and exploiting technological opportunities: The case of technology in Taiwan. *Renewable Energy* 35, 360-367.
59.   European Patent Office. http//ep.esp@cenet. Last accessed 7/31/2010.
60.   U.S. Energy Information Administration [EIA]. Last accessed May 1010. http://www.eia.doe.gov.
        a) For natural gas: http://tonto.eia.doe.gov/dnav/ng/ng_prod_sum_dcu_NUS_a.htm
        b) For oil: http://tonto.eia.doe.gov/dnav/pet/pet_crd_crpdn_adc_mbbl_a.htm
        c) For renewable energy sources biomass, solar, hydroelectric, geothermal, wind, biofuel, wood and total renewables: http://www.eia.doe.gov/emeu/aer/renew.html
        d) For uranium: http://www.eia.doe.gov/emeu/aer/nuclear.html
        e) For fossil fuel and nuclear: http://www.eia.doe.gov/emeu/aer/overview.html
        f) For coal: http://www.eia.doe.gov/emeu/aer/coal.html and http://www.eia.doe.gov/cneaf/coal/page/coal_production_review.pdf.
61.   Walpole, R.E., Myers, R.H. and Myers, S., 1998. Probability and Statistics for Engineers and Scientists, Sixth Edition. Prentice-Hall Inc., NJ.
62.   Miller, I., Freund, J.E. and Johnson, R.A., 1990. Probability and Statistics for Engineers, Fourth Edition. Prentice-Hall, Inc., NJ.
63.   *Engineering Statistics Handbook*; www.itl.nist.gov/div898/handbook /eda/ section 3/autocopl.htm. Last accessed May 2010.
64.   United States Geologic Survey. http://minerals.usgs.gov/minerals/pub/commodity/myb/; *Minerals Yearbook*, (2007). Last accessed 7/31/2010.
65.   Guseo, R. and Guidolin, R., 2010. Technological Forecasting and Social Change, Article in Press, 2010.
66.   Beretta, E., Gandolfi, A. and Sastri, C.C.A., 2008. Innovation in Materials Science, eds. J. A. Sekhar and J. D. Dismukes; Vol. 380, TransTech Publication, Key Engineering Materials, Switzerland, pp. 3-14.
67.   Betz Frederick, 2003. Managing Technological Innovation: Competitive Advantage From Change, 2nd Edition, John Wiley & Sons, New York, NY, ISBN# 0-471-22563-0.
68.   Schumpeter, J.A., 1942. Capitalism, Socialism and Democracy. Harper & Brothers Publishers, New York.
69.   Schumpeter, J.A., 1934. The Theory of Economic Development: An Inquiry Into Profits,

Capital, Credit, Interest and the Business Cycle. Harvard University Press, Cambridge, Mass.

70. Schumpeter, J.A., 1939. A Theoretical, Historical and Statistical Analysis of the Capitalistic Process. McGraw-Hill, New York, NY.
71. United States Geologic Survey. usgs.gov.
72. M. C. Connelly, J. D. Dismukes and J. A. Sekhar; New Relationships between Patents and Technological Innovation: Modeling Patent Activity as a Driver of Innovation, 2011, TFSC, Vol. 76, issue1, pg 308-318,
73. M. C. Connelly and J. A. Sekhar: U. S. Energy Production Activity and Related Patents. Submitted for Publication 2010.

MICRO PATTERNING OF DIELECTRIC MATERIALS BY USING STEREO-LITHOGRAPHY AS GREEN PROCESS

Soshu Kirihara, Naoki Komori, Toshiki Niki and Masaru Kaneko

1) Joining and Welding Research Institute, Osaka University
2) Graduate School of Engineering, Osaka University
1) Ibaraki / 2) Suita, Osaka, Japan

ABSTRACT

Dielectric micro patterns composed of ceramic/resin composites were fabricated to control terahertz waves effectively by using stereolithography. In this process, the photosensitive resin paste with titania particles dispersion was spread on a substrate with 10 µm in layer thickness by moving a knife edge, and two dimensional images of UV ray were exposed by using digital micro mirror device with 2 µm in part accuracy. Through the layer by layer process, the periodic structures composed of micro polygon tablets were formed. The electromagnetic wave properties of these samples were measured by using a terahertz spectroscopic analyzer. The band gap formation and localization behavior of terahertz waves in dielectric micro patterns will be reported. The electromagnetic waves in a terahertz frequency range with micrometer order wavelength are expected to be applied for novel sensors to detect gun powders, drugs, bacteria in foods, micro cracks in electric devices, cancer cells in human skin. The extremely thin terahertz wave devices could be fabricated successfully by using non-heating green processes.

INTRODUCTION

Photonic crystals with periodic variations in dielectric constants exhibit forbidden gaps in electromagnetic wave transmission spectra [1-4]. Spatially arranged patterns of grasses or ceramics can reflect right waves or microwaves perfectly in the specific wavelength regions comparable to the artificial periodicities through Bragg diffraction. Introduced cavities into the periodic structures to resonate the electromagnetic waves strongly can form transmission peaks in the band gaps at the specific wavelengths comparable to the defect sizes [5-8]. In our investigation group, the photonic crystals with diamond lattice structures have been fabricated successfully by using stereolithography [9-12]. This micro fabrication equipment of a computer aided design and manufacturing (CAD/CAM) system was newly developed through the collaboration projects [13-18]. The formed micro ceramic lattices of silica, alumina or titania could exhibit perfect band gaps in terahertz frequency ranges to prohibit the electromagnetic waves incidents form all directions. The resonation modes of transmission peaks were formed in the photonic band gaps by introducing air cavity defects. In this research, micro patterns of the dielectric materials were processed to diffract effectively and resonate strongly the terahertz waves though the photonic crystals theories. The extremely thin devices could be processed successfully by using the micro stereolithography of the non-heating green processes. In near future industries, electromagnetic waves in the terahertz frequency range with micrometer order wavelength will be expected to apply for various types of novel sensors which can detect gun powders, drugs, bacteria in foods, micro cracks in electric devices, cancer cells in human skin and other physical, chemical and living events [19-14]. Filtering effects of the electromagnetic waves for the perpendicular direction to the micro dielectric patterns were observed through time domain spectroscopic (TDS) measurements. These micro geometric patterns of extremely thin devices with a high dielectric constant were designed to concentrate the electromagnetic energies effectively through a finite difference time domain (FDTD) method.

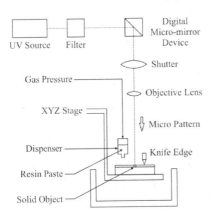

Figure 1 A schematically illustrated free forming system of a micro-stereolithography machine by using computer aided design and manufacturing (CAD/CAM) processes. (D-MEC Co. Ltd., Japan, SI-C 1000, http://www. d-mec.co.jp).

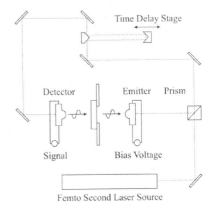

Figure 2 The schematically illustrated measuring system of a terahertz wave analyzer by using a time domain spectroscopic (TDS) detect method (Advanced Infrared Spectroscopy, Co. Ltd. Japan, J-Spec 2001, http://www. aispec.com).

EXPERIMENTAL PROCEDURE

The micro dielectric pattern was designed as the periodic structure composed of micro square tablets of $240 \times 240 \times 100$ μm in dimensions at intervals of 45 μm. These micro tablets of $9 \times 9 = 81$ in numbers were arranged to form the extremely thin dielectric device of $2520 \times 2520 \times 100$ μm in whole dimensions. The real sample was fabricated trough the micro stereolithographic

Figure 3　A dielectric micro pattern composed of titania particle dispersed acrylic resin fabricated by using the micro-stereolithography.

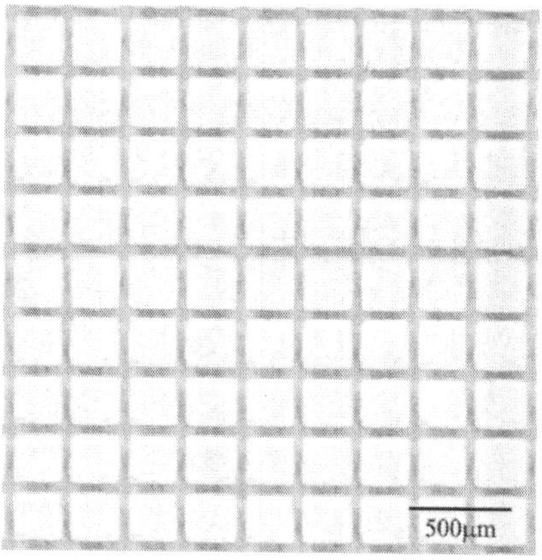

Figure 4　Microstructure in an acrylic resin bulk with titania particle dispersion observed by using a scanning electron microscope (SEM).

system. A designed graphic model was converted for stereolithography (STL) data files and sliced into a series of two dimensional layers. These numerical data were transferred into the fabrication equipment (D-MEC: SIC-1000). Figure 1 shows a schematic illustration of the fabrication system. As the raw material, nanometer sized titania particles of 270 nm in average diameter were dispersed into a photo sensitive acrylic resin at 40 volume percent. The mixed slurry was squeezed on a working stage from a dispenser nozzle. This material paste was spread uniformly by a moving knife edge. Layer thickness was controlled to 5 μm. Ultra violet lay of 405 nm in wavelength was exposed on the resin surface according to the computer operation. Two dimensional solid patterns were obtained by a light induced photo polymerization. High resolutions in these micro patterns had been achieved by using a digital micro mirror device (DMD). In this optical device, square aluminum mirrors of 14 μm in edge length were assembled with 1024×768 in numbers. Each micro mirror can be tilted independently, and cross sectional patterns were dynamically exposed through objective lenses as bitmap images of 2 μm in space resolution. After stacking and joining these layers through photo solidifications, the periodical arrangements of the micro dielectric tablets were obtained. A bulk sample of the titania dispersed acrylic resin with the same material composition was also fabricated to measure the dielectric constant of the composite tablets. A terahertz wave attenuation of transmission amplitudes through the micro pattern were measured by using a terahertz time domain spectrometer (TDS) apparatus (AISPEC: Pulse-IRS 1000). Figure 2 shows the schematic illustration of the whole measurement system. Femto second laser beams were irradiated into a micro emission antenna formed on a semiconductor substrate to generate the terahertz wave pulses. The terahertz waves were transmitted trough the micro patterned samples perpendicularly. The dielectric constant of the bulk samples were measured through a phase shift counting. Diffraction and resonation behaviors in the dielectric pattern were calculated theoretically by using a transmission line modeling (TLM) simulator (FLOMERICS: Micro-stripes Ver. 7.5) of a finite difference time domain (FDTD) method.

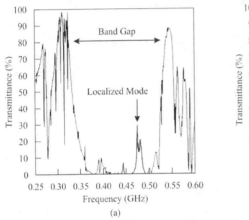 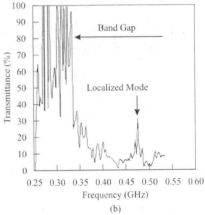

(a)　　　　　　　　　　　　　　　(b)

Figure 5 Transmission amplitudes of the terahertz wave through the dielectric micro pattern. The spectra (a) and (b) are measured and calculated properties by using the terahertz wave time domain spectroscopy (TDS) and a transmission line modeling (TLM) methods, respectively. In both transmission spectra, localized modes of transmission peaks are formed at specific frequencies in band gap regions.

RESULTS AND DISCUSSION

The dielectric micro patterns with the periodic arrangement of the acrylic tablets with the titania particles dispersion was fabricated successfully by using the micro stereolithography system of the non-heating green process as shown in Figure 3. Dimensional accuracies of the fabricated micro tablets and the air gap widths were approximately 0.5 percent in length. The nanometer sized titania particles were verified to disperse uniformly in the acrylic resin matrix as shown in Figure 4 thorough a scanning electron microscope (SEM) observation. The dielectric constant of the composite material of the titania dispersed acrylic resin was measured as 40. Figure 5-(a) and (b) show transmission spectra measured and simulated by using the TDS and TLM methods, respectively. The measured result has good agreement with the calculated one. Opaque regions were formed in both spectra form 0.33 to 0.57 THz approximately. Maximum

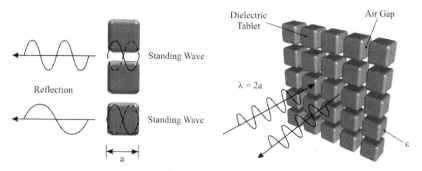

Figure 6 An intensity profile of electric field on a cross sectional plane in the dielectric micro pattern calculated at the peak frequency of the localized mode by using the TLM simulation of a finite difference time domain (FDTD) method.

attenuation was measured as about -20 dB in the transmission amplitude, and the minimum transmittance showed below 1 percent. The two dimensional photonic crystals with periodic arrangement with the lower dielectric contrasts were well known to open the band gaps limitedly for the parallel directions to the plane structures. However, through the theoretical simulations, it was verified that the micro patterns with the periodically arranged square tablets above 30 in dielectric constant could exhibit the clear forbidden bands in the transmission spectra toward the perpendicular direction to the plane structures. The fabricated dielectric pattern is considered to totally reflect the terahertz wave at the wavelength comparable to the optical thickness as schematically illustrated in Figure 6. Two different standing waves vibrating in the air and the dielectric regions form the higher and the lower edges of the band gap. The gap width can be controlled by varying geometric profile, filling ratio, and the dielectric constant of the tablets. As shown in Figure 5, the localized mode of a transmission peak was observed at 0.47 THz in the band gap. Figure 7 shows a simulated distribution of electric field intensities in the micro pattern at the localized frequency. The white area indicates that the electric field intensity is high, whereas the black area indicates it is low. The incident terahertz wave is resonate and localized along the two dimensionally arranged dielectric tablets at the specific frequency. The amplified terahertz wave can transmit through the micro pattern. Therefore, the transmission peak of the localized mode should be formed clearly in the photonic band gap frequency range.

CONCLUSIONS

Micro square tablets of acrylic resin with titania particles dispersions were arranged periodically in two dimensions by using micro stereolithography of a non-heating green process. Fabricated micro pattern was verified to be able to exhibit a forbidden band of opaque region. A localized mode of a transmission peak was clearly formed at a specific frequency to concentrate electromagnetic energies in the periodic arrangement of the dielectric constant. The terahertz waves are well known to resonate with various types of protein molecules, and expected to control the biological material syntheses by using frequency excitements through characteristic resonation effects. The fabricated micro pattern can include various types of solutions into their air gaps between the square tablets as illustrated in Figure 8, therefore, it will be applied for novel micro reactors to create useful biological materials.

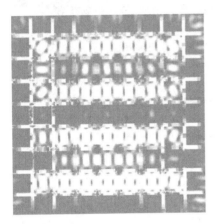

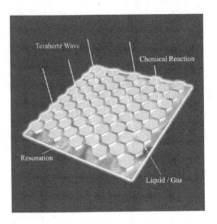

Figure 7 The schematic illustrations of formation mechanisms of the electromagnetic band gap through the dielectric micro pattern by Bragg diffraction. Incident direction of the electromagnetic wave is perpendicular to the dielectric arrangement with the plane structure.

Figure 8 A schematic illustration of photonic crystal micro reactor with the dielectric tablets arrangement fabricated by the stereo lithography of a green processing. The micro pattern can include various types of solutions into their air gaps between the square tablets

ACKNOWLEDGMENTS

This study was supported by Priority Assistance for the Formation of Worldwide Renowned Centers of Research - The Global COE Program (Project: Center of Excellence for Advanced Structural and Functional Materials Design) from the Ministry of Education, Culture, Sports, Science and Technology (MEXT), Japan.

REFERENCES

[1] E. Yablonovitch, Inhabited Spontaneous Emission in Solid-state Physics and Electronics, *Physical Review Letter*, **58**, 2059-2062 (1987).

[2] S. John, Strong Localization of Photons in Certain Disordered Dielectric Superlattices,

Physical Review Letter, **58**, 2486-2489 (1987).

3 K. M. Ho, C. T. Chan, and C. M. Soukoulis, Existence of a Photonic Gap in Periodic Dielectric Structures, *Physical Review Letter*, **65**, 3152-3165 (1990).

4 B. Temelkuran, Mehmet Bayindir, E. Ozbay, R. Biswas, M. M. Sigalas, G. Tuttle,

5 and K. M. Ho, Photonic Crystal-based Resonant Antenna with Very High Directivity, *Journal of Applied Physics*, **87**, 603-605 (2000).

6 S. Kirihara, Y. Miyamoto, K. Takenaga, M. W. Takeda, and K. Kajiyama, Fabrication of Electromagnetic Crystals with a Complete Diamond Structure by Stereolithography, *Solid State Communications*, **121**, 435-439 (2002).

7 S. Kirihara, M. W. Takeda, K. Sakoda, and Y. Miyamoto, Control of Microwave Emission from Electromagnetic Crystals by Lattice Modifications, *Solid State Communications*, **124**, 135-139 (2002).

8 S. Kanehira, S. Kirihara, and Y. Miyamoto, Fabrication of TiO_2-SiO_2 Photonic Crystals with Diamond Structure, *Journal of the American Ceramic Society*, **88**, 1461-1464 (2005).

9 H. Takano, B. S. Song, T. Asano, and S. Noda, Highly Efficient in-Plane Channel Drop Filter in a Two-Dimensional Heterophotonic Crystal, *Applied Physics Letters*, **86**, 241101-1-3, (2005).

10 S. Kirihara and Y. Miyamoto: Terahertz Wave Control Using Ceramic Photonic Crystals with Diamond Structure Including Plane Defects Fabricated by Micro-stereolithography, *The International Journal of Applied Ceramic Technology*, **6**, 41-44 (2009).

11 S. Kirihara, T. Niki, and M. Kaneko, Terahertz Wave Behaviors in Ceramic and Metal Structures Fabricated by Spatial Joining of Micro-stereolithography, *Journal of Physics*, **165**, 12082-1-12082-6 (2009).

12 S. Kirihara, T. Niki, and M. Kaneko: Three-dimensional Material Tectonics for Electromagnetic Wave Control by Using Micoro-stereolithography, *Ferroelectrics*, **387**, 102-111 (2009).

13 S. Kirihara, K. Tsutsumi, and Y. Miyamoto: Localization Behavior of Microwaves in Three-dimensional Menger Sponge Fractals Fabricated from Metallodielectric Cu/polyester Media, *Science of Advanced Materials*, **1**, 175-181 (2009).

14 W. Chen, S. Kirihara, and Y. Miyamoto, Fabrication and Measurement of Micro Three-Dimensional Photonic Crystals of SiO_2 Ceramic for Terahertz Wave Applications, *Journal of the American Ceramic Society*, **90**, 2078-2081 (2007).

15 W. Chen, S. Kirihara, and Y. Miyamoto, Three-dimensional Microphotonic Crystals of ZrO_2 Toughened Al_2O_3 for Terahertz Wave Applications, *Applied Physics Letter*, **91**, 153507-1-3 (2007).

16 W. Chen, S. Kirihara, and Y. Miyamoto, Fabrication of Three-Dimensional Micro Photonic

Crystals of Resin-Incorporating TiO2 Particles and their Terahertz Wave Properties, *Journal of the American Ceramic Society*, **90**, 92-96 (2007).

[17] W. Chen, S. Kirihara, and Y. Miyamoto, Static Tuning Band Gaps of Three-dimensional Photonic Crystals in Subterahertz Frequencies, *Applied Physics Letters*, **92**, 183504-1-3 (2008).

[18] H. Kanaoka, S. Kirihara, and Y. Miyamoto Terahertz Wave Properties of Alumina Microphotonic Crystals with a Diamond Structure, *Journal of Materials Research*, **23**, 1036-1041 (2008).

[19] Y. Miyamoto, H. Kanaoka, S. Kirihara Terahertz Wave Localization at a Three-dimensional Ceramic Fractal Cavity in Photonic Crystals, *Journal of Applied Physics*, **103**, 103106-1-5 (2008).

[20] M. V. Exter, C. Fattinger, and D. Grischkowsky, Terahertz Time-domain Spectroscopy of Water Vapor, *Optics Letters*, **14**, 1128-1130 (1989).

[21] D. Clery, Brainstorming Their Way to an Imaging Revolution, *Science*, **297**, 761- 763 (2002).

[22] K. Kawase, Y. Ogawa, Y. Watanabe, and H. Inoue, Non-destructive Terahertz Imaging of Illicit Drugs Using Spectral Fingerprints, *Optics Express*, **11**, 2549-2554 (2003).

[23] R. M. Woodward, V. P. Wallace, D. D. Arnone, E. H. Linfield, and M. Pepper, Terahertz Pulsed Imaging of Skin Cancer in the Time and Frequency Domain, *Journal of Biological Physics*, **29**, 257-259 (2003).

[24] V. P. Wallace, A. J. Fitzgerald, S. Shankar, N. Flanagan, Terahertz Pulsed Imaging of Basal Cell Carcinoma ex Vivo and in Vivo, *The British Journal of Dermatology*, **151**, 424–432 (2004).

[25] Y. Oyama, L. Zhen, T. Tanabe, and M. Kagaya, Sub-Terahertz Imaging of Defects in Building Blocks, *NDT&E International*, **42**, 28-33 (2008).

Author Index